Zoophysiology and Ecology
Volume 9

Coordinating Editor: D.S. Farner

Editors:
W.S. Hoar B. Hoelldobler H. Langer M. Lindauer

E. B. Edney

Water Balance in Land Arthropods

With 109 Figures

Springer-Verlag
Berlin Heidelberg New York 1977

Dr. Eric B. Edney
University of California
Laboratory of Nuclear Medicine and Radiation Biology
900 Veteran Avenue
Los Angeles, CA 90024/USA

The cover design depicts adult male and female desert cock-roaches *(Arenivaga investigata)* superimposed on histograms showing changes in the water content of these insects as they dehydrate and (in the case of females) rehydrate by absorbing water vapour. Fig. 97 in the text is relevant.

ISBN 3-540-08084-8 Springer-Verlag Berlin Heidelberg New York
ISBN 0-387-08084-8 Springer-Verlag New York Heidelberg Berlin

Library of Congress Cataloging in Publication Data. Edney, E. B. Water balance in land arthropods. (Zoophysiology and ecology; v. 9). Includes index. 1. Arthropoda—Physiology. 2. Osmoregulation. I. Title. QL434.72.E35.595'.2.77-424.

Typesetting, printing and bookbinding: Zechnersche Buchdruckerei, Speyer.
2131/3130-543210

Preface

Writers on arthropod water relationships range from bio-physicists and biochemists to population ecologists—a fact that gives cause to wonder whether the field is already too heterogeneous to be written about in a single book by a single author. I have partly avoided the problem by concentrating largely on physiological mechanisms and by omitting most aspects of behavioural regulation and most aspects of heat balance and body temperature, except when these impinge directly on water balance.

Even within this limited field there has been a lot of work during the past twenty years, as a result of which some problems have been solved (or at least more clearly defined), and many others have been opened up. On the whole there has been a welcome change to a more rigorous experimental approach and it is now possible for water balance people to state their problems in physiological terms.

Good progress has been made towards understanding the mechanisms involved in nearly all avenues of water uptake and loss, although problems indeed remain. The cuticle has yielded part of its secrets to electron micrography, but exploration by means of lipid biochemistry among other techniques is necessary for a real understanding of cuticle permeability. Recognition that water exchange through the cuticle is nearly always a two directional process has come with the introduction of isotopically labelled water to study the component movements. Work on respiratory water loss has led to the discovery of intermittent carbon dioxide release and to a much better understanding of spiracular control mechanisms and of their effects. But several aspects of the relationship between oxygen uptake and water loss are still unexplored.

The most important means of water uptake are usually by feeding and drinking, and problems concerning the control of these processes have proved to be interesting and quite obstinate. Perhaps the most intriguing of all uptake mechanisms is that of water vapour absorption. Here, there is much more information about rates and limiting conditions, and a strong suggestion that the site of uptake in some instances at least is the rectum or mouth rather than the cuticle, but the central mechanism itself still eludes description and poses an attractive

v

problem for cooperation between biologists and biophysicists. The significance of metabolic water production is better appreciated, and some integration of this with other avenues of gain and loss in field conditions has been attempted.

Great strides have been made in understanding what is perhaps the central problem: the internal mechanisms for osmotic and ionic control. Elegant experiments have yielded much information about the details of Malpighian tubule and rectal function at the organ level, and this in turn has led to an attack on the mechanisms of transepithelial transport, in both iso-osmotic and in contra-osmotic situations, in terms of fine structure. Exploitation of this area demands the cooperation of general physiologists with arthropod physiologists, and the results so far have been exciting and encouraging. In this connection, the need to eliminate excess water, e. g. by blood suckers and plant sap feeders, has become recognised and forms a useful counterweight to the usual emphasis on water conservation mechanisms. The water balance of eggs has received relatively little attention, although some important and interesting problems seem to be involved.

Most of the recent work has been concerned with the exploration of particular mechanisms, and there is now both a need and an opportunity to take a wider view and attempt to synthesise such knowledge into an understanding of water balance in whole animals in natural situations. Work along these lines should be innovative and productive; and would encourage a useful exchange of ideas between laboratory physiologists and field ecologists. I have tried to enlarge on this idea in the final chapter of the book.

Los Angeles, 1977   E. B. EDNEY

Acknowledgements

Much of this book was written while I was on sabbatical leave at the Imperial College of Science Field Station at Silwood Park in England, and my thanks are due to the academic and administrative staff for their generous provision of facilities and for many stimulating discussions.

The following people kindly read various parts of the manuscript, and I am most grateful to them for their useful comments: Dr. M.J. Berridge, Professor E. Bursell, Mr. P. Cooper, Mr. P. Franco, Dr. B.L. Gupta, Dr. D.R. Howton, Professor A.D. Lees, Dr. J.P. Loveridge, Dr. S.H.P. Maddrell, Dr. P.S. Nobel, Professor K. Mellanby, Dr. K.A. Nagy, Dr. J. Nevenzell and Professor Sir Vincent Wigglesworth. Any errors of omission or commission are of course my own responsibility.

I have benefited greatly from discussions with those named above, and from discussions or correspondence with the following among many other people: Dr. E.A. Bernays, Professor T.O. Browning, Dr. C.S. Crawford, Professor D.W. Ewer, Dr. J.D. Gee, Dr. N.F. Hadley, Dr. W.J. Hamilton III, Professor H.E. Hinton, Dr. J.E. Moorhouse, Dr. A.C. Neville, Dr. J. Noble-Nesbitt, Dr. J.E. Phillips, Dr. J.E. Treherne, Dr. B.J. Wall.

Several people have kindly lent original photographs, and these are acknowledged in the appropriate place. Thanks are also due to the following for permission to refer to unpublished work: Dr. E.A. Bernays, Mr. B.C. Bohm, Mr. M. Broza, Dr. T.M. Casey, Professor D.W. Ewer, Dr. N.F. Hadley, Dr. W.J. Hamilton III, Dr. J.M. O'Donnell, Dr. T.G. Wolcott.

The Intext Publishing Company, and Dr. F.J. Vernberg (as Editor) gave permission to reprint parts of a chapter I wrote for their book on *Physiological adaptations to the environment*. I thank them and all those authors and publishers who gave permission to reproduce figures or tables from journals or books.

The book is dedicated to Sir Vincent Wigglesworth—a pioneer in this and many other fields.

Contents

Chapter 1

Introduction

A. General

Because of its physical properties, water is uniquely suitable as a basic material for life, and forms a major constituent of all living cells. Water is generally abundant on the earth's surface, but it may be locally scarce, and the reflexion of these facts upon the structure, function and distribution of terrestrial arthropods, is the concern of this book.

Needless to say, the treatment is not exhaustive. We shall consider several important areas of interest concerning water uptake, water loss, and the regulation of water content. We shall inevitably be concerned with osmotic and ionic regulation. Body temperature will be considered only insofar as water affairs are involved: thus we shall consider evaporation as a component of body temperature determination, but refer only briefly to other aspects of body temperature. In general, attention will be directed mainly to work during the last two decades. Earlier investigations are referred to by Andrewartha and Birch (1954, 1960), Edney (1957) and Browne (1964).

Since then several reviews, usually of particular parts of the general field, have been published, including those by Beament (1961, 1964, 1965) on the structure and function of the insect cuticle, Bliss (1968) on terrestrial crustaceans, Bursell (1974a, b) on the effects of temperature and humidity, Cloudsley-Thompson (1970) in the course of a review on temperature regulation in invertebrates, Ebeling (1974) on cuticle permeability, Edney (1974) and Cloudsley-Thompson (1975) on desert arthropods, Hackman (1971, 1974) on insect cuticle structure in general and its more chemical aspects respectively, Locke (1974) on the structure and formation of the insect integument, including its mechanical properties, Maddrell (1967, 1971), Cochran (1975) on the mechanisms of excretion in insects, Rapoport and Tschapek (1967) on soil water in relation to arthropods, Riegel (1971) on excretion in arthropods, Shaw and Stobbart (1972) on osmoregulation in locusts and other desert arthropods, and Stobbart and Shaw (1974) on salt and water balance in insects. Finally, several aspects of Neville's (1975) review of arthropod cuticle are relevant to the present book.

Several of the reviews referred to appear in Rockstein's *The Physiology of Insecta*, and additional relevant material will be found in other chapters of that standard work. Other reviews of particular subjects will be referred to below in the appropriate places.

B. Terrestrial Arthropods

According to one widely accepted view (Tiegs and Manton, 1958; Manton, 1964, 1973) arthropods represent an organisational level rather than a monophyletic taxon. It seems that arthropodization from unknown invertebrate ancestors may have occurred at least three times, giving rise respectively to (1) the Crustacea, (2) the Chelicerata (xiphosurans, spiders, scorpions, Acarina, etc.), and (3) the Uniramia. Present day Uniramia are represented by (1) onycophorans, (2) a myriapodous group including millipedes, centipedes, pauropods and symphylans, and (3) a hexapodous group, including several non-winged groups (proturans, collembolans, thysanurans and diplurans) besides the pterygotes or winged insects.

Each of the major groups has terrestrial representatives, but judged by the number of species or by the range of environments inhabited, insects and arachnids are much more successful on land than are the rest. Insects today are very largely terrestrial, with a few secondarily aquatic freshwater forms, a few that live in the marine littoral zone, and one genus (*Halobates*—a gerrid) that lives on the ocean surface (Cheng, 1974). Spiders are also predominantly terrestrial with no fully marine forms, and the same is true of ticks, but there are many marine mites. The Xiphosura are entirely marine, but all the other remaining arachnids are terrestrial, as are the myriapodous groups: centipedes, millipedes and pauropods. Crustaceans are predominantly marine, but several malacostracans have become terrestrial to different extents, a few amphipods are entirely terrestrial (though highly cryptozoic) (Hurley, 1959, 1968). Several crabs, notably the coconut crab *Birgus*, *Coenobita*, *Gecarcinus* and others, are almost entirely terrestrial, returning to water only to breed (Bliss, 1968), but land isopods (woodlice or sowbugs) are the most fully terrestrial of all crustaceans (Edney, 1954, 1968 a), showing many degrees of adaptation to that mode of life.

C. Arthropod Structure

I. Size

Detailed consideration of the structure and function of particular organs will be delayed until later. Here we draw attention to some of the basic attributes of arthropods, as a grade of organisation, that are relevant to a discussion of their water relationships.

By vertebrate standards, terrestrial arthropods are small. The range of size is very great: about eleven orders of magnitude, from 10^{-8} g for some tarsonemid mites that are only 50 μm long to about 10^3 g for the land crab *Birgus*; but the large majority of terrestrial forms weigh less than 10^{-1} g. (The range for mammals: from 2.0 g for a pygmy shrew to 7×10^6 g for an African elephant is much narrower.) Since $S = kW^{2/3}$, where S is surface area, W is weight, and k is a constant that depends on units and shape, the surface: volume ratio of animals becomes progressively greater the smaller the size. Small size is, therefore, an important attribute of land arthropods, with great significance for water and heat exchange—matters to which we shall return later.

II. Integument

A second feature of general importance concerns the skeleton. In soft-bodied larvae such as caterpillars and grubs, the skeleton is largely hydrostatic; but in most adult forms (and in many larvae too) the integument forms a more or less rigid exoskeleton, allowing for muscle attachments and rapid, leveraged locomotory and other movements. The advantages of an essentially tubular exoskeleton for small animals include the fact (remarked upon by Locke, 1974) that for equal amounts of material, a hollow cylinder is three times more resistant to bending than is a rod. The mechanical functions of arthropod exoskeletons are varied, and their mechanical properties vary correspondingly, often between instars (Hepburn and Joffe, 1974), or between different parts of the cuticle in one instar, as in *Tenebrio* adults (Delachambre, 1975). For example, the cuticle may be highly elastic where kinetic energy storage is advantageous as in wing hinges, or flexible, as in soft-bodied larvae or at the arthrodial membranes between sclerites. Elasticity, where this occurs, is achieved by the inclusion of a rubber-like protein, termed "resilin" by Weis-Fogh (1960), in the material of the cuticle (for a review see Anderson and Weis-Fogh, 1964). In other structures, such as the wings of insects, the cuticle must be light and fairly rigid, but for the mouthparts of a beetle or a locust (Anderson, 1974) a hard, resistant cuticle is necessary. For a discussion of the mechanical properties and the chemical constitution of insect cuticle see Locke (1974), and Hackman (1971, 1974), and for a comprehensive review of the structure and functions of arthropod cuticle see Neville (1975). The adaptive significance of major differences in musculo-skele-tal architecture has been well described in a series of papers by Manton (refs in Manton, 1973).

But apart from its importance as a mechanical exoskeleton, the arthropod integument provides an interface between the organism and the environment. As such the integument must generally resist the efflux of water but permit the limited influx of certain molecules, of light and of mechanical information, for sensory perception. It must resist mechanical abrasion, and bear a variety of structures and microstructures (hairs, scales, sense organs) and colors, for protec-tion or advertisement, all of which are reflected in a remarkable versatility of shape and form.

Owing to inevitable wear and tear, and also to the need for growth in immature stages, an exoskeleton has to be replaced from time to time by moulting. This process involves a series of highly complex, beautifully integrated events; and although it is under the control of centrally produced hormones, most of the complex biological machinery associated with moulting is built into the integument, particularly into the epidermal cells which produce the cuticle. The process is carried out without undue loss of water or dry materials.

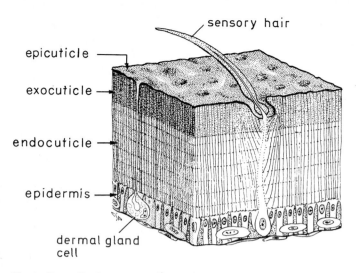

Fig. 1. Generalized structure of an arthropod cuticle. An epidermis (hypodermis), one cell thick, produces all the other structures before or after ecdysis. Dermal gland cell ducts run to the outer surface. The bulk of the cuticle consists of an inner, laminate, chitin-protein endocuticle and an outer, hardened and darkened exocuticle. External to these lies the thin epicuticle, a complex layer whose detailed structure is considered in Chapter 3. The cuticle is perforated by numerous fine, often twisted, ribbon-shaped pore canals, which terminate in even finer epicuticular canals through the inner epicuticle. (Redrawn after Wigglesworth, 1972)

Despite the great variety of form and function, all arthropod integuments conform to a basic common structure (Fig. 1)—a fact that is all the more remarkable in a group of animals that is believed to be polyphyletic. In all there is a single-layered epidermis (hypodermis) which secretes (and may re-absorb parts of) the cuticle itself. The latter is composed largely (so far as bulk is concerned) of the procuticle, which is a complex of protein and chitin intimately associated, perhaps as mucopolysaccharides. Chitin itself is a nitrogenous polysaccharide, identical with the material of the cell walls of fungi and probably having the form of long chains of acetylated glucosamine residues, as shown in Figure 2. Much of the procuticle is laminated, each lamina being composed

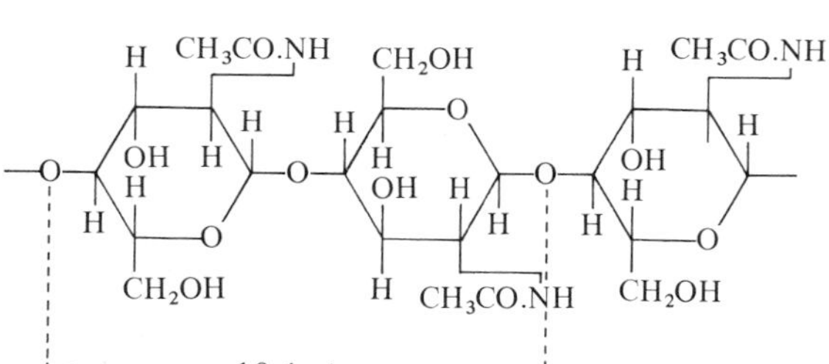

Fig. 2. Chitin is a nitrogenous polysaccharide identical with fungine from the cell walls of fungi. It is probably composed of long chains of acetylated glucosamine residues linked as shown here with a repeating pattern of 10.4 Å. (From Wigglesworth, 1972)

4

of several lamellae. In many arthropods (refs in Neville, 1975), laminae are laid down in daily cycles, and since the preferred direction of microfibrils (chitin crystallites) in each lamella within the daily cycle is rotated with respect to the previous lamella, the resulting appearance in oblique section is a series of parabolic curves (Fig. 3). This concept, originally due to a mathematician, Bouligand (1965), has since been confirmed and amplified by Neville and others for many arthropods as well as for animals in several other phyla (Gubb, 1975). Full references and discussions are given by Neville (1975).

Fig. 3. Electron micrograph of a section of exocuticle of *Astacus* (a freshwater crayfish), × ca. 7,600. The photograph shows the parabolic patterns created by the changing direction of chitin crystallites in successive lamellae. Ribbon-like pore canals (P.C.) rotate through 180° in traversing each lamella, forming crescentic areas in oblique section that conform to the parabolic pattern. Each canal contains a central bundle of filaments (F). (Photograph kindly furnished by A. C. Neville)

5

The second component of all arthropod cuticles is a thin, outer, epicuticle, containing protein, lipids and polyhydric phenols, but no chitin. The structure of this important region is complex and we shall refer to it later when experimental evidence necessary for its interpretation has been considered (Chap. 3.B.VI.).

The procuticle is traversed by fine pore canals, which are of the order of 100 nm in diameter, ribbon-shaped, showing an anti-clockwise twist (away from the observer) corresponding with the anti-clockwise shift in the preferred direction of microfibrils in the lamellate regions (Neville *et al.*, 1969; Neville, 1970) (Fig. 4)

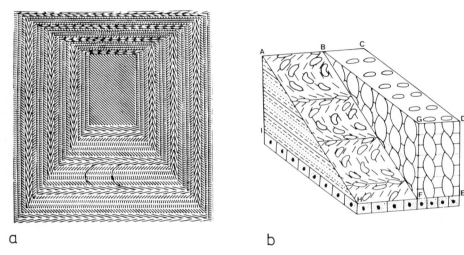

a   b

Fig. 4. (a) Parabolic patterns in cuticle appear in oblique section because the chitin microfibrils are arranged parallel to each other in any one lamella, but their orientation changes systematically between lamellae. The diagram (after Bouligand, 1965) represents a truncated pyramid of chitin seen from above, and two parabolic series of microfibrils are indicated on the lower oblique face. (b) An interpretation of Bouligand's model by Neville *et al.* (1969) showing how pore canals in the form of twisted ribbons with a pitch of 180°/lamella, in phase with the changing direction of the chitin microfibrils (seen in vertical section in planes BGF or DEFG) generate parabola-shaped artifacts in oblique section (seen in plane ABFH). Helically constructed pore canals would have to rotate by 360° in each lamella to provide a similar effect. Neville *et al.* (1969) prefer the twisted ribbon alternative

and each containing a central bunch of straight filaments. The outer part of the procuticle is made hard and resistant by quinone tanning, a process that involves oxygen cross links between the protein chains, to form "arthropodin." This hardened, darkened region, when it occurs, is termed the exocuticle, to distinguish it from the untanned and more flexible endocuticle. In some insects a middle, hard but still light coloured mesocuticle is also present, perhaps as a result of tanning by a different process (Anderson, 1974).

Cuticulin, a form of stabilised lipid, is present in parts of the procuticle as well as in the epicuticle (Wigglesworth, 1970) and confers a degree of rigidity on the cuticle before tanning. The precise nature of the lipid of cuticulin is

6

not known, but from the evidence available, Wigglesworth (1970) suggests that it may be "a multiple polyester, consisting of hydroxylated fatty acid chains cross linked by ester bonds between carboxyl and hydroxyl groups" (rather similar to the cutin of plants). It will be referred to further below in connection with the epicuticle.

Quinone tanning probably occurs in most other arthropods (Blower, 1951; Jones, 1954; Krishnan, 1951; Sewell, 1955; other refs in Richards, 1958), but in some crustaceans and in millipedes, hardening is additionally effected by the incorporation of calcium salts (Drach, 1939; Subramoniam, 1974). Conservation of these materials at moult time is ensured by re-absorption from the old and redeployment in the new cuticles, although the brief description of moulting in millipedes by Rajulu (1969) suggests that in these animals as in *Peripatus*, (Robson, 1964) the whole cuticle is shed and there is no moulting fluid.

The cuticle of the onycophoran, *Peripatus*, is aberrant in some respects: pore canals and dermal glands are absent, for example. But the cuticle is essentially arthropodan in structure (Robson, 1964), and in chemical composition (Hackman and Goldberg, 1975), albeit of a primitive kind. In all peracarids, including isopods, moulting occurs in two halves in sequence, and the calcium salts are absorbed and transported between the antherior and posterior regions of the body as appropriate.

The structure and moulting of crustacean cuticle has received less attention. In general it is fairly similar to that of insects, although in decapods the lower layers of the cuticle are deposited later in the moult cycle (ecdysis occurs when the new epicuticle is present), and the complex lipid and other layers of the insect cuticle have not been observed in crustaceans. For details of the moulting cycle see Drach (1944) and for the moulting process see Passano (1960), and Travis (1960, 1963). The cuticle of arachnids is in general rather similar to that of insects (Lees, 1946a, 1947; Jones, 1954). See Beadle (1974) for the fine structure of *Boophilus* cuticle, Henneberry et al. (1965) for the spider mite *Tetranychus*, Wharton et al. (1968) for the mite *Laelaps* and Sewell (1955) for the spider *Tegenaria*.

For further details of cuticular structure, see Dennell and Malek (1956 and other papers), Richards (1951, 1958), Wigglesworth (1948 a, b, c) as well as the more recent works cited above.

III. Body Cavity and Blood

In arthropods, the main body cavity is blastocoelic in origin and known as a haemocoele, in contrast to the intra-mesodermal coelom of vertebrates. The evolutionary origin of the haemocoele, which may have arisen in proto-arthropods partly in response to burrowing, is discussed by Manton (1970).

Arthropods, unlike vertebrates, have the main body cavity continuous with the blood vascular system, and the latter is said to be "open." This means that the blood of arthropods (often referred to as "haemolymph") is not usually at a significantly higher pressure than that of the other tissues, so that formation of urine by ultrafiltration under pressure, as in the vertebrate kidney, is impossible. Further, in insects as in other arthropods, save crustaceans and some arachnids, blood vessels are represented only by a tubular heart and a short median dorsal artery. This has the important consequence that the blood flow is relatively sluggish, and blood does not serve as the main means of respiratory gas transport. Instead there is in insects, and to a lesser extent in other non-crustacean terrestrial arthropods, a system of internal ramifying tubules, the tracheae, which convey oxygen mainly by diffusion directly to all metabolizing tissues. We shall be concerned with the effect of this on water loss.

On the other hand, where oxygen transport is not involved, a constant blood volume is not necessary as long as there is sufficient circulation to provide for the transport of food materials, metabolites, hormones and other substances over the relatively short distances involved. Thus the haemocoele and the blood it contains can function as a water store, and water can be removed from the blood to maintain the water content of the tissues and vice-versa, (as Mellanby first suggested in 1939) apparently without any dire osmotic repercussions (see Chap. 6).

IV. Excretion and Osmoregulation

The main organs concerned with the excretion of excess materials, particularly nitrogenous wastes and inorganic salts, are the Malpighian tubules, whose structure and function have been much studied in insects (see review by Maddrell, 1971). Such organs also occur in spiders, scorpions, acarines, centipedes and millipedes. Their form is very various, but they are always blind tubules (or loops) opening proximally into the hind gut. Commonly the walls of the hind gut, particularly the rectum, are specialised for water and salt transport (Wall and Oschman, 1975), so that the Malpighian tubules and rectal walls function in concert.

In arachnids the so called "coxal glands," and the salivary glands in ixodid ticks (see Chap. 6.C.III), also function as organs of osmoregulation (Lees, 1946b). The crustaceans differ from other arthropods in lacking Malpighian tubules, and excretion and osmoregulation are associated with antennal or maxillary glands. The nature of the nitrogenous endproduct varies, and the significance of this, as well as details of the excretory machinery will be considered below.

This concludes our brief survey of the basic attributes of arthropods that are relevant to water affairs, and we turn now to consider the properties and distribution of water itself from the same standpoint.

D. Water

I. The Properties of Water

There are several reasons why water is essential for life, but above all, this is because it serves as a material in which diffusion of solutes can occur, and related to this, as the solvent in which many biochemical reactions take place. Additionally, water is a liquid at most biologically significant temperatures, and poorly compressible, so that it can function as a flexible hydrostatic skeleton. It has a high heat of vaporisation and thermal capacity, and thus functions well as a means of heat distribution and temperature control. Water also takes part directly in basic metabolic processes. In plants it is the source of oxygen evolved in photosynthesis, and of hydrogen for carbon dioxide reduction. In animals, water is the result when oxygen is used as the final electron acceptor, and production of the common energy currency, ATP from ADP, involves dehydration of the latter, and the release of the components of water.

Useful descriptions of the biophysical properties of water are given by Nobel (1974) and House (1974), and the following brief account owes a great deal to these works.

Many of the important physical properties of water can be traced to its molecular structure (Fig. 5). The strong intermolecular forces in water which

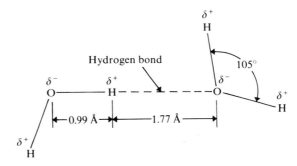

Fig. 5. Schematic structure of water molecules, indicating the hydrogen bonding resulting from the electrostatic attraction between the net positive charge on a hydrogen ($\delta +$) in one molecule and the net negative charge on an oxygen ($\delta -$) in a neighbouring water molecule. (From Nobel, 1974)

result in its high melting and boiling points (compared with other substances of similar electron structure) are due to attraction between the two hydrogen atoms of one water molecule and the oxygen atoms of two neighbouring molecules (hydrogen bonding), and involve an energy store of about 4.8 kcal mol^{-1} of H-bonds. In ice, nearly all the water molecules are hydrogen bonded, and about 15% of these bonds must be broken before melting occurs.

Disruption of some of these bonds as ice melts is responsible for the heat of fusion of ice (1.44 kcal mol^{-1} of water), and further energy is required when water evaporates, amounting to 10.5 kcal mol^{-1} at 25°C, and 9.71 kcal mol^{-1} at 100°C, as the remaining hydrogen bonds are broken. Energy is also required to overcome van der Waal's forces and for expansion, so that hydrogen bonds are not the sole repositories of energy.

The water molecule is polar, and like other such substances has a high dielectric constant (D), compared with that for non-polar liquids such as hexane. D for water is about 80.2 and for hexane about 1.87: some 43 times lower. Ions which result from the dissociation of electrolytes (e. g. Na$^+$ and Cl$^-$) attract polar water molecules and thus form a screen around themselves, so that their attraction for each other (which would lead to recombination) is much reduced,

9

and many more of them are able to exist in solution. In brief, water is an excellent solvent for polar particles.

Another important property of water is its very high surface tension which amounts to 72.8 ergs cm^{-2} at 20°C, and may be thought of as the energy necessary to expand the surface of an air-water interface. To move water molecules into the air-water interface from the interior of the aqueous phase calls for the breaking of hydrogen bonds and hence for the use of energy. Or, to put it another way, attraction acting on a water molecule below the surface is equal in all directions, while a molecule at the surface is attracted inwards and laterally only. Hence a laterally oriented attraction or "skin" is present at the surface.

When a liquid is contained within a solid (when water is in a glass vessel for example), the behaviour at the line of contact depends on the surface tension of the liquid (itself dependent on cohesive intermolecular forces) and adhesive forces between the liquid and the solid. For water and a clean glass or a proteinaceous surface, the contact angle (measured in the liquid) is small, the meniscus, if the solid surface is vertical, is bent upwards, and the liquid will rise in a capillary situation. If the solid surface is lipid, it will be hydrophobic, and the behaviour will be the reverse of the first situation.

The extent of capillary rise is expressed by the relationship

$$h = \frac{2\sigma\cos\alpha}{r\varrho g} \tag{1}$$

where h is the height in cm, σ is the surface tension in dynes cm^{-1}, α is the contact angle, r is the radius of the capillary in cm, ϱ is the density of the liquid in g cm^{-3}, and g is the acceleration due to gravity, about 980 cm s^{-2}.

If the liquid is water, the surface tension is 72.8 dynes cm^{-1} at 20°C, and if the capillary is strongly hydrophilic (clean glass or exposed polar groups), the contact angle approaches 0° and $\cos\alpha$ approaches 1. Then Eq. (1) gives,

$$h_{(cm)} = \frac{0.149 \text{ cm}^2}{r_{(cm)}} \tag{2}$$

from which it follows that water would rise in such a capillary with r = 20 μm to a height of 74.5 cm. We shall revert to this discussion below in relation to cuticular capillaries. For further quantitative treatment see Nobel (1974).

Meanwhile it is easy to see the significance of surface tension for small aquatic arthropods: if the surface of the animal is strongly hydrophobic (the legs of gerrids for example) α will be very large and at equilibrium the animal will be suspended on the surface film. But if the surface is strongly hydrophilic (as in aquatic crustaceans), the equilibrium position is below the surface, and insects, such as mosquito larvae, or hydrophilid beetles, that are usually submerged but need atmospheric air, would find it hard to break the surface film were it not for the strongly hydrophobic hairs surrounding the tip of the siphon or the elytra respectively.

II. Water in the Environment and in Animals

Water exists in the biosphere as ice, liquid water or water vapour, but in the present context we shall be concerned mainly with the liquid and vapour phases. The rate of movement, and the forces causing such movement of water from one place to another (between an animal and its environment, or across the rectal wall of an animal for example) are central to the discussion, so that it is necessary to consider how to deal with the concepts and quantities concerned.

The difference in free energy, or chemical potential, between water in different states determines (1) the direction in which water will tend to move, and (2) the work required to effect the movement if this is against a potential gradient. The full expression for the chemical potential of a substance j, in energy units per mole is given by:

$$\mu_j = \mu_j^* + RT \ln a_j + \bar{V}_j P + z_j F E + m_j g h \tag{3}$$

where μ_j^* is a constant reference level, $\bar{V}$ is partial molal volume in $cm^3 mol^{-1}$, a_j is chemical activity and depends on concentration in mol $liter^{-1}$, P is hydrostatic pressure in bars, $z_j F E$ is a component resulting from electrical potential, and $m_j g h$ is a gravitational component. In the case of water $z_j F E$ is zero and may be neglected.

Now osmotic pressure is related to activity[1] by the equation:

$$R T \ln a_w = - \bar{V}_w \pi \tag{4}$$

where π is osmotic pressure in bars, so that Eq. (3) may be written

$$\mu_w = \mu_w^* - \bar{V}_w \pi + \bar{V}_w P + m_w g h \tag{5}$$

and by identifying $m_w / \bar{V}_w$ as ϱ_w, the density of water, we can write:

$$\frac{\mu_w - \mu_w^*}{\bar{V}_w} = P - \pi + \varrho_w g h = \Psi \tag{6}$$

which is an evaluation of Ψ, or water potential, a quantity which is increasingly being used by plant physiologists. It takes into consideration hydrostatic pressure (e.g. cell turgor), solute activity (osmotic pressure) and height above a standard level. Values of Ψ are negative because the reference state, pure water, is considered to have a water potential of $\Psi_w^* = 0.0$, and Ψ_w is either equal to or less than this.

When air and water are in contact in an isothermal, isobaric closed system, molecules of water evaporate, and others condense until at equilibrium the two rates are identical, and the air is said to be saturated with water vapour.

---

[1] In the following discussion, a_w is used to refer to the chemical activity or activity of water either when the liquid phase is concerned or when both liquid and vapour phases are concerned; while a_v is used to refer to the vapour phase alone.

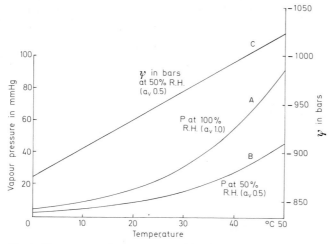

Fig. 6. The amount of water contained in air at any one relative humidity (R.H.) or water vapour activity (a_v), increases exponentially with temperature, as shown by curves A and B. Curve C shows the effect of temperature on water potential (Ψ) at 0.5 a_w, based on the relationship, $\Psi_w = \dfrac{RT}{\bar{V}_w} \ln a_w$. This relationship is not truly linear since $\bar{V}_w$ varies slightly with temperature

If now the temperature rises, a new equilibrium is reached in which a larger amount, or higher partial pressure, of water is present in the vapour phase, the increase being exponential with temperature, as shown in Figure 6. In natural environments, air is hardly ever saturated with water vapour, and if P is the actual partial pressure of water vapour, and P* the potential saturation vapour pressure, then P* − P (or ΔP) is a measure of vapour pressure difference sometimes known as "saturation deficit," and $\dfrac{P \cdot 100}{P*}$ is relative humidity (R. H.). Furthermore, if we assume that water vapour behaves as in ideal gas, then the activity of water vapour is,

$$a_v = \frac{P}{P*} = \frac{\% \, R.H.}{100}. \tag{7}$$

Relative humidity is a ratio, and as Figure 6 shows, for air at any one $a_v \left(\text{or } \dfrac{R.H.}{100}\right)$, the vapour pressure difference (P* − P, or ΔP in pressure terms), is greater at higher than at lower temperatures. In other words, a given amount of water in a sample of air will represent higher relative humidities at lower temperatures.

Water tends to move away from regions of higher to regions of lower partial pressure (P), but it follows from the above that this does not necessarily correspond with the direction of falling R.H. gradient. For example, if air at the bottom of a scorpion's burrow is at 15°C and 90% R.H. ($a_v = 0.9$), its vapour pressure (P) will be 11.5 mm Hg or 15.3 mbar. If at the same time, air outside the burrow

12

is at 28°C and only 50% R.H. ($a_v = 0.5$), P will be 14.2 mm Hg or 18.9 mbar, and water vapour will move into the burrow against the a_v gradient (Fig. 7).

This principle is probably of general ecological importance. It was recognized many years ago by Williams (1924, 1954) in relation to the microclimate of desert caves (Fig. 8); and similar reasoning underlies Lowry's (1969) important

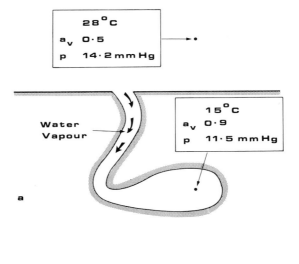

a

b

Fig. 7. (a) Although the relative humidity inside a burrow may be higher than that outside, if the temperature is sufficiently low inside, water vapour will move inwards. (b) Although air at the surface of a lake is saturated with water vapour, its vapour pressure may be lower than that of unsaturated air above, and water vapour will condense on the lake

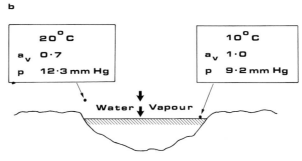

point that "vapour pressure deficit" of a sample of air is not by itself an indication of drying power. The temperature of the surface from which evaporation may occur must be known in order to determine the vapour pressure of water there, and thus to derive the effective vapour pressure *difference*.

If a solute is present in water, the water potential Ψ_w and activity a_w of the solution are reduced, and (again if the system is assumed to be closed) this results in a proportional reduction in P and a_v in the air in equilibrium with it. The extent of this reduction in dilute solutions is determined by the molar fraction (N_w) of water in the solution, i.e., the ratio of the number of moles of water to the number of moles of all substances in the system. When the solute is not an ideal non-electrolyte (i.e., it ionises in solution) its molar concentration must be expressed as osmoles per liter.

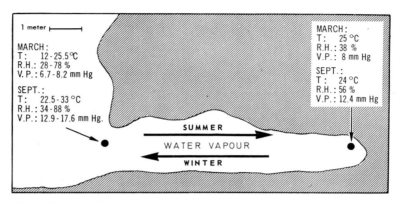

Fig. 8. Diagram of an imaginary cave to illustrate real measurements made by Williams (1954) near Cairo, Egypt. There is a net movement of water into the cave during the summer (Sept.) and out during the winter (March), even though the R.H. in the cave is higher in summer than in winter. T: temperature; R.H.: relative humidity; V.P.: vapour pressure. (From Edney, 1974)

For example, the activity of a 1.0 osmolar solution of NaCl (about 35 g liter^{-1}) is $a_w = \dfrac{55.5}{55.5 + 1} = 0.982$, which is approximately the activity of sea water. The osmolar concentration of human blood is about 0.3 and that of insect blood about 0.3–0.5 osmol liter^{-1}.

Osmotic concentrations are often expressed in terms of freezing point depression, $\Delta °C$, Δ_i and Δ_0 being used for internal and external media respectively. A 1.0 osmolar solution has $\Delta 1.86°C$, and this is also the approximate value for sea water. A 1.0 osmolar solution has an osmotic pressure of about 24.37 bars at 20°C. Thus a solution whose $a_w = 0.95$, which is in equilibrium with air at $a_v = 0.95$, would have a solute concentration, c_s, given by, $0.95 = \dfrac{55.5}{55.5 + c_s}$, from which $c_s = 2.92$ osmol liter^{-1} and an osmotic pressure (π) at 20°C of $\pi = 2.92 \times 24.37 = 71.2$ bars or 72.1 atmospheres, if the solution behaved as an ideal one.

Several conclusions of biological interest follow from the foregoing. Firstly, since the body fluids of most arthropods have $a_w \simeq 0.995$ (the equivalent of 300 mosmol liter^{-1}), unless the air around them is nearly saturated, a steep gradient (equivalent to 150 bars at $a_v = 0.9$) will tend to cause loss of water by evaporation.

Secondly, doubling (for example) the concentration of an animal's body fluid will have a negligible effect on the gradient, and hence on the rate of evaporation of water from it. Thirdly, any net inward movement of water (when

14

a tick absorbs water vapour for example) must take place against a very steep overall gradient (about 340 bars at $a_v = 0.8$, or 80% R.H.), and once again, any biologically reasonable decrease in a_w will not reduce the gradient significantly.

Animals such as teleost fish, with $\Delta_i = 0.8\,°C$ ($a_w = 0.992$) that live in sea water where $\Delta_0 = 1.86\,°C$ ($a_w = 0.982$), are subject to a less dehydrating environment than animals that live in air below 98% R.H. In other words, the ocean is not a physiological desert, as is sometimes said of it.

III. Evaporation of Water

Net water loss through the cuticle or from the respiratory surfaces of arthropods are both forms of evaporative loss (unlike excretory or faecal water loss) and, therefore, depend on P in the surrounding air. The extent of such loss, in terms of the proportion of total body water initially present, will be greater in smaller than in larger animals, even though the rate of evaporation per unit surface area is the same. For example, a 1 mg animal has a surface: volume ratio 100 times that of a 1 kg animal of similar shape. Again because weight specific metabolic rates are usually higher in smaller animals, ventilation, and hence weight specific respiratory water loss, may be expected to increase with decrease in size.

According to Fick's law, the diffusion of water vapour across a homogeneous membrane or through a layer of air, follows the relationship:

$$J = \frac{D\Delta C}{\Delta X} \tag{8}$$

where J is the rate of flux in $\text{mass} \times \text{time}^{-1} \times \text{length}^{-2}$; D is the coefficient of diffusion in $\text{length}^2 \times \text{time}^{-1}$ (usually $\text{cm}^2\ \text{s}^{-1}$); C is in $\text{mass} \times \text{volume}^{-1}$ (usually mol liter^{-1} or mg cm^{-3}); and ΔX is the length of the diffusion path (usually in cm).

For D (in $\text{cm}^2\ \text{s}^{-1}$)/$\Delta X$ (in cm), it is sometimes convenient to substitute $1/R$ (where R is resistance in s cm^{-1}), and then Eq. (8) may be rewritten:

$$J = \frac{\Delta C}{R} \tag{9}$$

where J is flux rate in $\text{mg cm}^{-2}\ \text{s}^{-1}$, C is concentration in mg cm^{-3}, and R is resistance in s cm^{-1}.

The extent to which we can use this relationship in the special case of transpiration across an arthropod integument will be considered below (Chap. 3).

To conclude this section we return to an earlier point—that water concentration can be treated in several ways. Water potential, Ψ_w, can be attributed to water in either liquid or vapour phases, and takes into account the osmotic concentration π of a solution and its gravitational and hydrostatic circumstances, although the last factor applies only to the liquid phase. Water activity (a_w) can also be used for either phase, and takes into account π in the liquid phase.

Water activity is a ratio, and is proportional to relative humidity in the vapour phase. Vapour pressure, P, may be attributed to either phase, and is an amount, expressed in a variety of units including mm Hg (equivalent to pressure), or any other suitable units such as atms., bars or tors. Here we shall generally use a_w rather than Ψ, because the former is currently in use by animal biologists (see however Galbreath, 1975) and (at constant temperature and pressure) Δa_w is proportional to the rate of evaporation while Ψ is not.

The quantities concerned are interconvertible according to the following approximations:

$$a_w = \frac{R.H.}{100} = \frac{P}{P*} \tag{10}$$

$$\Psi_w = \frac{RT}{\overline{V}_w} \ln a_w \quad \text{(neglecting hydrostatic pressure} \tag{11}$$
$$\text{and any gravity component)}$$

$$\frac{RT}{\overline{V}_w} = \begin{matrix} 1243 \text{ bars at } 0°C \\ 1332 \text{ bars at } 20°C \\ 1454 \text{ bars at } 50°C. \end{matrix}$$

Chapter 2

Water Content

A. Total Body Water

Over 99% of all the molecules in a living arthropod are water molecules, but because of the very high molecular weights of the large organic molecules of animal tissues, water content expressed as a precentage of total weight is very much less than its mole fraction—about 70%. In fact, as Table 1 shows, water content is very variable, both within and between species of arthropods, and we may briefly consider the reasons for such variability.

Table 1. Selected data on water contents of arthropods[a]

Taxon	Stage	Percent water content	Remarks	Reference
Insecta				
Apterygota				
Thysanura				
Thermobia domestica (= *Lepismodes inquilinus*)	A	70	normally fed	Okasha (1971)
Pterygota				
Orthoptera				
Chortophaga	L	75–78	growing	Bodine (1921)
Chortophaga	L	61–67	hibernating	Bodine (1921)
Chortophaga	A	73–77	growing	Bodine (1921)
Chortophaga	A	62–69	old	Bodine (1921)
Melanoplus	L	75	active	Bodine (1921)
Melanoplus	L	65	hibernating	Bodine (1921)
Locusta migratoria	A	74	young adults	Loveridge (1973)
Locusta migratoria	A	62	20-day old adults	Loveridge (1973)
Dictyoptera				
Arenivaga investigata	L	63–71	desert cockroach	Edney (1966)
Periplaneta americana	L	68.5	domestic cockroach	Edney (1968b)
Psocoptera				
Liposcelis	A	66	book louse	Knülle and Spadofora (1969)
Isoptera				
Hodotermes mossambica	A	56	newly emerged alate ♂♂	Hewitt *et al.* (1971)
Hodotermes mossambica	A	49	newly emerged alate ♀♀	Hewitt *et al.* (1971)
Hodotermes mossambica	A	76	10-day old pairing ♂♂	Hewitt *et al.* (1971)
Hodotermes mossambica	A	67	10-day old pairing ♀♀	Hewitt *et al.* (1971)

[a] A: adults, P: pupae, L: larvae.

Table 1 (continued)

Taxon	Stage	Percent water content	Remarks	Reference
Homoptera				
Aphis fabae	A	64–73	unflown alates, % total body weight	Cockbain (1961b)
Aphis fabae	A	72–76	unflown alates, % fat-free body weight	Cockbain (1961b)
Coleoptera				
Blaps gigas	L	76		Marcuzzi (1960)
Calandra granaria	A	46–47	grain weevils	Robinson (1928) Buxton (1932a)
Limonius sp.	L	58	wireworms	Jones (1951)
Popillia japonica	L	78–81	flea-beetles	Ludwig (1936)
Popillia japonica	P	74	flea-beetles	Ludwig (1936)
Popillia japonica	A	67	flea-beetles	Ludwig (1936)
Phyllophaga	L	77	20°C or 0.5°C	Sweetman (1931)
Phyllophaga	A	77	feeding	Sweetman (1931)
Phyllophaga	A	78	after 2 months fast	Sweetman (1931)
Pimelia angulata	A	62	10.0% fat content	Marcuzzi (1960)
Pimelia angulata	L	68	4.3% fat content	Pierre (1958)
Tenebrio molitor	L	58	13.7–16.7% fat content	Hochrainer (1942) Marcuzzi (1960)
Tenebrio molitor	L	58	on low water content bran	Mellanby (1958)
Tenebrio molitor	L	75	low water bran + water ad lib.	Mellanby (1958)
Tenebrio molitor	L	62	mature, 0.133 g 20°C	Marcuzzi (1960)
Tenebrio molitor	L	54	mature, 0.133 g 28°C	Marcuzzi (1960)
Tenebrio molitor	L	68	juvenile, 0.058 g 20°C	Marcuzzi (1960)
Tenebrio molitor	L	59	juvenile, 0.058 g 28°C	Marcuzzi (1960)
Tenebrio molitor	L	56	R.H. 10%	Marcuzzi (1960)
Tenebrio molitor	L	61	R.H. 80%	Marcuzzi (1960)
Tenebrio molitor	P	64	7.5% fat content	Marcuzzi (1960)
Tenebrio molitor	A	62	7.2% fat content	Marcuzzi (1960)
Tenebrio molitor	A	63	R.H. 10%	Marcuzzi (1960)
Tenebrio molitor	A	67	R.H. 95%	Marcuzzi (1960)
Tenebrio molitor	A	64	11% fat	Schulz (1930)
Tenebrio molitor	A	54	18% fat	Schulz (1930)
Tenebrio molitor	A	61	6.3% fat	Hochrainer (1942)
Tribolium confusum	L	70–83	no difference between 10–24°C and 15–75% R.H.	Dreyer (1935)
Tribolium confusum	A	45	? conditions	Marcuzzi (1960)
Lepidoptera				
Bombyx mori	A	64–69	silk worm	Buxton (1932a)
Bombyx mori	L, P	77–79	silk worm	Buxton (1932a)
Bombyx mori	A	70	♂♂	Saccharov (1930)
Bombyx mori	A	79	♀♀	Saccharov (1930)
Cirphis	L	87–89		Robinson (1928)
Euxoa segetum	L	84	active	Saccharov (1930)
Euxoa segetum	L	71–75	hibernating	Saccharov (1930)
Telea polyphemus	L	92		Robinson (1928)

Table 1 (continued)

Taxon	Stage	Percent water content	Remarks	Reference
Hymenoptera				
Formica exsecoides	A	63–82	no effect of season, temp. or hibernation	Dreyer (1938)
Perga dorsalis	L	77.6	saw-flies	Seymour (1974)
Diptera				
Phlebotomus papatasi	A	65–70	fat 5%, active, sand-flies	Theodore (1936)
Phlebotomus papatasi	A	52–56	fat 15%, hibernating	Theodore (1936)
Glossina morsitans	A	75–77	fat-free, newly emerged	Jackson (1937)
Glossina morsitans	A	63	(65% of fat-free) before feeding	Brady (1975)
Glossina morsitans	A	65	(72% of fat-free) 2 days after feeding	Brady (1975)
Glossina morsitans	A	66	(69% of fat-free) 5 days after feeding	Brady (1975)
Opifex fuscus	L	85.2	in artificial sea-water	Nicolson and Leader (1974)
Siphonaptera				
Xenopsylla brasiliensis	Pre-P	70	before hydration	Edney (1947)
Xenopsylla brasiliensis	Pre-P	75	after hydration	Edney (1947)
Xenopsylla cheopis	L	76	70% R.H.	Knülle (1967)
Xenopsylla cheopis	L	78	93% R.H.	Knülle (1967)
Crustacea				
Isopoda				
Oniscus asellus	A	66		Bursell (1955)
Porcellio scaber	A	65–76	after 24 h in R.H. 79–100%	Den Boer (1961)
Arachnida				
Araneida				
Oxyopes salticus	A	77	five different species	Vollmer and
Achaearanea tepidariorum	A	85	within this range. No correlation with dryness of habitat	MacMahon (1974)
Dugesiella hentzi	A	73	fat content ≃ 10.3% body wt.	Stewart and Martin (1970)
Acarina				
Acarus siro	A	66	flour mites, R.H. 75%	Solomon (1966)
Acarus siro	A	70	flour mites, R.H. 95%	Solomon (1966)
Bryobia praetiosa	A	66–68	fasted in 0–93% R.H. for 2 days	Winston and Nelson (1965)
Echinolaelaps echidnina	A	75	spiny rat mites, ♀♀	Wharton and Devine (1968)
Dermatophagoides farinae	L	59	house dust mites	Arlian and Wharton (1974)
Dermacentor variabilis	L	57	92% R.H.	Knülle and Devine (1972)

Table 1 (continued)

Taxon	Stage	Percent water content	Remarks	Reference
Scorpionidea				
Hadrurus arizonensis	A	66.7		Hadley (unpub.)
Centruroides sculpturatus	A (males)	64.2		Hadley (unpub.)
Centruroides sculpturatus	A (gravid females)	71.4		Hadley (unpub.)
Vejovis confusus	A	69.3		Hadley (unpub.)
Paruroctonus mesaensis	A	69.0		Hadley (unpub.)
Paruroctonus mesaensis	A	70.5		Yokota (unpub.)
Paruroctonus mesaensis	immature	73.0		Yokota (unpub.)
Myriapoda				
Diplopoda				
Orthoporus	—	54	varies seasonally	Crawford (unpub.)

As the water content of arthropod cuticle is relatively low [in the range 13–70% (Richards, 1951)], those animals in which the integument makes up a large proportion of the body weight, such as adult beetles, will have lower water contents than others. Thus, the saturniid larva *Telea polyphemus* has 92% water while the adult beetle, *Calandra granaria* has a mere 46% (Buxton, 1932a). Developmental stages of a species may have different water contents, as do the eggs, larvae, pupae and adults of the silkworm *Bombyx mori* (where there is adult sexual dimorphism) (Buxton, 1932a), and the beetle *Limonius* (Jones, 1951); and age within an instar may also have an effect (Marcuzzi, 1960). A well-documented example of this is provided by Bullock and Smith (1971) who found that in several groups of caterpillars including noctuids and tortricids, proportional dry weight increases exponentially with size ($y = a \cdot x^{1.1094}$, where y is dry weight, x is wet weight, and a is a constant that varies with species).

Differences are also associated with physiological condition. For example, the amount of fat present will profoundly affect the overall water content expressed as a percentage of body weight, although it is without significance so far as the immediate water status of the animal is concerned. For this reason Jackson (1937), Jack (1939) and Bursell (1959a), in their work on tsetse flies, preferred fat-free dry weight as a basis for expressing water content. Fat content may account for at least some of the differences between stages. In *Tenebrio molitor* for example, larvae with about 58% water have between 16.9 and 13.7% fat, while pupae, with 64% water, have only 7.5% fat and adults with about 61% water have 6.3 to 7.2% fat (Marcuzzi, 1960). Expressed on a fat-free basis, the larvae would have between 67 and 70%, the pupae 69% and the adults 65.1 to 66.5% water—a somewhat narrower range (Table 2). Hagvar and Ostbye (1974) found that in several carabid and curculionid beetles the percentage

Table 2. Water and fat content of *Tenebrio molitor*. (Based on data from Hochrainer, 1942;[1] Marcuzzi, 1960[2]) Numbers are percentages of total wet weight, except for the last line

	Larvae		Pupae	Adult	
	(1)	(2)	(2)	(1)	(2)
Total dry	41.9	41.7	36.2	39.0	38.3
Fat	13.7	16.9	7.5	6.3	7.2
Fat-free dry	28.2	24.8	28.7	32.7	31.1
Water	58.1	58.3	63.8	61.0	61.7
Total	100	100	100	100	100
Fat-free total	86.3	83.1	92.5	93.7	92.8
Water as % of fat-free total	67.3	70.2	69.0	65.1	66.5

of water present varies inversely and quite strongly with total weight: e. g. in the carabid *Calanthus*, a weight range from 3 to 7 mg corresponds with a range in water content from 80 to 60%, and this may well be due, at least in part, to the tendency to lay down more fat in larger individuals.

Differences in fat content also largely account for apparent differences in water content between larvae and adults in *Pimelia angulata*, but in others, fat accounts for only part of the difference, as Marcuzzi (1960) found in adult *Tenebrio molitor*. Similarly, in *Locusta*, water content (percent of live weight) falls from nearly 75% after moulting to about 62% by the 18th day, and subsequently rises slightly; while the fat content, low at first (about 5%) rises to 10% at 15 days. These are real changes, not artefacts resulting from their expression as percentages of total wet weight (Loveridge, 1973).

Ambient humidity may affect water content in several ways: by absorption of water hygroscopically into the cuticle, by true absorption of water vapour, or perhaps indirectly by encouraging greater activity and thus greater respiratory water loss. *T. molitor* adults have about 63% and 67% water content in a_v's of 0.1 and 0.95 respectively (Marcuzzi, 1960). Ambient humidity has little effect on water content in the mite *Bryobia* (Winston and Nelson, 1965), but in other species water vapour is absorbed to an equilibrium that depends on a_v, as in *Acarus siro* (Solomon, 1966) and *Xenopsylla cheopis* larvae (Knülle, 1967). In the latter, water contents are 73%, 76% and 78% in a_v's of 0.43, 0.70 and 0.93 respectively (see Chap. 9).

Temperature may also affect water content (Marcuzzi, 1960), and seasonal effects are apparent in the sand-fly *Phlebotomus* (Theodor, 1936) and in the grasshopper *Melanoplus* (Bodine, 1921). But neither temperature nor season appears to affect water content in the ant *Formica exsecoides* (Dreyer, 1935, 1938).

In several arthropods, e. g. the grasshopper *Chortophaga*, but not in all, e. g. the bug *Leptocoris* (Hodson, 1937), the water content of hibernating individuals is low. But the widespread belief that low water content confers a resistance to freezing has been strongly questioned by Salt (1961).

Diet affects water content in several ways. Unless precautions are taken, apparent water content may be affected by food in the gut (Bullock and Smith, 1971), and also by the nature of the food. According to Robinson (1928) plant feeders generally have a higher water content (82–90%) than others such as grain feeders (50–60%). However, the content may not be genetically determined, for Mellanby (1958) found that *T. molitor* increases in water content from 58% to 75% when given free water to drink in addition to dry food (see Chap. 7).

While fasting, water content remains approximately constant in *Tribolium* (Dreyer, 1935) as it does in *Tenebrio* (Buxton, 1932a) and *Phyllophaga* (Sweetman, 1931). However, fasting, like hibernation, leads to lower water content in *Melanoplus* (Bodine, 1921). Further discussion and examples will be found in Buxton (1932a), Hochrainer (1942), Ludwig and Anderson (1942), Mellanby and French (1958), Bullock and Smith (1971), and Bursell (1974b). Rapoport and Tschapek (1967) deal particularly with soil fauna and Schmidt (1955) with Coleoptera.

Clearly water content by itself does not throw much light on either the immediate water status or the water balance mechanisms of an arthropod. However, when seen in relation to questions of storage, water reserves, and the limits to tolerable depletion, water content does become more biologically significant.

B. Water Reserves

We shall consider first lower limits of total water content, and then the extent and distribution of water storage. For references to early work see Johnson (1942) and Wigglesworth (1972).

Once again there appears to be considerable variability among species as regards lower limits. Johnson rightly points out that if the relationship between rate of loss and vapour pressure deficit (ΔP) is linear and if survival is limited only by water loss, the curve of longevity on ΔP should be hyperbolic. He showed that in some cases (e. g. Ludwig, 1936; Ludwig and Landsman, 1937, for *Popillia*) this is approximated, and similar results have been found in other insects. But as Bursell (1974b) points out, such a relationship is not usually to be expected, for water loss is not usually proportional to ΔP (spiracular control, variable permeability and other things intervene—see Chaps. 3 and 4), and several factors may affect the onset of death by desiccation.

In some insects, including beetle larvae, whose normal fat free water content is about 75%, the lower limit is about 60% (Buxton, 1930, 1932b; Ludwig, 1936). Bursell (1959a) found that a normal water content of 76% may be reduced to between 61 and 64% (depending on species) before death occurs in tsetse flies *(Glossina)*. This represents a loss of about half the total water in optimally hydrated flies (Fig. 9).

When the water content of different stages differs, as it does in *Popillia japonica*, the lethal limit for weight loss (largely water) also varies widely, but the limit for water content is more consistent. In *Popillia*, tolerable limits for final weight as percent of original weight vary from 45 to 69% (early larvae to pupae) but water content is thereby reduced to between 59% (pre-pupae) and 66% (pupae) (Ludwig, 1936).

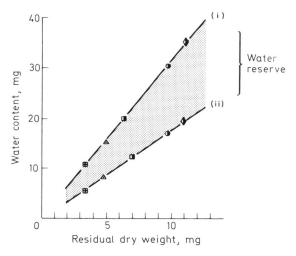

Fig. 9. The relationship between size and water reserves in different species of tsetse-flies. *Curve (i)* shows the water content of flies that have recently emerged from pupae maintained at high R.H.; *curve (ii)* shows the water content of flies that have been dehydrated to a point where they are no longer capable of righting themselves. The difference between the two curves shows the extent of the water reserves, which is about 45% of the water originally present. ⊞: *Glossina austeni*; △: *Glossina palpalis*; ◼: *Glossina pallidipes*; ◑: *Glossina brevipalpis*; ◆: *Glossina fuscipleuris*. (From Bursell, 1970)

We shall now consider a few of the available data in order to see the extent of variability in tolerable lower limits, and to see whether any pattern emerges (e.g. do arthropods from xeric areas tolerate lower water contents?).

Arenivaga investigata is a desert cockroach, whose normal water content is 67% which may be reduced to 60% in a week without lethal effects. Allowing for oxidation water this represents loss by transpiration of about 30% of the water originally present (Edney, 1966). Similarly, in the beetle *Carabus*, where different species have water contents from 75 to 80.5% of total weight, the tolerable water loss is 17–22.5% of original water present, corresponding with a total weight loss of 22–29% (Schmidt, 1955).

Abushama (1970) found that a desert grasshopper *Poecilocerus*, tolerates a water loss of 35% of its total weight, while *Anacridium*, a more mesic species dies after losing only 26%. In another grasshopper, *Chortophaga*, Ludwig (1937, 1950) found that after dehydration or fasting for a week or until death, the water content remains substantially the same at about 70%, even though the insects lose 22% or 40% respectively of their original water, for dry weight is lost at the same proportional rate.

23

Even greater water losses have been recorded in *Thermobia* (Sahrhage, 1953), *Tineola*, and certain psocids and mites. *Dermatophagoides* can lose 52% of its original water, thus reducing the water content to 59% (Arlian and Wharton, 1974) and *Liposcelis* may lose as much as 67% of original water, to give a content of 41% (Knülle and Spadafora, 1969) (Table 3). Both these arthropods can live in rather dry surroundings without any free water.

Table 3. Tolerance of water loss in certain arthropods

A. *Liposcelis* (data from Knülle and Spadafora, 1969)

	At equilibrium in 75% R.H.		After 11 days in 33% R.H.				Near death			
	Wt. in μg	% of total weight	Wt. in μg	% of total weight	Loss as % of original component	total	Wt. in μg	% of total weight	Loss as % of original component	total
Water	53.5	66	28.1	53	47	31	17.5	41	67	44
Dry matter	27.2	34	25.2	47	7	2	25.2	59	7	2
Total	80.7	100	53.3	100	34	34	42.7	100	47	47

B. *Dermatophagoides* (data from Arlian and Wharton, 1974)

	At equilibrium in 75% R.H.		After 14 h over P_2O_5			
	Wt. in μg	% of total weight	Wt. in μg	% of total weight	Loss as % of original component	total
Water	9.71	75	4.68	59	52	39
Dry matter	3.28	25	3.28	41	0	0
Total	12.99	100	7.96	100	39	39

Orthoporus toxicolens, a millipede from mesic habitats, loses water rather slowly, and tolerates a total weight loss of 40% before death in 17 days. Most of this loss is water, judged by the low metabolic rate: 65.6 μl O_2 g^{-1}h^{-1}. *Pachydesmus crassicutis*, from hygric regions, tolerates 65% weight loss (again mostly water), but survives for only 1 day owing to rapid dehydration (Stewart and Woodring, 1973).

Vollmer and MacMahon (1974) found that the water content of spiders varies considerably with species: from about 77% in *Oxyopes salticus* to 85% in *Achaearanea tepidariorum*, among those they measured. The extent of tolerable dehydration also varies, but not necessarily in parallel with water content or with dryness of habitat. Earlier work in this area is also referred to by Vollmer and MacMahon (1974).

Four species of geophilomorph centipedes, one lithobiomorph and one scolopendromorph, all survive until they have lost about 40% of their initial body weight, although they take different times to do this (Lewis, 1963). The desert scorpion *Buthotus minax*, has been shown to tolerate a loss of 40% of its original wet weight over six weeks (Cloudsley-Thompson, 1962) and other desert species survive a loss of 30% (Hadley, 1974).

At present there is no evidence to suggest a generally greater tolerance of low water contents in arthropods from xeric areas. Instead they are protected against rapid water loss, as will appear below.

A few extreme cases are known where almost complete dehydration leads only to cryptobiosis, a virtually non-living state, but one from which recovery is possible. Cryptobiosis is defined by Hinton (1960a, 1971) as a state where metabolic activity comes reversibly to a standstill, and Crowe (1971) uses the term "anhydrobiosis" to refer to a particular form of cryptobiosis associated with dehydration. Anhydrobiosis is well known in tardigrades (Crowe, 1971; Crowe and Cooper, 1971), and nematodes (Cooper and Van Gundy, 1971), and it has been observed in the larva of a chironomid midge *Polypedilum vanderplancki* (Hinton, 1960). This remarkable insect, which lives in temporary rock pools in Africa can withstand repeated dehydration to 8% water content, and when dehydrated to less than 3%, it can withstand 102°C for one minute or liquid air at −190°C for many hours. *Polypedilum dewulfi* is another species that shows anhydrobiosis (Miller, 1970), and the stratiomyid larva *Cyrtopus*, which occurs in similar habitats, tolerates drying of its environment by dropping its metabolic rate to about 1/100 of normal (aestivation) though it does not become truly anhydrobiotic. Other chironomid larvae are known to aestivate, but not to become anhydrobiotic. *Paraborniella tonnoiri*, which lives in Australian rock-pools provides a good example (Jones, 1975). This larva can survive six months drought by aestivating in dry mud cocoons, and while in this state it loses very little water, although water loss is still very high if the larva is removed from its cocoon. It also produces no urine, and tolerates temperatures up to 58°C, compared with an upper limit of 45°C in the active state. (For blood volume and osmotic effects see Chap. 6.C.V.2.)

Anhydrobiosis as a means of avoiding the otherwise disastrous effects of hot dry conditions has been reported by Poinsot-Balaguer (1976) for the collembolans *Folsomides variabilis* and *Brachystomella parvula*. These animals survive either as eggs or as adults, while other collembolans that apparently do not go into anhydrobiosis, survive only as eggs. For reviews of the biological significance of cryptobiosis and anhydrobiosis see Hinton (1971) and Crowe (1971) respectively.

Dehydration beyond the limits appropriate to each species results in death. However this is not an all or none process, and sublethal dehydration is known to effect various biological processes adversely, particularly during development. Examples of such effects include depression of oxygen uptake and rate of development in *Melanoplus* (Bodine, 1933), *Sitona* and *Lucilia* (Evans, 1934) and others (refs in Wigglesworth, 1972). On the other hand, large changes in water content may occur before any effect is observed, as Buck (1965) found for chironomid larvae whose rate of metabolism remains unaffected by a loss of 50% of the insects' normal water content.

The immediate cause of death by dehydration is not apparent; however, in order to understand the situation more clearly we need to know at least where the water that is lost comes from. Water loss from a specific water store would have no immediate harmful effect, while loss of the same amount from living tissues could be dangerous. Thus questions of water storage within the body are raised.

C. Location of Water Reserves

The most obvious compartment for water storage in arthropods would seem to be the blood, as Mellanby (1939) first suggested; and there can be little doubt that this is so. In order to function most efficiently in this way, it should be possible to move water alone from the tissues to the blood, and in the opposite direction, according to the needs of the tissues. This implies a capacity to tolerate large changes in both volume and osmotic concentration of the blood. However, the blood could also serve as a water store, to replace water lost from the tissues, provided that the excess electrolytes left as a result of transpiration are removed by excretion or inactivated in some way.

Several studies have in fact revealed considerable changes in blood volume in cockroaches as a result of radiation and other factors (Munson and Yeager, 1949; Yeager and Munson, 1950; Wheeler, 1963; Wharton et al., 1965 a, b). Lee (1961) working with the locust Schistocerca observed a fall in blood volume without a proportional fall in tissue water when the insects were fed dry grass, and Edney (1968 b) found a decrease in blood volume in Arenivaga and Periplaneta as a result of dehydration. In these insects tissue water decreases nearly proportionately, at least in the initial stages. As dehydration progresses, however, blood volume decreases to vanishing point. In Carausius, Nicolson et al. (1974) have also found that dehydration may reduce the total water content by 50%, and that the loss is sustained almost entirely by the blood, whose volume decreases to one seventh of normal (see Chap. 6.C.V.2.).

A short term reduction in blood volume of 13% following a full meal of fresh grass was observed by Bernays and Chapman (1974a) in Locusta migratoria, and this seems to represent movement of water alone, for the blood osmotic pressure increases appropriately from 350 to 392 mosmol liter^{-1}. Other similar fairly small changes in blood O.P. have been reported by Djajakusumah and Miles (1966) after dehydration in the grasshopper Chortoicetes, and by Browne and Dudzinski (1968) in blowflies.

The conclusion is that water seems to be moved between blood and tissues, and in this sense blood acts as a water store. However, there are exceptions. For example, Verrett and Mills (1975) found that in Periplaneta there is a cyclic change in the total body water and in blood volume associated with the six-day oothecal development cycle, but they believe that the true water storage regions are the integument and associated tissues, the fat body, and the gut. In their interpretation, changes in blood volume simply reflect an intermediate stage in the passage of water from the storage area to the oocytes. The extent to which osmoregulation also occurs in association with water storage will be considered more fully in Chapter 6.

Other, perhaps simpler, water stores have been reported in various arthropods. The first example to be discovered was that of the cotton stainer *Dysdercus fasciatus*, by Berridge (1965a). In larvae of this insect the gut is discontinuous and the contents of the rectum come only from the Malpighian tubules. In the fifth (last) larval instar, which lasts for eight days, *Dysdercus* feeds for five days during which time it eliminates a large quantity of urine and the blood volume falls. For the remaining three days, dilute (i.e. hypo-osmotic) urine is stored in the rectum and may be called upon to replace that lost by cuticular transpiration and to restore the blood volume (Fig. 10). The rectal wall cannot

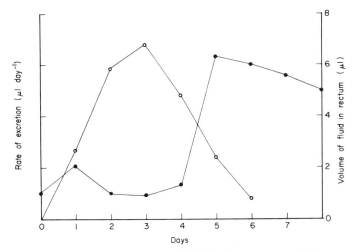

Fig. 10. Rate of excretion *(open circles)* and accumulation of fluid in the rectum *(solid circles)* by fifth instar *Dysdercus*. (From Maddrell, 1971; data from Berridge, 1965b)

absorb water against an osmotic gradient, but some 5.8 μl of iso-osmotic fluid may be absorbed before the rectal fluid becomes hyper-osmotic, and this represents more than 40% of the normal blood volume. Presumably the strongly hypo-osmotic urine is produced by reabsorption of potassium and perhaps other ions in the proximal region of the tubules after the distal region has secreted an iso-osmotic fluid, as happens in *Rhodnius* (Maddrell and Phillips, 1975a).

In the termite *Hodotermes mossambica*, Hewitt *et al.* (1971) found that newly emerged alates, when confined in pairs, increase their water content from 56% to 76% (males) and from 49% to 67% (females) by drinking, so that a colony-forming pair starts operations in good water status. The water is stored in large salivary gland reservoirs. After drinking, sodium and potassium concentrations

Table 4. Ionic concentrations, in meq $100 \, g^{-1}$, in the salivary sac fluid of *Hodotermes mossambica*. (Data from Hewitt *et al.*, 1971)

	Blood		Sac fluid	
	Males	Females	Males	Females
Sodium	12.00	14.85	1.75	1.69
Potassium	1.45	2.53	0.40	0.19

are much higher in the blood than in the salivary sacs, as shown in Table 4. Watson *et al.* (1971) went on to show, by the use of radioactive colloidal gold, that water enters the sacs not directly but via the gut, the precise path between gut and sacs being so far undetermined.

Larvae of the blowfly *Lucilia* also drink copiously and store water in their salivary glands which may expand to nearly one-fourth of the total body weight. The water then serves as a reservoir during pupal diapause (Mellanby, 1938).

In the cockroach *Leucophaea maderae*, salivary reservoirs also function as water stores (Laird *et al.*, 1972), maintaining the blood O.P. fairly constant during dehydration. The sac fluid, which is cell-free, represents 5–10% of total body weight in hydrated insects and has an O.P. of about 34 mosmol liter^{-1}. Again the route of movement of water into and out of the sacs has not been established. The salivary sacs of *Periplaneta* contain liquid similar to that of *Leucophaea* (Wall, 1970), and in both cases, storage involves movement of water against an overall osmotic gradient from blood to salivary sac lumen.

In the isopod *Porcellio scaber*, however, it seems that the gut itself may act as a water store (Lindqvist and Fitzgerald, 1973). The evidence is that in normal animals osmotic concentration in the blood and in fluid expressed through the mouth is about 313 and 244 mM respectively, while after 30 min dehydration blood concentration remains virtually unchanged while the gut fluid rises to about 390 mM. Again the path of movement of the water is uncertain.

It seems likely that water storage may be rather widespread in arthropods, and further exploration should be fruitful. Elimination and storage are of course opposite processes, but the physiological mechanisms involved may be similar in so far as water and ion transport are concerned. In this connection it is interesting that the salivary glands of certain ticks, whose walls are microstructurally suitable for ion transport, have been shown to function as excretory organs. Those so far investigated include *Boophilus* (Tatchell, 1967, 1969), *Dermacentor* (Kaufman and Phillips, 1973a, b, c), and *Amblyomma americanum* (Sauer and Hair, 1972) and the process is discussed further in Chapter 6.C.III.

D. Resistance to Desiccation

The concept of "resistance to desiccation" has been used in two ways. In the first it means tolerance of large water losses and low water contents, and in the second it refers to the restriction of water loss in one way or another. In desert arthropods both of these abilities would seem to be useful, but although such animals are demonstrably well adapted to conserve water by restricting loss, there is little evidence of tolerance of low water concentrations in their living tissues (Edney, 1974), and it seems that in general, prevention of loss from the living tissues rather than tolerance of its effects is characteristic of mesic and xeric land arthropods. In one way or another (mostly by the use of water stores), these animals can withstand water losses equal to 30% or so of their weight. Man can tolerate only 10–12% loss if he is to recover from dehydration (Schmidt-Nielsen, 1964). Instances of anhydrobiosis referred to in Chapter 2.B are obvious exceptions to the above generalization.

Chapter 3

Water Loss—Cuticular

A. Introduction

There are several avenues for water loss from arthropods: the general cuticle, the respiratory surfaces via the spiracles (if these are present), the excretory system (of which the hind gut is usually but not always a part) and the mouth, through which material from the salivary glands and part of the alimentary canal may be discharged. We shall consider the extent, and the possibilities for control, of net outward flow of water through each of these channels. The present chapter is concerned with water loss through the cuticle, and in this regard considerations of size are of the greatest importance. This has been referred to before, and is illustrated in another context below (Chap. 5.A.).

B. Transpiration and Integumental Structure

In much of this chapter we shall refer to the "integument" or to the "cuticle" as if it were homogeneous in structure over the whole animal. For certain purposes this is a justifiable simplification, but it is as well to bear in mind that the structure, and probably the permeability, of different parts of the cuticle (arthrodial membranes, sensory structures, gland ducts, muscle insertions, Slifer's patches, for example) vary considerably; and that the surface through which transpiration occurs is not only very difficult to measure but may vary from time to time in the same insect as its shape (e. g. the position of the appendages) changes.

Our present knowledge of cuticular structure and function in relation to water balance has been derived from several lines of enquiry, including measurement of the effects of environmental factors (particularly temperature and humidity) on transpiration, measurement of the effects of moulting, direct microscopical observation including much recent work on fine structure, and biochemical analysis, particularly of the cuticular lipids.

I. Cuticular Water Loss

Information about cuticular water loss may be expressed in various ways, each of which is appropriate to certain enquiries:

1. As a proportion of the total weight of the individual animal concerned. This is particularly useful if the original water content percent of total weight and the limits of tolerance to depletion are known, and if the biological effects of water loss are in question. But it tells little about the relative permeabilities of integuments (except under different conditions in the same animal).

2. As a rate of transpiration per unit surface area. This is satisfactory for comparative purposes provided that all measurements are made at the same temperature, humidity and air velocity, but again it tells little about the specific properties of the integument, or about probable biological effects, unless we know the size of the animal, and the value for k in the surface–weight relationship $S = kW^{2/3}$, from which we can derive weight-specific water loss.

3. As a rate of transpiration per unit area and vapour pressure difference. This, as in 2, permits the calculation of weight-specific loss only if the surface/weight coefficient is known; but, if transpiration obeys Fick's law we should be able to calculate a permeability coefficient for each kind of integument—a very useful comparative value. The question is whether transpiration does obey Fick's law, and to this we shall now turn.

II. Permeability of the Integument

The factors concerned in transpiration across the integument of arthropods are in fact extremely complex.

As mentioned in Chapter 1.D.III., Fick's law, [Eq. (8)],

$$J = \frac{D\Delta C}{\Delta X}$$

is applicable to the diffusion of a substance only through a homogeneous membrane where the media on each side of the membrane are identical. If this were true for arthropod integuments, Eq. (8) could be rewritten as

$$J = \frac{C_i - C_o}{R_c + R_a} \tag{12}$$

where J is the flux rate of water in mg cm^{-2} s^{-1}; C_i and C_o are concentrations of water vapour inside the cuticle and outside the boundary layer of air respectively, in mg cm^{-3} (this involves the simplifying assumption that water passes through the cuticle as water vapour); R_c and R_a are resistances of the cuticle and the air boundary layer respectively, in s cm^{-1}.

Most of the recent work on integumental water loss has been reported in terms of mg cm^{-2} h^{-1} mm Hg^{-1}, using vapour pressure difference (ΔP) in mm Hg as a measure of the force tending to move water across the cuticle, and this is a useful measure of permeability. If a value for the resistance of the cuticle or its reciprocal, permeability in comparable units, is required, then

Eq. (12) is appropriate, where concentrations rather than partial pressures are used, and R is in $s\,cm^{-1}$.

Unfortunately the real situation is not so simple, as Beament (1961) has pointed out. In the first place, the cuticle itself (apart from the epidermis) is multilayered, and it is entirely possible that the permeability of one or more of the layers may change according to its water content [as do many organic materials (King, 1944)]. If so, the rate of diffusion of water through the whole cuticle would depend partly on the absolute water content of one or more of its components; in other words D is a function of C in Eq. (8).

Secondly, if R_c is very low, J will be greatly affected by R_a, and air movement becomes increasingly important. Different R_a's in such cases may induce differences in a_w in the cuticle, and hence changes in R_c. Measurement of transpiration when R_a is high compared with R_c will thus be affected greatly by changes in R_a and give little information about R_c. However, if R_c is high compared with R_a, i.e. the integument is (loosely) very impermeable, and if much of this resistance is conferred by epicuticular lipids and unchanged by water absorption, then measurement of a permeability coefficient for the integument is more reasonable.

Nobel (1974) gives the thickness and resistance to water vapour diffusion of unstirred boundary layers of air adjacent to leaves of different sizes, in air moving at various speeds. For a leaf 5 cm long in air at 20 °C moving at $10\,cm\,s^{-1}$, the unstirred boundary layer is about 0.25 cm thick and $R_a = 1\,s\,cm^{-1}$. For an unstirred layer 0.025 cm thick, $R_a = 0.1\,s\,cm^{-1}$. Since R_{total} for most arthropods is very much greater than this (see below) we need not be very concerned in field situations with errors caused by the air resistance component of the total resistance. (For plant cuticles, R_c generally varies from about 20–200 s cm^{-1}.) In laboratory situations, in nearly still air, R_a may be more important.

Thirdly, it is possible that epidermal cell membranes contribute significantly to the total resistance. Fourthly, it has been suggested that the cuticle is asymmetric as far as water permeability is concerned; and lastly, most experiments so far have measured net outward movement of water, but recent work using labelled water molecules has shown that gross movement in both directions occurs and may be greater than net flow (Arlian and Wharton, 1974; and other papers discussed in Chap. 9).

The last three points will be discussed below; but, it is already clear that an arthropod cuticle should not be expected to behave like a simple homogeneous membrane with a constant permeability coefficient. For further discussion of the principles involved and of the evidence till then, see Beament (1961), and for a recent treatment of the problem mainly in relation to plants, see Nobel (1974).

III. Temperature and Transpiration

Interest in the effects of temperature on transpiration was originally stimulated by Gunn (1933), who found that water loss from living *Blatta orientalis* increases with temperature more rapidly above 30°C than below, and ascribed this to the onset of active ventilatory movements which set in at that temperature. In the light of later work showing a large effect of ventilation on water loss (Church, 1960; Weis-Fogh, 1967; Loveridge, 1968a), Gunn's interpretation seems reasonable. However, such ventilation does not account for all of the increase in water loss, because Ramsay (1935), working with *Periplaneta*, found a similar increase in transpiration rate at about 30°C even in dead insects with sealed spiracles, and even when increasing ΔP at higher temperatures is taken into account. The precise interpretation of these results may be debatable (Edney, 1957), but unquestionably Gunn's and Ramsay's work drew attention to the cuticle as an avenue for water loss in insects and opened up a field of investigation which has been active ever since.

Another important step occurred when Wigglesworth (1945) found that transpiration varies considerably between species, usually in a manner appropriate to their ecological situation; and that, for each species, while temperature has little effect below a "critical" or "transition" point, above this level transpiration increases rapidly with rising temperature. Slight abrasion of the surface of the epicuticle, demonstrable only histochemically, leads to rapid transpiration even at low temperatures and abolishes the transition effect.

At the same time Beament (1945) confirmed and amplified these results in vitro. He extracted chloroform soluble materials from the cast skins (largely epicuticles) of the same species as those used by Wigglesworth, and deposited the lipids on inert membranes of tanned gelatine or cleaned butterfly wings. Such membranes (particularly the latter) then behave in essentially the same way as whole dead insects: they show comparably low transpiration rates, transition phenomena at similar temperatures, and are rendered much more permeable by superficial abrasion. In 1947 Lees extended these findings to ticks, and showed that their cuticles are essentially similar in structure and properties to those of insects. A selection of these early temperature/transpiration curves is shown in Figure 11.

The main impact of these now classical papers was to confirm and explain earlier indications by Wigglesworth (1933), Ramsay (1935), Bergman (1938), Pryor (1940), Alexander *et al.* (1944) and others, that epicuticular lipids are crucially important in determining the properties of arthropod cuticles in relation to water transport. At this time Wigglesworth (1945) and Beament (1945) proposed that permeability may be controlled not so much by the whole epicuticle as by a monolayer of oriented lipid molecules organised at the interface between a superficial lipid layer and the underlying epicuticle, with the polar (hydrophilic) ends of the lipid molecules inwards and the non-polar (hydrophobic) hydrocarbon chains outwards. Transpiration, they proposed, increases at a particular temperature, when a transition occurs in the oriented monolayer leading to a less ordered state and increased permeability. Beament (1964) subsequently proposed that at lower temperatures, the lipid molecules are oriented at an angle of 65° with respect to the hydrophilic substrate surface, rather than normal to it, so that the repeating groups of atoms along the carbon chain interdigitate between chains which are in the "tilt-packed" position, thus forming an almost crystalline structure and blocking the passage of water molecules effectively.

33

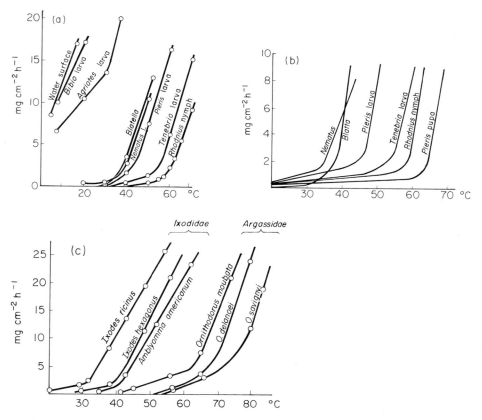

Fig. 11a–c. The relationship between air temperature and the transpiration of water from arthropods—a selection of some of the early data. (a) dead insects (after Wigglesworth, 1945); (b) artificial membranes prepared by depositing waxes extracted from insect exuviae on cleaned butterfly wings (after Beament, 1945); (c) dead ticks (after Lees, 1947). The point at which transpiration begins to increase rapidly with rising temperature is referred to as the "critical" or (preferably) "transition" temperature. This is clearly more abrupt in some cases than in others. The effect of increasing vapour pressure deficit (ΔP) at higher temperatures is not taken into account, so that some of the increase in transpiration rate must be ascribed to this rather than to a change in cuticle permeability

Above the transition temperature the free (CH_3) end of the molecule is supposed to vibrate around the fixed polar (OH) end, thus opening intermolecular channels to water molecules.

An alternative to this model (Davis, 1974a) is referred to below (Chap. 3.B.VII.), but whatever the correct molecular interpretation may be, the evidence from these and other papers points clearly to a superficial layer in the epicuticle as being an important barrier to transpiration. This evidence may be summarised as follows.

1. Light surface abrasion leads to a large increase in transpiration. (Wigglesworth, 1945; Beament, 1945; Lees, 1947).

2. Adsorption by inert dusts of the soft, grease-like surface lipids of cockroaches leads to increased transpiration (Wigglesworth, 1933; Ebeling, 1961; and other references in Ebeling, 1974).

3. The acquisition of impermeability by the new cuticle after ecdysis coincides with the deposition (or incorporation) of surface lipids (Wigglesworth, 1947, 1948b).

4. Lipids extracted from the epicuticle have melting zones close to their respective transition temperatures (Beament, 1945).

5. Such lipids, when deposited on inert membranes, produce impermeability and transition phenomena (Beament, 1945).

Most of the early measurements referred to above were made in dry air, and reported in terms of rates of water loss per unit area. They did not, therefore, take account of the increasing vapour pressure gradients occasioned by higher temperatures—increases which would have been responsible for some of the increased transpiration. Since then a bewildering array of data has been obtained by many workers on the relationship between temperature, transpiration and cuticle structure. Much of this work tends to confirm the existence of transition phenomena at least in some circumstances and to varying extents—although transition had at one time been questioned (Holdgate and Seal, 1956; Mead-Briggs, 1956; Edney, 1957). (See Beament, 1961 for a survey of relevant work up to then.) Here we shall attempt to consider some of the more significant themes that have emerged, and to suggest reasons for discrepancies where they occur.

Very probably many of the apparent discrepancies in results that have appeared are due to the use of different techniques and of animals in different conditions. This is all the more likely because as we now know, the integument itself is complex and cannot be treated as a simple homogeneous membrane.

Some of the results do not require a sharp transition temperature for their interpretation (Holdgate and Seal, 1956; Mead-Briggs, 1956; Delye, 1969; Edney and McFarlane, 1974; Gilby, pers. comm.), but many others seem to show

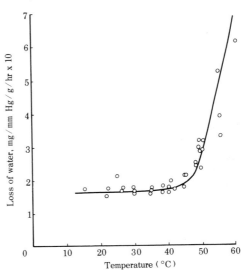

Fig. 12. The effects of temperature on the permeability of the cuticle of *Locusta*. Air, not cuticle, temperatures are shown. Most of the effect shown in this curve is due to changes in cuticle permeability, for changes in ΔP have been allowed for. (From Loveridge, 1968b)

35

that even after allowing for the effect of vapour pressure gradients, transpiration increases rather slowly with temperature in the lower range, and much faster above a transition point or region.

Beament's (1958, 1959) measurements in carefully controlled conditions certainly show this in several insects and ticks (nymphs of the locust *Schistocerca* apparently being exceptional), and further examples reported by other workers include *Melanoplus* and other grasshoppers (Chefurka and Pepper, 1955) *Gastrimargus* (Koidsumi, 1934), *Locusta* (Loveridge, 1968 b) (Fig. 12), tenebrionid beetles (Ahearn, 1970a), and spiders (Humphreys, 1975).

One source of discrepancy may be the curiously vulnerable nature of the epicuticular lipid barrier. When measuring transpiration in adult *Schistocerca*, Beament (1959) found that the effect of such handling of the insects as is necessary to block the spiracles, leads to higher apparent rates of water loss in blocked insects than in those with free spiracles, and to a relatively smooth temperature/transpiration curve. This he attributed to mechanical disruption of the surface lipids. Other experiments, however, including the early ones of Wigglesworth (1945), Lees (1947), Beament (1945), and the more recent ones of Loveridge (1968 b), do show transition effects even though the insects were handled before experiment. In nature, abrasion of the cuticle, if this occurs, can presumably be repaired by the secretion of new wax, for this is known to occur in vitro. (For recent information on this repair, see Chap. 3.B.VI.)

In arthropods with rapid transpiration rates, the surface of the cuticle may be at a lower temperature than that of the ambient air (Edney, 1951), and Beament (1958) pointed out that this effect might lead to false impressions about the temperature/transpiration relationship. By measuring cuticle temperature and rate of transpiration simultaneously in *Periplaneta* he found that if cuticle temperature rather than ambient temperature is used as the independent variable a sharp increase in the rate of transpiration appears at the transition temperature as transpiration increases without a parallel rise in cuticle temperature.

Edney and McFarlane (1974) made continuous recordings of weight changes in dead *Periplaneta* and the desert cockroach *Arenivaga*, in conditions where the cuticle temperature was also known, but they observed continuous rises in both cuticle temperature and transpiration in continuously rising ambient temperature, and were thus unable to confirm a step-like break in the temperature/transpiration curve (Fig. 13). They recorded small irregular perturbations in weight loss which they attributed to changes in the geometry of the insect as the head moves on the body, legs fold and unfold and so forth, and to opening of the spiracular valves; but the effects were comparatively minor. Such movements were observed in dead insects and are perhaps caused by differential contraction of arthrodial membranes and of underlying muscles as these structures dry out. Gilby (pers. comm.) also was unable to find a break in the *Periplaneta* curve in very similar conditions.

The significance of the "evaporative cooling" effect thus remains uncertain. It will seriously affect the temperature/transpiration relationship only when the transition point is sharp and permeability is very high (an unlikely combination), but its effect should be measured when possible, or at least allowed for.

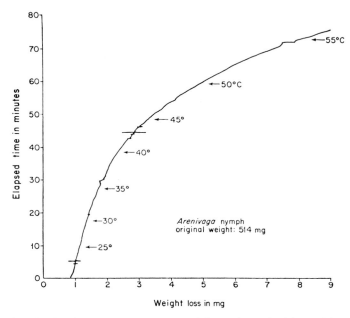

Fig. 13. Continuous recording of weight loss by an *Arenivaga* nymph in continuously rising ambient temperature (indicated at intervals by *arrows*). Cuticle temperature (not shown) also rose without any break. (From Edney and McFarlane, 1974)

An interesting sidelight is the observation by Galbreath (1975) in drawing attention to the reverse situation, where the cuticle of an arthropod is warmer than the adjacent air. He found that although the water potential (Ψ) of the blood of a soil beetle larva *Costelytra* is a little larger than that of its environment (-8.67 bars in the third stage larva), nevertheless, water leaves the insect for the soil because the insect's surface temperature is about 0.2 °C higher.

Other points arising from Beament's (1958, 1959) work are (1) in insects such as *Rhodnius*, *Pieris* and *Tenebrio* with hard waxes, subjection to temperatures above the transition range results in relatively higher rates of transpiration when the insects are returned to low temperatures, and in abolition of transition phenomena; (2) the transition temperature varies between individual insects within a species (though not so much as between most species) and during a moult cycle within a stadium, and (3) in several insects with highly impermeable cuticles, one transition temperature is shown immediately after a moult, but a second one, at a higher temperature appears later, correlated, perhaps, with deposition of the cement layer above the original wax, or perhaps with the formation of a second wax layer by migration. The more recent work of Nelson and Camin (1967), Hafez *et al.* (1970) and particularly of Davis (1974a, b) is very relevant and is discussed below and in Chap. 3.B.VII.

An alternative explanation for the double break transpiration/temperature curves has been suggested by Locke (1974) who proposed that the double peak in specific electrical resistance shown by an artificially formed phospholipid bilayer (Thompson, 1964) may be correlated with the two discontinuities in the permeability/temperature curves of insects such as *Rhodnius* and the pupae of *Pieris* and *Tenebrio* (Beament, 1959). However, differences between the two systems, and lack of direct evidence, render such an explanation tentative.

Arthropods other than insects have also been studied in this regard. In isopods, Edney (1951b) found that transpiration increases with temperature, but only to the extent that ΔP increases, so that a transition point is absent. Transpiration is rapid at low temperatures compared with *Blatella* (Wigglesworth, 1945), but is slower than in *Blatella* at higher temperatures, a fact which suggests that there may be a resistance to water transport in the isopod cuticle that is not subject to change at higher temperature. Bursell (1955), however, observed two transition points in *Oniscus asellus* and one in *Porcellio scaber*, and proposed that the lipids responsible lie deep in the cuticle, since superficial abrasion has little effect. Warburg (1965, 1968a) observed transition points in both *Armadillidium* and *Venezillo* (the latter is an isopod from xeric areas), but all these authors used different techniques, and the issue remains unclear. For further consideration of transpiration in isopods see Edney (1968a).

In millipedes transpiration appears to increase in proportion to ΔP as temperature rises (Cloudsley-Thompson, 1950, 1959; Edney, 1951b) and the same is true of the uropygid *Mastigoproctus* (Crawford and Cloudsley-Thompson, 1971), although Ahearn (1970b), using smaller animals, observed a marked increase in transpiration at 37.5°C. Ahearn's animals had been kept away from soil for some time before experimentation so that abrasion (if this should occur in soil) may have been repaired.

In centipedes, including *Lithobius* (Mead-Briggs, 1956), *Scolopendra* spp., *Rhysida* and *Ethnostigmus* (Cloudsley-Thompson, 1959; Cloudsley-Thompson and Crawford, 1970), no transition temperature has been observed. On the other hand, arachnids, including *Tegenaria*, two species of *Zilla*, *Meta* and *Lycosa* (Davies and Edney, 1952), *Ciniflo* (Cloudsley-Thompson, 1957), *Geolycosa* (Humphreys, 1975), *Pandinus*, *Androctonus*, *Scorpio* and *Euscorpius* (Cloudsley-Thompson, 1956) do show transition phenomena, and have been credited with a discrete lipid layer in the epicuticle.

Ticks have been referred to above, and more recent work confirms the presence of a sharp transition temperature in *Haemaphysalis* (Nelson and Camin, 1967), *Hyalomma dromedarii* and *Ornithodorus savignyi* (Hafez et al., 1970). The last named species shows hardly any increase in permeability between 30° and 60°C (this is about $2 \, \mu\text{g cm}^{-2}\text{h}^{-1}\text{mm Hg}^{-1}$), above which there is a rapid increase to $60 \, \mu\text{g cm}^{-2}\text{h}^{-1}\text{mm Hg}^{-1}$ at 75°C, a temperature which is, of course, well above the lethal temperature of the animal. On the other hand moulting abolishes the transition effect and raises permeability at low temperatures for several days.

The effect of developmental stage and of feeding on cuticle permeability has been well documented by Davis (1974a, b) in the rabbit-tick *Haemaphysalis leporispalustris*. Engorged nymphs of this animal show progressively decreasing permeability, and increasingly abrupt transition phenomena during the first 14 days after feeding. On the first day there is nothing that could be called a transition point and water loss is relatively rapid, while after 14 days, water

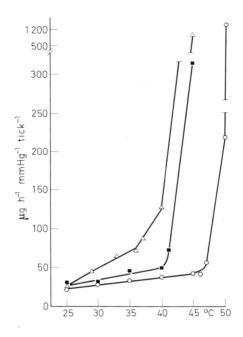

Fig. 14. The effect of ambient temperature on rate of transpiration from the rabbit tick, *Haemaphysalis leporispalustris*. △—△: engorged nymphs 1 day after drop off; ■—■: the same, 4 days after drop off; o—o: the same, 15 days after drop off. Stage of development strongly affects both permeability and the abruptness of the transition temperature in a single species. (Redrawn from Davis, 1974b)

loss is much slower up to 47 °C, above which it increases abruptly (Fig. 14). This work and that of Hafez *et al.* (1970) shows that in one species transition may be very gradual or abrupt according to the state of the animal. Adult teneral male ticks have a significantly lower permeability than females. Engorged adult females have a much lower transition temperature (35 °C) than fasted nymphs (45 °C), and also a much higher permeability at biologically significant temperatures. Unfortunately the data are given in terms of rates per tick, so that comparisons with other arthropods in terms of permeability cannot readily be made.

IV. The Effect of Hydration and Other Factors on Permeability

A complicating factor in the interpretation of temperature/transpiration data is the fact that the rate of water loss from arthropods in experimental situations is not usually constant, but is faster at first than later. This has been observed in isopods (Edney, 1951b; Bursell, 1955), land crabs (Herreid, 1969a, b), myriapods (Perttunen, 1953; Crawford, 1972), scorpions (Cloudsley-Thompson, 1956, 1967; Hadley, 1970b; Crawford and Wooten, 1973), spiders (Humphreys, 1975) and in many species of insects including beetles (Edney, 1971; Schmidt, 1955), locusts (Loveridge, 1968b), *Pieris* and *Tenebrio* pupae (Beament, 1959) and *Rhodnius* nymphs. It is apparently not true of the onycophoran *Peripatopsis moseleyi* (Dodds and Ewer, 1952) in spite of earlier reports to the contrary (Manton and Heatley, 1937). However, the phenomenon is certainly widespread and is probably general.

Bursell's work on *Oniscus* shows that the first very high rate of transpiration is due to loss of water from the outer layers of the cuticle. Subsequently there is a further decrease in the transpiration rate which is related to a drop in total (not cuticular) water content. This he suggests, is due to an increase in electrolyte concentration of the blood affecting the protein constituents of the cuticle in such a way as to cause contraction, leading to increased resistance to water movement. In *Porcellio*, according to Salminen and Lindqvist (1972) and Lindqvist *et al.* (1972), the water content of the cuticle is maintained at a constant level during transpiration, either by pumping from within or, additionally, by a spreading over the cuticle of water regurgitated from the mouth (Lindqvist, 1971, 1972). Lindqvist (1968, and subsequent papers) also found that in *Porcellio* antennectomy leads to higher cuticular water content and to reduced transpiration, while leg amputation increases transpiration as does hydration; but the mechanisms involved are unknown.

Cuticular water is certainly not constant in land isopods. In addition to changes that may be occasioned by strong general dehydration, Mayes and Holdich (1976) found an approximately 6.5-h cycle for water content of the cuticle in *Oniscus asellus* kept in saturated air. As they point out, this may be associated with the periodic release of gaseous ammonia which occurs in these animals (Chap. 6.B.II.; Wieser *et al.*, 1969).

Decrease in transpiration rate with time could give misleading results. If transpiration from one animal is measured at different temperatures, beginning at the lowest temperature, then an increase which should result from higher temperatures could be cancelled by a decrease as a result of dehydration, or even lead to apparently lower rates of transpiration at higher temperatures (below the transition point), as it may do in *Cryptoglossa* (Ahearn, 1970a). Similarly, if one animal is exposed to a series of different humidities, beginning with the lowest, a high initial loss (perhaps of cuticular water) may give the impression of an actual decrease in permeability (increase in R_c) as ambient humidity rises. This may account for Vannier's (1974a) observation that the cuticle of the collembolan *Tetradontophora* appears to have a higher R_c at 12.5% R.H. than in dry air.

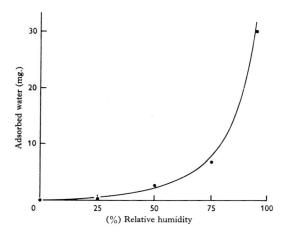

Fig. 15. The amount of hygroscopic water absorbed by the dry cuticle of a whole dead locust that originally weighed 0.86 g. (From Loveridge, 1968b)

40

Loveridge (1968 b) made a careful examination of these questions in *Locusta* and concluded that a high initial rate of loss is ascribable to loss from the cuticle, for the latter was shown to be quite strongly hygroscopic (Fig. 15). However, he also observed a subsequent slow decrease in transpiration representing a true increase in R_c, the precise cause of which is uncertain. (For further discussion of possible mechanisms, see Loveridge, 1968b.) However, this does have the very important result that over long periods of time, evaporative water loss through the cuticle is reduced at lower external humidities and the extent of this is shown in Figure 16.

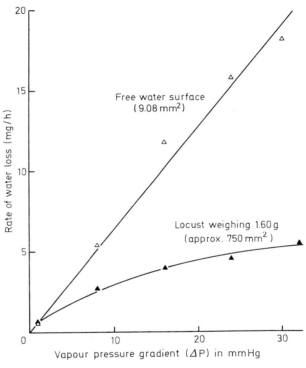

Fig. 16. Evaporation from a free water surface of 9.8 mm² and from a standard locust (surface area of head, thorax and abdomen about 750 mm²) in the same apparatus. Evaporation from the free water surface is linear with *Δ*P but in a locust the rate falls off in drier air. (From Loveridge, 1968 b)

In ticks, Hafez *et al.* (1970) also found that the rate of transpiration is approximately halved as a result of losing water equal to 15% of original wet weight, and the change in cuticular permeability is apparently due to a decrease in total body water rather than to dehydration of the cuticle, for dehydrated ticks kept in high a_v for 3 h to hydrate the cuticle, also lose water slowly. In the spider *Geolycosa geoffroyi*, Humphreys (1975) found that ambient relative humidity affects the rate of loss even when *Δ*P has been allowed for. Here, as in Loveridge's locusts, a change in cuticle permeability may be the explanation.

41

V. The Nature of Cuticular Lipids

There have been several good reviews of work on cuticular lipids (Gilby, 1965; Jackson and Baker, 1970; Hackman, 1971, 1974; Neville, 1975) so that we shall not go deeply into the matter here except insofar as cuticle permeability to water is concerned. For a summary of the available comparative data see Table 5.

Periplaneta, which has a greasy lipid layer, also has a higher proportion of hydrocarbons (75–77%) (Gilby and Cox, 1963; Gilbert, 1967a, b; Beatty and Gilby, 1969) than any of the insects with hard waxes, but hydrocarbons are nevertheless the major component of most. There is a remarkably close correspondence (in hydrocarbons present) between two species belonging to different genera of cockroaches (*Blatta* and *Leucophaea*) (Tartivita and Jackson, 1970), although significant differences exist between the sexes in *Periplaneta australasiae* and *Periplaneta fuliginosa*, where cis-9-tricosene is a major constituent in males only (Jackson, 1970). In *Sarcophaga* too, the cuticular lipids (about 100 μg per fly) are mainly hydrocarbons, n-alkanes and mono-methyl alkanes predominating. But here there are important differences in the nature and amounts of constituents between young and old flies (Jackson *et al.*, 1974). There are seasonal and topological differences in the relative proportions of hydrocarbons in an ant, and between sexes of blowflies and houseflies (refs in Hackman, 1974). In the fire ants *Solenopsis invicta* and *S. richteri*, hydrocarbons (all saturated) predominate, but there are large differences in their composition between developmental stages (Lok *et al.*, 1975). Unfortunately the biological significance, if any, of these chemical differences is unknown.

As regards other classes of lipids, fatty acids usually form a substantial contribution, but alcohols are apparently absent, except for *Tenebrio* larval exuviae, where they represent 55% by weight of all lipids. Alkyl (wax) esters, aldehydes, sterols, sterol esters and glycerides are also present in variable amounts in all cuticles. Phospholipids are absent, except for a trace in *Lucilia* puparia and 2% in *Tenebrio* larvae.

A high proportion of hydrocarbons is not always found. In the desert scorpion *Paruroctonus mesaensis*, thin layer chromotography shows that hydrocarbons represent only 18% of total chloroform-extractable lipids. These are all saturated n-alkanes whose chain length varies from C 21 to C 33 or more—the higher numbers usually representing a branched chain molecule. Triglycerides, free fatty acids, cholesterol and alcohols form the other major constituents (Hadley, pers. comm.).

The grease of cockroaches is mobile while on the insect but hardens to a wax when removed, and this was ascribed at first to evaporation of solvents (Beament, 1955). However, the major hydrocarbon of *Periplaneta* cuticle is a C_{27} cis,cis-6,9-heptacosa diene, which belongs to a group of compounds known to undergo autoxidation in air. Atkinson and Gilby (1970) believe that the grease remains mobile while on the insect owing to the action of polyhydric phenols which prevent the formation of free radicals responsible for autoxidation. These polyphenols are also believed to be the precursors of the orthoquinones that effect tanning of the exocuticular proteins after ecdysis.

Table 5. Composition of insect cuticular lipids[a]

Lipid class	Anabrus simplex abdominal sclerites (1)	Bombyx mori exuviae (2, 3)	Adults (4) ♂	Adults (4) ♀	Puparia (4)	Lucilia cuprina Puparia (5) Free lipid	Puparia (5) Bound lipid	Pupal and moulting membranes (5) Free lipid	Pupal and moulting membranes (5) Bound lipid	Peri-planeta americana nymphs (6, 7)	Pteronarcys californica Adults (8)	Pteronarcys californica Naiads (8)	Tenebrio molitor larval exuviae (9)
Hydrocarbons	48–58	+	59.2	65.7	33.3	55.0	35.7	51.0	59.9	75–77	12	3	10
Fatty acids	15–18	+	8.8	5.4	15.6	14.9	21.5	3.4	8.4	7–11	49	12	5
Alcohols	—	—	—	—	—	—	—	—	—	—	—	—	55
Alkyl (wax) esters	9–11	+	15.8	16.1	25.6	0.2	2.3	3.1	0.6	3–5	4	1	13
Aldehydes	—	—	—	—	—	—	0.5	—	—	8–9	—	—	—
Sterols	2–3	—	—	—	1.8	1.8	3.6	3.2	3.3	<1	18	1	—
Sterol esters	—	—	—	—	—	7.3	0.7	14.2	8.0	—	—	—	—
Glycerides	—	—	+	+	+	2.4	0.3	6.7	0.7	—	7	78	—
Acidic resins	12–14	—	—	—	—	—	—	—	—	—	—	—	—
Phospholipids	—	—	—	—	—	—	0.8	—	—	—	—	—	2

[a] From Hackman (1974).
[b] +: present but amount not stated.
[c] Reference: (1) Baker et al., 1960; (2) Bergman, 1938; (3) Amin, 1960; (4) Goodrich, 1970; (5) Gilby and McKellar, 1970; (6) Beatty and Gilby, 1969; (7) Gilby and Cox, 1963; (8) Armold et al., 1969; (9) Bursell and Clements, 1967.

43

In an interesting paper comparing the surface lipids of the adult (aerial) and larval (aquatic) stages of the stone-fly *Pteronarcys*, Armold *et al.* (1969) report that the adult form has twice as much surface lipid as the larva, and the chemical composition also differs considerably: the bulk of the adult lipids are free fatty acids (49%), sterols (18%) and hydrocarbons (12%), while the larva has a preponderance of triglycerides (78%) and free fatty acids (12%). These authors believe that the surface lipids of the adult would have a higher melting range than those of the larva, and also be less permeable to water. The larval lipid mixture is oily at room temperatures, but they believe it would be solid in field conditions and may play a part in water and ion transport. Impermeability is particularly valuable for a terrestrial insect that probably does not eat, and may not drink.

Another recent example of work relating composition to function is that of Davis (1974a, b) on the rabbit tick *Haemaphysalis* which shows quite striking changes in transpiration phenomena in different physiological states (Chap. 3.B.III.). Davis estimated the proportions of various solvent classes of epicuticular lipids by thin layer chromotography and found that as permeability changes occur, so do the presence and proportions of different lipids. There are, for example, different kinds of sterols in ticks with or without abrupt transition points. Unfortunately the particular lipids were not identified, so that the results, suggestive as they are, need amplification.

Wigglesworth's (1975) recent study of the incorporation of lipids into the epicuticle of *Rhodnius* is referred to below (Chap. 3.B.VI.), and the nature of the epicuticular water barrier is discussed in Chapter 3.B.VII. For a discussion of arthropod lipids other than those in the cuticle, see Gilbert and O'Connor (1970).

VI. The Structure of the Epicuticle

The epicuticle of arthropods is a complex structure of great biological importance. It has been extensively studied and the information about it has recently been reviewed by Locke (1974) and Neville (1975). Unfortunately the complexity of the structure and the technical difficulty of studying it have led to confusion with regard to the nomenclature of its various components.

Much information about the epicuticle has come from the study of moulting and its significance for water affairs, and we begin by considering this avenue of approach.

The process of moulting is known to lead to a certain amount of extra water loss in several arthropods, including *Rhodnius* (Wigglesworth and Gillette, 1936) (Fig. 17), *Tenebrio* (Buxton, 1930), isopods (Webb-Fowler, quoted by Edney, 1960a) and scorpions (Toye, 1970). But in view of the fact that nearly all the old cuticle is digested and reabsorbed before it is shed, the extra amount of water lost seems remarkably small, and this of course again directs attention to the outermost layers.

The early work of Wigglesworth (1947, 1948b) on moulting and its significance for water balance in *Rhodnius* and *Tenebrio* provides a good example of the

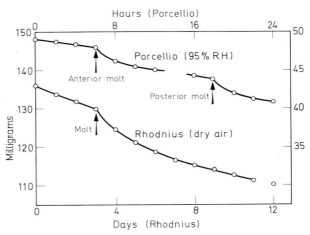

Fig. 17. The effect of moulting on transpiration in a land isopod *(Porcellio scaber)* and an insect *(Rhodnius)*; in neither is the extra water loss excessive. There is an interval of 2–3 days between the posterior and anterior moults in land isopods. [Data for *Porcellio* from Webb-Fowler, 1955 (quoted by Edney, 1960a); for *Rhodnius* from Wigglesworth and Gillette, 1936]

effective combination of physiological experiment with histochemical methods. The information then obtained has had to be modified in the light of improved techniques (Wigglesworth, 1970, 1975) (see below), but the essentials remain, and we shall use them as a starting point.

As the time for ecdysis approaches little besides the old epicuticle is left, and at this time, if the old cuticle is artificially removed, the underlying new cuticle is highly permeable to water and its surface is hydrophilic and argentaffin (i.e., it reduces silver nitrate). A short time later, immediately before ecdysis, removal of the old epicuticle exposes a hydrophobic surface, and the new cuticle is impermeable. At this stage, treatment of the new cuticle with a cold lipid solvent exposes the hydrophilic, permeable, lower layer; but a little later still, more vigorous treatment with hot chloroform is necessary to produce the same effect. These observations were explained by supposing that, shortly before ecdysis lipid material is laid down over the new epicuticle after passage through the pore canals, that the innermost layer of lipids forms the impermeable oriented monolayer, and that shortly after ecdysis, a protective cement layer is secreted over the surface of all, via the dermal gland ducts.

The material forming the main body of the epicuticle was termed cuticulin by Wigglesworth (1947), and the term is so used by later authors such as Weis-Fogh (1970) and others. Cuticulin was originally defined by Wigglesworth (1933) as a refractile, non-chitinous component of the cuticle and epicuticle, probably a lipid polymer. Subsequently, the term was restricted to the body of the epicuticle, since "sclerotisation" of the procuticle was shown to be effected by quinone tanning of protein.

After re-examination by a new technique, Wigglesworth (1970) demonstrated the widespread presence of cuticulin not only in the epicuticle (where it is present in the "cement layer" and in the inner and outer epicuticle proper) but also in the main body of the cuticle and elsewhere in the skeletal structures of several species of insects. Wigglesworth suggested that oenocytes are the source of these lipids, and this was confirmed by Diehl (1973) who used radiotracers to show that 40%–60% of the cuticular lipids are formed by the oenocytes of the peripheral abdominal fat body.

Locke (1960, 1965, 1974), Rinterknecht and Levi (1966), Filshie (1970a, b) and Noble-Nesbitt (1970a), restrict "cuticulin" to an outermost layer, about 50 to 150 Å thick, which is the first to be formed in the development of a new cuticle. In tracheoles this layer first appears as twin dense lines about 85 Å thick. Locke prefers "dense homogeneous layer" or "inner epicuticle" for the main body of the epicuticle.

Wigglesworth (1975) has recently completed a revision of the structure and function of the epicuticle of *Rhodnius*, and it will be convenient to refer to this now before trying to compare the various interpretations and nomenclatures used by other authors for other insects.

The outer part of the fully formed nymphal abdominal cuticle of *Rhodnius* is represented diagramatically in Figure 18 and in Figure 19 an attempt is made to show the order in which the various events connected with moulting occur. The beginning of the moulting process is signaled by apolysis (separation of the epidermal cells from the old endocuticle). This is followed by the secretion of the new outer epicuticle and the new inner epicuticle in that order. Both these layers contain abundant lipid material but the outer (about 12 to 17 nm) in *Rhodnius* is much thinner than the inner (about 300 to 400 nm). The protein component of the inner layer becomes strongly stabilised by tanning at a later stage, and the lipids of both layers are strongly stabilised, perhaps by polymerisation. Both layers contain fine, permeable, lipophilic channels, the epicuticular channels, about 20 to 25 nm wide, through which, at this stage, the moulting fluid, containing enzymes for the digestion of the old cuticle, may pass.

After producing the epicuticular layers, the epidermal cells begin to lay down the components of the main, or procuticle, leaving spaces, or pore canals, which are continuous apically with the fine epicuticular channels already referred to and which provide pathways for the passage of materials between the epidermis and the outermost regions of the epicuticle.

Shortly before ecdysis, the epicuticular channels contain silver binding material which is discharged mainly at the junction of the inner and outer epicuticles, from where it spreads laterally and inwards, impregnating the greater part of the inner epicuticle. A small amount of this material also appears as a very fine membrane outside the epicuticle. This is the material that Wigglesworth (1947) previously described as polyphenolic, and as spreading over the outer surface of the epicuticle. Its precise nature is unknown, but it may well be tyrosine, or a protein rich in tyrosine, as a forerunner of the polyphenols that do appear at a later stage.

Shortly thereafter, a lipid material which is seen first in the outer regions of the pore canals, is extruded through the epicuticular channels as small droplets that spread and coalesce to form a continuous, stainable, lipid layer. As the time for ecdysis approaches, the space between the epicuticle and the old cuticle becomes dry and filled with air, and at the same time the surface lipid becomes

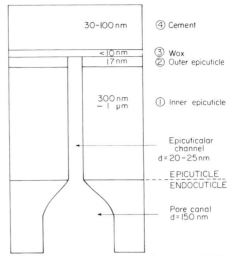

Fig. 18. The structure of *Rhodnius* epicuticle, based on a description by Wigglesworth (1975). *1*: The *inner epicuticle* has been known as the cuticulin layer (Wigglesworth, 1972), and the dense layer or protein epicuticle (Locke, 1961). It contains highly tanned protein and abundant lipid (Wigglesworth, 1975). *2*: The *outer epicuticle*, also known as the resistant layer (Wigglesworth, 1947). This layer is rich in lipids, but the precise composition is unknown (Wigglesworth, 1975). *3*: Silver binding material, combined with lipid, both probably exuded from the pore canals, harden to form the *wax layer*. This is hydrophobic and probably corresponds with Filshie's (1970a) outer epicuticle or his superficial layer (1970b), and with the surface monolayer of oriented lipid of Locke (1966) and Gluud (1968). *4*: The *cement layer*, produced by the dermal glands, is a mucopolysaccharide, later impregnated with wax. Layers 3 and 4 in electron micrographs together look like a triple layer with lipid staining mostly in the upper and lower regions

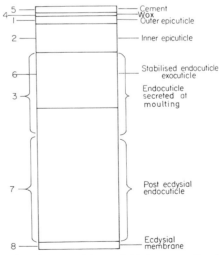

Fig. 19. The sequence in which various components of an insect cuticle are laid down or altered in the course of a moulting cycle. The cement is produced by dermal glands; all the rest are the products of epidermal cells. Pre-ecdysial endocuticle is also known as "procuticle." Post ecdysial endocuticle continues to be laid down in layers (often with a daily cycle) after ecdysis. (Partly after Locke, 1974)

47

less stainable as it permeates the silver binding material outside the epicuticle (referred to above), and becomes strongly stabilised and unreactive. Thus a hard, thin, resistant, hydrophobic surface, which Wigglesworth terms the "wax layer," is formed.

Finally, soon after ecdysis, a material is exuded from the dermal glands and spreads over the whole "wax layer," forming a protective cement, about 30 to 100 nm thick, closely associated with the wax layer on its under surface. This cement material, which may be a mucopolysaccharide, is hardened, but not darkened, by tanning, and may also become secondarily impregnated with further wax. Surface abrasion is known to increase the permeability to water of the epicuticle dramatically, but if death by dehydration is prevented by keeping the abraded insect in high humidity, repairs are effected and impermeability is restored. It seems that in this situation the cement layer is not reformed but secretion of lipid material in a carrier protein is hardened to form a resistant almost glassy layer, with a well marked dark staining region (about 25 to 50 nm thick) along the inner face, and another, much more tenuous dark staining line on the outer face.

The foregoing account, which largely follows Wigglesworth's (1975) description, may be summarised and compared with the situation as described by other workers as follows:

The *inner epicuticle* (1 in Fig. 18) is the "cuticulin layer" of Wigglesworth (1947), the dense layer, or protein epicuticle of Locke (1961) and the inner epicuticle of Weis-Fogh (1970). It varies in thickness up to 1 μm and is rich in lipids impregnating a highly tanned protein component. The *outer epicuticle* (2 in Fig. 18), is Wigglesworth's (1947) resistant layer, Dennell and Malek's (1955) paraffin layer, Locke's (1961) cuticulin layer and Weis-Fogh's (1970) outer epicuticle. It is a highly resistant, lipid-rich material, about 17 nm thick in *Rhodnius*.

Running through the whole epicuticle there are simple, straight epicuticular channels about 20 nm in diameter. Locke (1960, 1961, 1965) has described the structure of the epicuticle of several insects, including the caterpillars of *Calpodes* and *Galleria*, the beetle larva *Tenebrio* and the honey bee *Apis*. According to his observations the epicuticle contains wax canals which are usually multi-branched systems of fine canals, about 6 to 13 nm in diameter that spring from the outer ends of pore canals and contain wax filaments (Figs. 20 and 21). These probably correspond to the epicuticular channels of *Rhodnius* (Wigglesworth, 1975), although a branched appearance is not present in that insect.

The *wax layer* of *Rhodnius*, (3 in Fig. 18), is a thin layer, 10 nm or less, in which lipid material has impregnated silver-binding material to produce a hard, hydrophobic substance. It probably corresponds with Filshie's (1970a, b) "outer epicuticle" and "superficial layer." [Filshie uses "cuticulin" for the outer epicuticle (2 in Fig. 18)]. The wax layer also corresponds to the surface monolayer, or oriented lipid layer of Locke (1966).

The *cement layer* (4 of Fig. 18) is the tectocuticle of Richards (1953). It is a stabilised mucopolysaccharide secreted by the dermal glands, later impregnated with further wax. In electronmicrographs of *Rhodnius* this region appears as a triple layer some 30 to 100 nm thick, with a clear central region sandwiched between thinner upper and lower layers that stain deeply with lipid

48

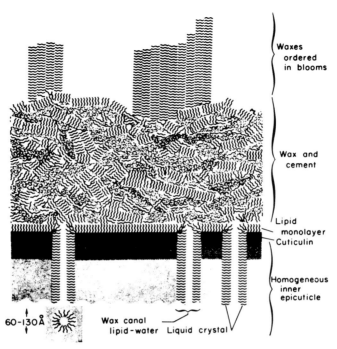

Fig. 20. The outermost layers of an insect cuticle (from Locke, 1974). The lipid monolayer as shown here is continuous with the wax in wax canals, and the latter is in the form of liquid crystals in the middle phase. There is (as Locke, 1974, points out) no direct evidence of this structure. Dimensions of the wax canals (60–130 Å) are somewhat smaller than those of the epicuticular channels, (Wigglesworth, 1975), in *Rhodnius* (200–250 Å)

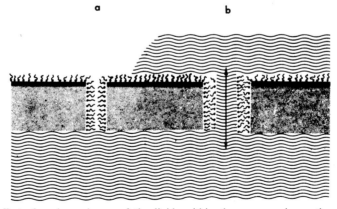

Fig. 21a and b. A possible effect of a phase change of the lipids within the wax canals on the permeability of cuticle to water. A wax canal occupied by a lipid water crystal (a) in the middle phase and (b) in the complex hexagonal phase. If water, or a high humidity in the environment, can induce this phase change then the apparent asymmetry in permeability to water could be accounted for. (From Locke, 1974)

49

stains. The lower of these two areas is probably the "wax layer" of Wigglesworth (1975) (3 in Fig. 18) and is referred to by Gluud (1968) and by Locke (1966) as the monomolecular lipid layer.

The larval cuticle of cyclorrhaphous flies (that of the last larval stage becomes the puparium) has always been considered somewhat aberrant. It lacks wax and cement layers according to Wolfe (1954a, 1955), although a wax layer is present in the adult cuticle (Wolfe, 1954b). It has been studied in detail by Dennell (1946). More recently Filshie (1970a) observed the formation of the outermost layer, which is resistant to concentrated acids, and confirmed that the material is not soluble in lipid solvents (although it occupies the same position as the lipid layers of other insects). It is laid down after the formation of the inner layers of the epicuticle by secretion from the epidermis and passage through the epicuticle to the outside via "epicuticular filaments" which are probably the homologues of the epicuticular channels of *Rhodnius*.

The lipid nature of the contents of the pore canals, and the superficial layer have been questioned by Filshie (1970a, b) on the grounds that these structures look identical in electron micrographs irrespective of whether the material has been subjected to strong lipid solvents or not. Locke and Krishnan (1971) responded by proposing that the filaments in question arise in fact as protein liquid crystals rather than lipid liquid crystals, and are subsequently tanned. The fact that the contents of pore canals and epicuticular channels may be proteinaceous (or tyrosine-like) at one time, lipid at another, may help to resolve this problem, but the manner in which materials move up and down these channels, often simultaneously, or at least in very rapid succession is still very puzzling.

VII. The Epicuticular Water Barrier

Does the recent work on transition phenomena, epicuticular lipids, and epicuticular structure, point to a conclusion about the water barrier?

The organization of lipid molecules into stable monolayers (the foundation of classical water barrier theory) has been called into question. The preponderant lipids from *Periplaneta* cuticle grease are hydrocarbons which are non-polar and are therefore unlikely to form stable monolayers, as Hackman (1971, 1974) points out. Alcohols, which do have polar moieties, and which would form monolayers, are apparently absent from cockroaches and all other insects so far investigated, save *Tenebrio* larvae. Hackman suggests that polymeric films formed from aldehydes or acidic resins may be more likely as candidates to explain the very high impermeabilities found in some insect cuticles.

Changes in the orientation of lipid monolayers, as well as changes in electrical properties of lipid bilayers, are well known to occur with phospholipids (Thompson, 1964), but once again, the latter are conspicuously absent from epicuticular lipids (except apparently for 2% in *Tenebrio* larvae). However, other classes of lipids also form stable monolayers, and of these, fatty acids are almost universally present in arthropod cuticles, probably in sufficient amounts to function in this way.

It does seem that epicuticular lipids are stabilised, perhaps polymerised, when they occur in the fully formed "wax" layer (Wigglesworth, 1975), or elsewhere in the epicuticle, but this does not conflict with the monolayer theory, for even a mixture of polymerised lipids (provided they are polar lipids) may form crystal-like oriented monolayers. On the other hand the fact that the lipids in question are dispersed throughout the "wax" layer, which also contains derivatives of tyrosine or polyphenolic substances, is perhaps a point against the monolayer theory.

The existence of monolayers has also been questioned for tick cuticles by Davis (1974b). Her work with *Haemaphysalis* confirms and amplifies the earlier finding of Beament (1945) and some later authors, that the shape of the transpiration/temperature curve is extremely variable, not only between developmental stages, but even in one individual in different physiological states. Davis refers to known phase transition phenomena for many phospholipids and sterol esters (Chapman and Leslie, 1970; Small, 1970), and points out that there are many similarities between these effects and the temperature transition phenomena seen in many arthropods. Setting this beside her own observations on amounts and classes of lipids present in tick epicuticles with different transpiration rates (Chap. 3.B.V.), Davis suggests (1) that the cuticular lipids responsible for impermeability are in a crystalline phase at lower temperatures, but undergo a phase change to become liquid crystals at higher temperatures; (2) that instead of permeability being conferred by a monolayer of the oriented hydrocarbon tails of polar lipids, this results from the packing of many molecules throughout the "cuticulin" layer (i.e. the outer epicuticle) and that changes in the packing structure of one kind of hydrocarbon chain may affect changes in the arrangements of many other nearby lipid molecules, leading to greater intermolecular spaces for the passage of water molecules; and (3) that greater or less homogeneity in the lipid species involved, and differences in proportions present, may account for different degrees of abruptness of transition and also for differences in general permeability levels.

Insofar as (3) above is true for "bulk" lipids it is also true for monolayers. Thus there is little cause for dispute about the heterogeneity of epicuticular lipids—the question that remains is whether or not the lipids are organised into monolayers. An advantage of the monolayer theory is that it is easy to envisage how impermeability of an arthropod cuticle could result from such a continuous crystalline layer. It is less easy to envisage the kind of organization that would confer very high impermeability according to the "bulk" lipid theory. However, lipids are distributed throughout the outer epicuticle as well as the "wax" layer and an oriented monolayer has never been unequivocally observed. Thus the evidence at present may be rather in favour of bulk dispersal (I believe it is), but the original monolayer theory of Beament has by no means been disproved. In either case it will be the task of future research to identify the nature and properties of the lipids concerned as these affect the permeability of the cuticle to water.

VIII. Asymmetrical Permeability of the Cuticle

Quite early on Hurst (1941, 1948) and Beament (1945, 1948) found that in vitro the permeability to water of several arthropod cuticles differs according to the direction of water movement. Differences of as much as 100:1 (Hurst, 1941) but usually in the region between 16:1 and 2:1 are measured (Richards *et al.*, 1953) when one side of a separated cuticle, or even a cast skin, is exposed to water, the other to air. Resistance (R_e) always appears to be less in the direction epicuticle to procuticle.

In the intact living cuticle also there is some evidence of asymmetry in the same sense. In those arthropods that absorb water vapour from unsaturated air, the process of absorption usually occurs much faster than does loss of water to dry air in the same species. For example, Knülle and Spadafora (1969) found that *Liposcelis* (a psocid) loses 50% of body water in 11 days in air at a_v 0.33, but recovers all the water lost in 6.7 h when transferred to a_v 0.58—a 38-fold difference in rate. In *Tenebrio molitor* the situation is similar (Locke, 1974). But this is not always the case, for Lees (1946a) found uptake by the tick *Ixodes* to be 1.9 mg cm^{-2} day^{-1} against a loss into dry air of 5.6 mg cm^{-1} day^{-1}. Where uptake is more rapid than transpiration into dry air, and if uptake occurs through the same pathway as loss, we are probably justified in concluding that asymmetry of resistance exists. However, a note of caution is introduced by the fact that in some arthropods at least, the integument is not the site of water vapour uptake (Noble-Nesbitt, 1970b, 1973; Rudolph and Knülle, 1974; O'Donnell, pers. comm.) (a topic to be discussed further in Chap. 9) in which case asymmetry in cuticle resistance is not necessarily involved.

It is not difficult to understand asymmetrical resistance to the passage of water in certain circumstances. As Hartley (1948) pointed out, for any bilaminar membrane in which resistance to the passage of water varies with the water content in one layer but not in the other (the pro- and epi-cuticles respectively in the present context), R_{total} is bound to vary according to which layer is in contact with water. But this does not help solve the present problem, for in vivo the arthropod procuticle is never in contact with air, and in any case, both in vivo and in vitro, $R_{in} < R_{out}$—the reverse of the effect that would be expected if $R_{procuticle}$ decreases when that layer is in contact with water.

Several proposals to account for asymmetrical behavior have been made, usually in the form of molecular valves (Hurst, 1941) or electrogenic ion pumps (Beament, 1961, 1964), but none so far has been generally accepted. We shall now briefly refer to two recent proposals—the second derived in part from the first—that have stimulated useful discussion, and are relevant to more general aspects of cuticular structure.

The first, proposed by Locke (1965, 1974) is suggested by the work of Luzzati and Husson (1962) and of Stoeckenius (1962), who demonstrated that certain lipids such as soaps and phospholipids, may exist as lipid water liquid crystals in various phases determined by temperature and by the proportion of water present. When such crystals are in the "middle phase," lipid molecules lie in a cylinder with their hydrophobic ends inwards, and their polar, hydrophilic ends outwards, water being on the outside (Fig. 20). In the "complex hexagonal phase" of these crystals, the molecules form a double cylinder (Fig. 21), with a central core of water; both external and internal surfaces of the cylinder wall of lipid molecules being hydrophilic, their hydrophobic tails lying within the wall itself. Filaments in cylinders of an appropriate size have been observed in some epicuticles particularly where wax secretion is copious *(Calpodes, Apis)*, but also in other areas where the integument is not so specialized [*Calpodes, Rhodnius, Galleria* and others, (Locke, 1974) Fig. 22]. These are the wax canals referred to above (Chap. 3.B.VI.). If such canals and their filaments extend to the outer surface, and if the phase of the liquid crystals is changed by the presence or absence of water, the observed asymmetry might be accounted for. However this proposal has its own problems: there is doubt as to the continuity of the wax canals with the surface, and indeed whether the contents are lipid (Filshie, 1970a, b). Furthermore the liquid crystalline form has not been demonstrated in hydrocarbons which are usually the preponderant lipids in epicuticles. Nevertheless, the proposal is a stimulus for further enquiry.

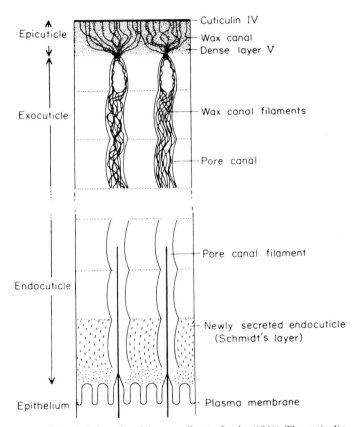

Fig. 22. Diagram of the structure of *Tenebrio* larval cuticle according to Locke (1961). The cuticulin layer and dense layer correspond to Wigglesworth's (1975) outer and inner epicuticles. The wax canals probably correspond with the epicuticular channels of *Rhodnius* (Wigglesworth, 1975), although the latter are simple, unbranched canals

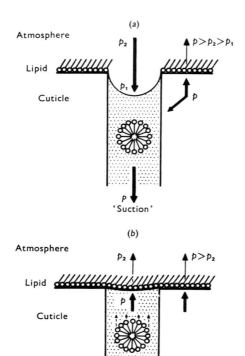

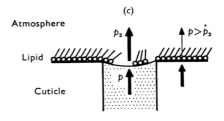

Fig. 23 a–c. Beament's microcapillary valve model, modified by Noble-Nesbitt. (a) suction from below produces a meniscus, breaking the impermeable lipid layer, and causing lowered water activity at its surface, so that water molecules from the air (of higher partial pressure, p_2) condense on the meniscus, where the partial pressure is lower (p_1). (b) suction from below decreases and the water rises to the surface, where it is covered by a lipid monolayer which prevents rapid evaporation. (c) the arthropod is dead and supposedly incapable of producing new lipid material to cover the exposed water surface, so that evaporation is rapid. (After Noble-Nesbitt, 1969)

The second and perhaps the most likely hypothesis so far is a modification of the fore-going due to Beament (1965) and further modified by Noble-Nesbitt (1969), and shown in Figure 23. This model proposes that fine, water-filled capillaries (of the order of tens of Å) open through the epicuticle to the surface, and that during normal transpiration, resistance is high owing to a continuous monolayer of oriented lipid molecules over the mouth of the capillary. (The existence of such layers has indeed been seriously questioned, but for the sake of the present discussion we will accept some form of mobile lipid barrier to water.) When water molecules move inwards, during vapour uptake (see Chap. 9), suction from below leads to the breaking of the surface lipid layer (the lipid molecules may perhaps form micelles within the capillary with their polar ends outwards), and impedance is consequently low. When uptake ceases, water in the capillary rises, the surface lipid is reformed, and remains intact owing perhaps to replenishment by further secretion from below. If the insect dies, the surface layer is disrupted and impedance to the outward movement is reduced. In nearly all cases, transpiration is indeed more rapid from dead than from living arthropods. This model was used by Noble-Nesbitt (1969) to account for water vapour uptake through the cuticle in *Thermobia*. The same author subsequently found that the cuticle is not the site of such uptake in that insect (Noble-Nesbitt, 1970b), nevertheless the model is useful insofar as it also helps to account for asymmetry in transpiration rates. Pores with the diameter of the order required are present in *Thermobia* (Noble-Nesbitt,

1967) and *Calpodes* (Locke, 1964) and other insects (Filshie, 1970b; Wigglesworth, 1975) but as we mentioned above, they may not be open to the air. The model is suggestive but by no means compelling.

Because of the theoretical difficulties involved in the concept of a rectifier system for water molecules and of the lack of rigorous proof that water vapour is absorbed through the cuticle of any arthropod, or that any such cuticle is asymmetrical in resistance to water movement, there is little to be gained from further speculation.

IX. Epidermal Cells as Water Barriers

A simplified view of the general situation concerning water barriers in the integument, proposed by Berridge (1970), is shown in Figure 24. Barriers to the movement of water are considered to be (1) the basal membrane of the epidermal cell, (2) the apical membrane, and (3) the epicuticle. Water activity is shown by

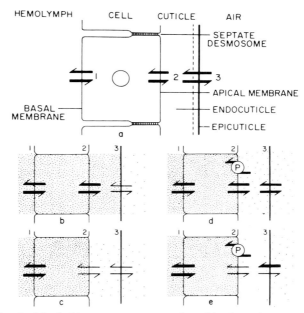

Fig. 24a–e. Possible mechanisms involved in limiting water movement from blood to air across an arthropod integument. The permeability barriers (a) are supposed to be *(1)* the basal and *(2)* apical cell membranes, and *(3)* the epicuticle. (b) the main resistance is in the epicuticle, and water activity in the endocuticle will be approximately in equilibrium with that in the epidermal cell and the blood. (c) the resistance is shared between apical membrane and epicuticle, and a_w in the endocuticle is less than that in the cell. If the resistance of the apical membrane is very much greater than that of the epicuticle, then a_w in the endocuticle would approach that of air. (d) a pump is supposed to move water from the endocuticle to the cell, and again a_w (endocuticle) is less than a_w (cell) and approaches a_v (air). (e) the pump is supposed to be capable of reducing a_w (cuticle) below a_v (air), a situation which would permit the absorption of water vapour. (From Berridge, 1970)

55

the degree of stipling. In Figure 24 b, the main barrier to transpiration is assumed to be the epicuticle. In this case water activity drops but little across the procuticle, and a_w in that compartment is essentially the same as that in the blood. Regulation of transpiration could be accomplished by changing R_e—perhaps (as Bursell suggested for *Oniscus*) by further dehydration leading to shrinkage and closer packing of water resistant elements.

In (c), the main resistance is presumed to be the apical cell membrane (R_{ap}), and a_w falls steeply across this barrier. If $R_{ap} > R_c$, water activity in the cuticular compartment will approximate that of the air outside. The initial rapid rate of water loss recorded in many arthropods, including isopods (Edney, 1951b; Bursell, 1955) and locusts (Loveridge, 1968 b) would be accounted for by water leaving the cuticular compartment—subsequent loss being dependent on R_{ap}. Regulation would then be achieved by altering R_{ap}, a process not impossible to envisage since that membrane is part of the living epidermal cell. In this connection, although the chitin present in the procuticle is crystalline (Neville, 1975) and therefore probably does not vary in water content, water in the protein component probably varies considerably. Possibly the apical cell membrane may be the site of a hormonal control of cuticular water loss that has recently been proposed by Treherne and Willmer (1975a, b) and referred to below in this section.

In (d), a pump of some kind is supposed to exist at the apical cell membrane, removing water from the cuticle, thus reducing transpiration by decreasing the activity gradient from cuticle to air. This system corresponds with the suggestion of Lees (1946 a) for ticks, Davies and Edney (1952) for spiders, and Winston and Nelson (1965) for the clover mite *Bryobia*. Winston (1967) found that a_w of both cockroach and locust cuticles was indeed lower than that of the blood, and he used this as evidence for a pump in the epidermis. However, a pump is not a necessary hypothesis, for low cuticular water activity would result from an apical membrane water barrier alone as in (c).

Finally, in (e), the apical membrane pump is assumed to reduce the activity of water in the cuticle below that of the surrounding air, so that a net absorption of water into the blood follows. The nature of the pump, if one is present, remains obscure.

The process of plasticization of the cuticle in feeding *Rhodnius* described by Maddrell (1966) and Reynolds (1974a, b, 1975), involves an increase in cuticular water content which might (but need not) be causally linked with changes in the function of the apical membrane. (For further discussion of plasticization see Chap. 6.D.) On the other hand, the abdominal cuticle of normal *Rhodnius* nymphs does not swell when placed in a solution similar in composition to *Rhodnius* blood (Reynolds, 1975). This suggests, but again does not prove, that unlike the situation in cockroaches and locusts (Winston, 1967), isopods (Bursell, 1955) or locusts (Loveridge, 1968 b), water activity in the normal *Rhodnius* cuticle is in equilibrium with that of the blood.

Recent thinking about the nature of plasma membranes has been usefully reviewed in relation to water transport by Oschman *et al.* (1974). Much of the membrane is now believed to be a bi-layer of lipid molecules, most of which are not enclosed in a protein sandwich (as used to be thought according to the original Danielli and Davson (1935) model, and water is thought to move directly through this bi-layer rather than through pores in it. Resistance to such movement would reside largely in the outer, ordered regions of the hydrocarbon chains, and there is evidence that this may be altered by a variety of factors, including the direct or indirect effect of hormone action. For further discussion see Oschman *et al.* (1974). What concerns us here is that if the accumulating evidence is correct (that cuticle resistance is variable and is affected by such factors as body water content, cuticle hydration, and ambient humidity) and if the recent report of Treherne and Willmer (1975b) that cuticle resistance is under hormonal control is verified, then an attractive, if tentative, hypothesis would be to suppose that information about both internal and external parameters may be fed through sensory channels to the CNS, there to affect the production and release of a hormone which in turn controls water balance by its action on the apical membrane of epidermal cells. However, experimental evidence is lacking. For further discussion see Chapter 6.D.

C. Ecological Implications

I. Transpiration and Habitat

Despite the great complexity of the cuticle comparative measurements of transpiration rates are valid for ecological purposes and for the compilation of water balance budgets. Doubtless measurements have been made in different ways and with animals in different physiological conditions. Nevertheless it is usually possible to arrive at an approximate figure for overall cuticular water loss and to compare species from different habitats in this respect. A selection of the large number of data available has been set out in this way in Table 6. Since permeability varies with temperature, humidity, age, stage, hydration, and probably other parameters, we cannot speak of a permeability constant for each species; however, we can speak of a permeability coefficient in any one set of conditions. In the following discussion permeability data are quoted in terms of $\mu g\ cm^{-2}\ h^{-1}\ mm\ Hg^{-1}$, followed (in parenthesis) by the value in cm s^{-1} $\times\ 10^{4}$.

Among the insects alone, permeability varies from 0.3 (0.9) for pupae of the tsetse fly *Glossina morsitans* to 190 (559) for caterpillars of the swift moth *Hepialus*. Eggs of insects such as *Rhodnius*, which may be laid in dry places, have very low permeabilities. These, as well as pupae in xeric environments, cannot replace lost water by feeding and consequently need to be relatively impermeable.

In fact, as Table 6 shows, it is possible to find good examples of relationships between permeabilities and habitats in all classes of arthropods. Even among terrestrial isopods, the integument of the xeric *Venezillo arizonicus* is some five times less permeable than that of the hygric *Porcellio scaber*. In land crabs also, overall water loss rates (percent of total body weight) are correlated with terrestrialness of habitat (Edney, 1961; Herreid, 1969a) and the rates are generally

Table 6. Arthropod transpiration[a]

Taxon	Habitat	Permeability		Resistance	Reference
		μg cm^{-2}h^{-1} mm Hg^{-1}	$10^4 \times$ cm s^{-1}[b]	s cm^{-1}	
Decapod crustaceans					
Uca annulipes	hygric	80	235	42.6	Edney (1961)
U. marionis	hygric	200	586	17.1	Edney (1961)
Isopod crustaceans					
Porcellio scaber	hygric	110	323	30.9	Edney (1951b)
Hemelepistus reaumuri	xeric	23	68	156	Cloudsley-Thompson (1956)
Venezillo arizonicus	xeric	32	94	106	Warburg (1965)
Insects					
Agriotes larvae	hygric	600[c]	1,760	5.68	Wigglesworth (1945)
Agriotes pupae	hygric	23	94	106	Wigglesworth (1945)
Hepialus larvae	hygric	190	559	17.9	Wigglesworth (1945)
Nematus larvae	hygric	20	58.8	179	Wigglesworth (1945)
Bibio	hygric	76	223	44.7	Wigglesworth (1945)
Blatta orientalis	mesic	48	141	70.9	Mead-Briggs (1956)
Calliphora erythrocephala	mesic	51	149	67.1	Mead-Briggs (1956)
Periplaneta americana	mesic	55	161.7	61.8	Mead-Briggs (1956)
Glossina palpalis	mesic	12	35	286	Mead-Briggs (1956)
Glossina morsitans	mesic-xeric	8	23.5	425	Bursell (1957a)
Arenivaga investigata	xeric	12.1	36.3	275	Edney and McFarlane (1974)
Centrioptera muricata[d]	xeric	6.3	18.5	541	Ahearn (1970a)
Cryptoglossa verrucosa[d]	xeric	8.4	24.6	406	Ahearn and Hadley (1969)
Eleodes armata[d]	xeric	17.2	50.4	198	Ahearn and Hadley (1969)
Eremiaphila monodi	xeric	8.0	24.0	417	Delye (1969)
Glossina morsitans (pupae)	xeric	0.3	0.9	$>10^4$	Bursell (1958)
Locusta migratoria	xeric	22	65	154	Loveridge (1968b)
Manduca sexta	xeric	39.6	116.4	86	Casey (unpub.)
Rhodnius prolixus	xeric	12.0	35.3	283	Holdgate and Seal (1956)
Tenebrio molitor (larvae)	xeric	5	14.7	680	Mead-Briggs (1956)
T. molitor (pupae)	xeric	1	2.9	3,401	Holdgate and Seal (1956)
Thermobia domestica	xeric	15	44	227	Beament (1964)
Chortoicetes terminifera	mesic	41	120.5	83	Lees (1976)
Insect eggs					
Phyllopertha	hygric	60	176	56.7	Laughlin (1957)
Lucilia	mesic	15	44.1	226	Davies (1948)
Rhodnius	xeric	3	8.8	1,136	Beament (1949)
Myriapods					
Glomeris marginata	hygric	200	588	17.0	Edney (1951b)
Lithobius sp.	hygric	270	792	12.6	Mead-Briggs (1956)
Orthoporus ornatus	xeric	7.9	23.1	433	Crawford (1972)

Table 6 (continued)

Taxon	Habitat	Permeability		Resistance	Reference
		$\mu g\,cm^{-2}\,h^{-1}$ $mm\,Hg^{-1}$	$10^4 \times$ $cm\,s^{-1\,b}$	$s\,cm^{-1}$	
Arachnids					
Ixodes ricinus (tick)	mesic	60	176	56.8	Lees (1947)
Lycosa amentata (spider)	mesic	28.3	83.2	120	Davies and Edney (1952)
Mastigoproctus (whip scorpion)	mesic	21.0[d]	62.0	161	Ahearn (1970b)
Pandinus imperator (scorpion)	mesic	76	223	44.8	Cloudsley-Thompson (1959)
Androctonus australis (scorpion)	xeric	0.8	2.4	4,167	Cloudsley-Thompson (1956)
Aphonopelma (tarantula)	xeric	3.56[d]	10.5	954	Seymour and Vinegar (1973)
Buthotus minax (scorpion)	xeric	0.98	2.89	3,465	Cloudsley-Thompson (1962)
Eurypelma sp. (tarantula)	xeric	10.3	31.2	321	Cloudsley-Thompson (1967)
Galeodes arabs (solpugid)	xeric	6.6	19.4	515	Cloudsley-Thompson (1961)
Hadrurus arizonensis (scorpion)	xeric	1.22	3.59	2,786	Hadley (1970b)
H. hirsutus (scorpion)	xeric	25.0	74.0	135	Cloudsley-Thompson (1967)
Leirurus quinquestriatus (scorpion)	xeric	1.27	3.75	2,667	Cloudsley-Thompson (1961)
Ornithodorus moubata (tick)	xeric	4.0	12.0	833	Lees (1947)
Ornithodorus savignii (tick) (♀ unfed)	xeric	2.0	5.9	1,700	Hafez *et al.* (1970)
Hyalloma dromedarii (tick) (♀ unfed)	xeric	12.0	35.3	283	Hafez *et al.* (1970)

[a] Transpiration in arthropods from various habitats, at temperatures between 20° and 30°C.

[b] Permeability in $cm\,s^{-1}$ is the reciprocal of resistance (R) in $s\,cm^{-1}$. R is derived from the relationship,

$$R = \frac{\varDelta C}{J}$$ where C is in $mg\,cm^{-3}$ and J is in $mg\,cm^{-2}s^{-1}$. Values for permeability and resistance are actually obtained from measurements that include a boundary layer of air. R_{air} is usually much lower than R_{cut}, but in certain circumstances R_{air} may become important (see Chap. 3.B.II.).

[c] This is for larvae whose integuments have been abraded by movement through the soil. It is not, therefore, a measure of the permeability of the intact cuticle.

[d] Derived from values in terms of percentages of total weight lost, and therefore very approximate.

much higher than in insects or vertebrates. The desert cockroach, *Arenivaga investigata*, 12.1 (36.3), loses water less than one fourth as rapidly as the domestic *Periplaneta americana*. Among arachnids, permeabilities of the mesic spiders *Lycosa*, 28.3 (83.2), may be compared with 6.6 (19.4) for the desert solfugid *Galeodes*. Similar correlations have been indicated by Norgaard (1951), Davies and Edney (1952), Duffey (1962) and Almquist (1971) for other spiders.

The black African scorpion *Pandinus imperator* from hygric rain forests has a permeability of 76 (223) while desert scorpions show some of the lowest recorded permeabilities for any arthropod, 0.8 (2.35) for *Androctonus australis*, 1.22 (3.59) for *Hadrurus arizonensis*.

Two species of termites, *Heterotermes aureus* and *Reticulotermes tibialis*, have distributions in Arizona that show a relationship between cuticular water loss and local climate. *R. tibialis* loses water about ten times as fast as *H. aureus*, and occurs only in relatively moist, cool uplands, while *H. aureus* is found in adjacent but lower desert and desert grassland habitats (Haverty and Nutting, 1976).

The very large caterpillars (horn-worms) of *Manduca sexta* are at first sight exceptional. These larvae live in true deserts (as well as elsewhere), yet, as Casey (unpublished) has shown, they transpire more rapidly than any other xeric arthropod (see Table 6). The explanation is that these insects eat large amounts of jimson weed leaves (*Datura* spp.) which contain a high proportion of water, so that cuticular transpiration is probably an advantage as a means of eliminating excess water.

In a somewhat different context, permeability in the rabbit tick *Haemaphysalis* (Davis, 1974a) is much higher in engorged females and nymphs than in fasted ones, and it has been suggested that this could aid in eliminating excess water. But Davis's data show that the amount of water lost in this way is negligible compared with excretion through the salivary glands, even when the transition temperature is low enough to effect an increased transpiration in ticks living at the body temperatures of their hosts.

In the millipedes *Cingalaborus* and *Aulacobolus*, according to Rajulu and Krishnan (1968), permeability varies with season, being lower in the dry part of the year (summer). A discrete wax layer they believe, is never present, but the whole epicuticle itself appears to be absent in the winter.

The differences are great and the picture on the whole is fairly consistent: animals or their developing stages which are small, unable to replace lost water, or which live in xeric sorroundings, have low permeabilities, and vice versa.

However, it is important to recognise that integumental permeability is by no means a complete guide to ecological adaptation with respect to water balance, because it says nothing about size and therefore about the proportion of water reserves lost in unit time for unit permeability. Animals that have both low permeability as well as large size (as in several desert scorpions whose weight may be several grams) are exceedingly well adapted to withstand desiccating conditions for long periods.

This point is further illustrated in Figure 25 where rates of water loss are compared in several Namib desert insects of various sizes (Edney, 1971). In Figure 25a, $\log_e$ (percent of original weight lost in five days) is plotted against $\log_e$ (wt in mg) for one species, *Gyrosis moralesi*, and a significant negative correlation is found where $y = 5.57 - 0.69x$, and $r = 0.84$. Similar relationships are found for other species of beetles. If the different species are compared by using the rate of loss appropriate to the mean size for each species, again there is a significant negative correlation between proportion of weight lost and mean size, both on a log scale: $y = 3.72 - 0.27x$, and $r = 0.88$ (Fig. 25b).

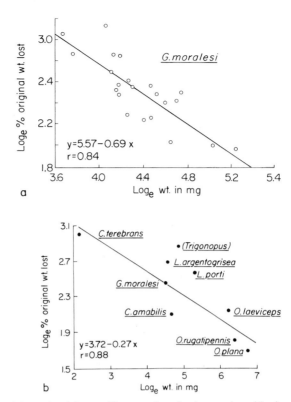

Fig. 25a and b. The relation between weight and weight specific water loss (both on a logarithmic scale) in desert beetles during 5 days at 27°C in dry air. (a) data for individuals of one species, *Gyrosis moralesi*, show a highly significant negative correlation, with a slope <1.0, so that water loss is proportional to a fractional power of weight; (b) a negative correlation also exists between log (mean weight) and log (water loss) in different species when these are compared. Differences between mean loss rates in species of about the same weight are probably due to differences in cuticular permeability or in spiracular efficiency. *Trigonopus* sp. is a mesic tenebrionid for comparison, *Ctenolepisma terebrans* is a thysanuran, and the remainder are all desert tenebrionids: *Lepidochora argentogrisea, Lepidochora porti, Gyrosis moralesi, Calosis amabilis, Onymacris laeviceps, Onymacris rugatipennis* and *Onymacris plana*. If the thysanuran in Figure 25b is omitted, the *y* intercept is 4.09, the slope is −0.33, and *r* is 0.77. (Redrawn from Edney, 1971a)

All the insects concerned are tenebrionid beetles save *Ctenolepisma*, which is a thysanuran. If the point for *Ctenolepisma* is omitted, the y intercept is 4.09, the slope is −0.33 and r is 0.77. Average weights, proportional weight losses and permeabilities are shown in Table 7.

The data illustrate several points. Firstly, they show an overall effect of size: the largest beetles *(Onymacris plana)* lose a smaller proportion of their original weight per day than the small ones when all are fasted in dry air at 27°C. Secondly some insects of similar size (e.g. *Lepidochora argentogrisea, Gyrosis moralesi, Calosis amabilis* which all weigh about 90 mg) show different weight-specific water loss rates, and such differences are probably due to differences

Table 7. Comparative rates of weight loss (presumed water) at 27°C in various species of Namib desert insects (and *Trigonopus*—a mesic beetle). (Data from Edney, 1971a)

Species	Mean wt (mg)	% wt lost in 5 days at R.H. 5%	Permeability $\mu g\,cm^{-2}h^{-1}$ mm Hg^{-1}	$10^4 \times$ cm s^{-1}	Resistance s cm^{-1}
Tenebrionid beetles					
Onymacris plana	838.5	5.6	1.53	4.50	2,222
O. laeviceps	483.9	9.1	3.41	10.02	998
O. rugatipennis	601.5	7.2	1.87	5.50	1,818
Gyrosis moralesi	85.2	12.8	2.24	6.59	1,517
Calosis amabilis	96.7	8.8	1.09	3.20	3,125
Lepidochora porti	187.5	14.2	3.49	10.26	975
L. argentogrisea	86.7	15.0	1.91	5.62	1,901
Trigonopus sp.	125.8	29.5	4.13	12.14	824
Thysanuran					
Ctenolepisma terebrans	8.7	21.1	0.68	2.00	5,000

in permeability. Thirdly, whatever the cause of differences in weight specific water loss, such differences certainly correspond with ecological facts. Thus *Onymacris plana*, which is active by day in the summer and runs on unprotected sand, has the lowest proportional water loss rate of all. Of the three intermediate sized species, *Calosis amabilis* is most desertic, being active by day in exposed areas. It might therefore be expected to have a low permeability, and indeed, as Table 7 shows, it is the least permeable of all the beetles studied. *Onymacris laeviceps*, active by twilight, has a higher proportional rate of loss, while the two species of *Lepidochora*, active largely by night, have the highest rates of loss except for *Ctenolepisma*, and this insect absorbs water vapour. The beetle *Trigonopus* sp. is a mesic tenebrionid found in pine woods, and was included in the above measurements for comparative purposes. Its proportional water loss is about twice as high as the highest desert species and its permeability is higher than all the rest.

Land crabs studied by Herreid (1969b) show a similar effect: weight specific weight loss decreases with weight from 7.0% h^{-1} for *Uca minax* weighing 0.2 g to 0.6% h^{-1} for *Cardisoma guanhumi*, weighing about 120 g, and the same is true for the scorpion *Diplocentrus spitzeri*, where log (% original wt lost h^{-1}) varies inversely with log (weight) (Crawford and Wooten, 1973).

If we consider arthropods that absorb water vapour (see Chap. 9), then the correlation between permeability and habitat is further complicated—a point illustrated by the work of Hair et al. (1975) on ticks, where size, rates of absolute and weight specific water loss, and critical equilibrium humidity (C. E. H.) (for water vapour uptake) differ in different species. The combined effect of all these parameters in the three species, *Amblyomma americanum*, *Amblyomma maculatum* and *Dermacentor variabilis*, plays a part in determining distribution. For example, *Dermacentor* lives in somewhat drier regions than the other two, is intermediate in size, has a lower absolute water loss rate than the others, and a lower C. E. H. (84–85%) than *A. maculatum* (92–93%).

Finally, it is interesting to compare the permeabilities we have been discussing with those measured, albeit by very different means, in plants. Plant physiologists think of "leaf resistance" to water vapour moving from the mesophyl cell wall to the outside air as being offered by two pathways in parallel: one consisting of the intercellular air spaces and stomatal pores in series, the other consisting of the cuticle. When the stomata are open, the total resistance is usually low (by arthropod standards), in the region of 2 to 10 s cm^{-1} for mesophytes and 5 to 20 s cm^{-1} for xerophytes. When the stomata close resistance increases, and the residual cuticular resistance is usually between 20 and 80 s cm^{-1} for mesophytes and 100 to 200 or more s cm^{-1} for xerophytes (Nobel, 1974). These values compare with about 10^4 s cm^{-1} for tsetse pupae, 3,125 s cm^{-1} for the tenebrionid *Calosis amabilis* to a minimum recorded resistance of 12.6 s cm^{-1} for the centipede *Lithobius*.

II. The Significance of the Transition Temperatures

The biological relevance of different cuticular permeabilities is clear. It is possible to think of a particular permeability characteristic as being the result of an interplay of various selective advantages and disadvantages (even though the selective advantage of high cuticular permeability can at present only be guessed at). Is it also true that the temperature and abruptness of transition is independently selected? At present this seems unlikely. Firstly, we cannot equate the onset of rapid transpiration with evaporative cooling because transition often occurs at temperatures above those normally encountered, or even above its possessor's lethal limit (about 50°C for *Tenebrio* larvae and *Rhodnius*). On the other hand transition appears to occur in some species at temperatures below the insects' preferred range. In *Thermobia*, for example, transition occurs at 28°C. There are indeed cases where lower permeability seems to correspond with higher transition temperature: for example Ahearn (1970a) found transition temperatures of 40°, 47.5° and 50°C for the desert tenebrionids *Eleodes armata*, *Cryptoglossa verrucosa*, and *Centrioptera muricata* respectively, and the beetles stand also in that order of decreasing permeability. *Periplaneta* has a low transition range and a high permeability, but there are many different shapes of temperature/ transpiration curves: sharp as in *Schistocera* adult (Beament, 1959), intermediate as in locusts (Loveridge, 1968b), hardly discernible as in *Schistocerca* nymphs (Beament, 1959), or highly variable within one individual in different physiological states, as in *Haemaphysalis* (Davis, 1974a). It has been suggested by Camin (quoted by Davis, 1974a) that in this case, a low transition range leading to rapid loss of water immediately before feeding may stimulate rapid engorgement, but this remains to be proved. Transition points may be single, in *Calliphora* (Beament, 1959) and *Arenivaga* (Edney and McFarlane, 1974), or double as in *Oniscus* (Bursell, 1955) *Rhodnius*, *Tenebrio* and *Pieris* pupae (Beament, 1959) and the scorpion *Hadrurus arizonesis* (Hadley, 1974). Unfortunately we do not yet know the molecular basis for these great differences in permeability and in permeability/temperature relationships. Perhaps the position of the transition region, when it exists, is a by-product of particular molecular structures determining permeability characteristics. Such characteristics are selected for, while the accompanying transition phenomena that necessarily follow are selectively neutral, or at least of minor importance.

D. Conclusions

It would be useful to be able to synthesise the foregoing sections into a consistent, acceptable model of integumental structure in relation to water loss. This is not at present possible, but we can at least try to assess progress so far and perhaps point to promising fields for future investigation.

In part the present muddle stems from the fact that people (the present author included) have not recognised the complexity of the system (the integument) with which they are working, and have not appreciated the biophysical stumbling-blocks along the path. (These points have essentially been made already by Beament, 1961.)

The data are indeed difficult to reconcile with each other; but in general they indicate the following: rates of transpiration, either per animal, or per unit surface, or per unit of water activity gradient, differ greatly, and this indicates differential permeability sensu stricto between and within species, although precise permeabilities cannot usually be measured because transpiration does not obey Fick's law (i.e. the resistance of component layers may vary with water concentration), and in any case, permeability varies in different regions. Resistance to the outward movement of water is perhaps provided by the apical membrane of epidermal cells, probably by structures (perhaps tanned or polymerised lipo-proteins) in the exocuticle and certainly by the structure of the epicuticle. The effect of temperature on permeability to outward movement of water varies greatly. In some arthropods (isopods, millipedes) this is negligible, in others (insects, arachnids) there is an effect, which may be slight or profound, and in these cases permeability undergoes a *progressive* increase with rising temperature, the onset of the change being gradual in some cases, but rather abrupt (giving a sharp transition point) in others. In all, however, transpiration continues to increase above the transition point or region, not only as a result of increasing ΔP but also as a result of progressively decreasing R_c. Changes in permeability are probably not due to a single change in phase of an oriented lipid monolayer, but may be due to progressive changes (perhaps phase changes from crystalline to liquid crystalline Chap. 3.B.VIII.) either in monolayers composed of mixtures of polar lipids or in multilayered, heterogeneous lipid or lipo-protein complexes. Transpiration may vary greatly with period of exposure, probably because water is lost first from the protein procuticle; and with degree of hydration. It is more rapid in dead than in living arthropods. In general (exceptions abound), arthropods with high transpiration rates have less well defined transition ranges than those that are more strongly water-proofed; high transition temperatures are often associated with low transpiration rates at biologically significant temperatures, and both transition range and basic transpiration rates vary appropriately with ecological distribution (Chap. 3.C.).

It is now fairly clear that if the immediate interest lies in the relationship of cuticle structure to water balance, and through this to arthropod biology, we shall not get very much further by continuing to measure evaporation from whole arthropods, even dead ones, because it is almost impossible to know what is measured. Much more profitable for this field of enquiry will be the measurement of the properties of isolated integuments, preferably live, and using autoradiography and labelled water for the accurate assessment of net movements. In this connection it is important to recognise that in many experimental systems water molecules move through the cuticle in both directions simultaneously, and that net flow (the parameter often measured) is simply the algebraic sum of possibly much greater gross fluxes in both directions. Work along these lines has already been started by Wharton, Knülle, Kanungo, and their associates and this will be considered later in connection with water vapour absorption (Chap. 9).

On the other hand, if immediate interest has an ecological slant it may be important to know the extent of cuticular water loss (1) in comparison with other sources of loss, and (2) in relation to questions of distribution and biological strategies for living in difficult habitats. For these purposes measurements of total transpiration and of the effect of temperature on this, are not only justifiable but essential, and useful work along these lines will no doubt continue. Such work will have increased validity if it turns out that the use of radioactive isotopes for the measurement of water fluxes and metabolic rates in free running animals proves to be as suitable for arthropods as it has been shown to be for vertebrates (Nagy and Shoemaker, 1975). Work employing such techniques is being undertaken in several laboratories and results may be expected before long (e. g. Bohm and Hadley, 1976). Such work, however, should not be expected to reveal the intricacies of integumental structure and function.

Chapter 4

Water Loss—Respiratory

A. Introduction

If necessary it is possible to reduce water loss through the cuticle of an arthropod to very low levels. However, respiratory membranes, through which oxygen is absorbed from the gaseous phase, are moist, probably because the rate of diffusion of oxygen through water is some 34 times more rapid than through "chitin" (Dejours, 1975). Respiratory surfaces, essential for life as they are, therefore constitute a potential site for dehydration in land arthropods, and the form and function of the whole respiratory system can usually be seen as a compromise between the need for adequate oxygen and the need to conserve water. For reviews of these matters see Waggoner (1967), Bursell (1970), Berridge (1970), Edney (1974), and for insect respiration in general, see Miller (1974).

The rate of oxygen consumption by living tissues is such that the surface area of any reasonably active arthropod larger than a mite is insufficient for gas exchange and has to be augmented in various ways, usually by the extension of the integument inwards in the form of tubes or plates.

B. The Tracheal System of Insects

Probably the most efficient form of respiratory equipment in arthropods is the tracheal system of insects. Essentially this is a system of tubes, branching inwards from paired, segmentally arranged openings at the surface. The larger, more superficial trunks are usually interconnected, and they may be expanded to form large, relatively limp air sacs. Further away from the spiracles, the tracheal branches become finer, and eventually end blindly in very fine tubules, less than 1 μm in diameter, known as tracheoles. The luminal surface of the whole system including the tracheoles is supported by helical or annular stiffenings, the taenidia (Richards and Korda, 1950; Whitten, 1968). In the larger branches the lining is similar to the external cuticle, but in the finer branches the surface probably represents a very fine epicuticle, where cuticulin [a stabilised lipid, probably associated with a protein (Wigglesworth, 1970)] is present. Whether or not a wax layer is also present in the finer tracheoles is doubtful (Beament, 1964).

Developmentally, the tracheal system is formed by invagination of the integument; its basal (non-luminal) cell layer is continuous with the epidermis below the cuticle, and the finest branches of the system (the tracheoles) are intracellular and are not shed at a moult (Wigglesworth, 1973).

Figure 26 shows diagrammatically the structural components of an insect tracheal unit. The opening into the spiracle is often protected by sieve-like plates

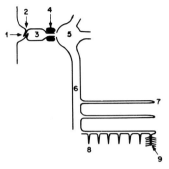

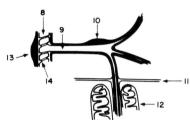

Fig. 26. The structural components of an insect tracheal system. *1*: spiracle; *2*: filter; *3*: atrium; *4*: spiracular valve; *5*: tracheal manifold; *6*: primary trachea; *7*: secondary trachea; *8*: tertiary trachea; *9*: tracheole; *10*: tracheal end-cell; *11*: sarcolemma; *12*: mitochondrion; *13*: tracheal epithelium; *14*: taenidium. (From Bursell, 1970)

or lips. Internal to this are the spiracular valves worked by muscles which usually cause closing, opening of the valves occurring by elasticity. Figure 27 is an electron micrograph to show the fine structure of a tracheole.

The diffusion constants for oxygen through air and through living tissue are 11×10^4 and 0.14 ml cm^{-2} min^{-1} atm$^{-1} \mu^{-1}$ respectively (calculated from data in Krogh, 1919). For this reason tracheolar surfaces must lie close to the sites of oxygen use. In a very active tissue such as a dragon-fly wing muscle during flight, Weis-Fogh (1964) has calculated that even with oxygen at a partial pressure of 142.5 mm Hg in the tracheoles (close to the 150 mm Hg in the atmosphere), any mitochondrion that is more than 10μ from a tracheole surface would not receive an adequate oxygen supply. Consequently, tracheoles indent the sarcolemma of muscle fibres, and thus become functionally intracellular, but they are not known to penetrate this or any other cell membrane.

Weis-Fogh (1964) has confirmed earlier calculations by Krogh (1920) that diffusion can account for most of the transfer of oxygen from outside to the tissues. However, in many insects, particularly in larger, active ones, the diffusion path is shortened either by tidal ventilation or by one-way flow of air in the main tracheal trunks and air sacs. Such ventilation of course increases the rate of water loss in dry air, and we shall refer to this again below.

The tracheoles of some insects contain water to an extent that depends on the local oxygen demand. Thus capillary forces tend to pull water into the lumen, but this is resisted [probably indirectly via the colloidal properties

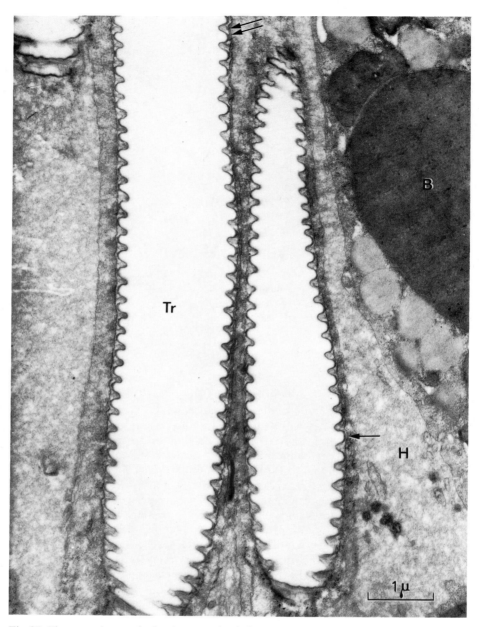

Fig. 27. Electron micrograph showing a tracheole in the developing leg of *Sarcophaga*. Note the presence of taenidia and the very thin layer of cytoplasm *(arrows)* between tracheolar lumen *(Tr)* and haemocoele *(H)*. *B*: hemocyte. An unpublished micrograph by the late Dr. Whitten similar to one in Whitten (1968); by permission of Dr. Robert C. King

of the tracheolar wall (Wigglesworth, 1972)] by local osmotic concentration, and since the latter is higher in the absence of oxygen (Wigglesworth, 1938), more tracheolar surface is exposed to air when the need for oxygen is greater (Fig. 28). Other tracheoles, e.g. those in the epidermis of *Rhodnius*, are always filled with gas (Wigglesworth, 1954).

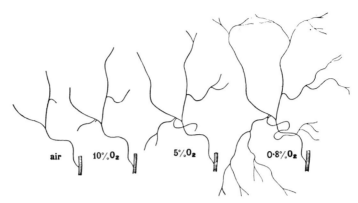

air   10%O₂   5%O₂   0·8%O₂

Fig. 28. The extent of gas in a group of tracheoles in the abdomen of a flea at rest, exposed to different concentrations of oxygen. The lower the oxygen concentration the greater the area of tracheolar lumen exposed by withdrawal of fluid. (From Wigglesworth, 1972)

It has been suggested, or implied, that the rate of water loss is in some way reduced by the fact that the moist respiratory membranes are situated at the ends of long tracheae (Edney, 1957) or are protected in other ways. In particular, the elytra of desert beetles are believed to reduce water loss by creating high humidities in the subelytral cavity into which some of the spiracles open (Ahearn, 1970a). And the sunken, hair-covered spiracles of xeric buprestid beetles are believed to have the same effect, (Bergold, 1935; Hassan, 1944; Hadley, 1972; Wigglesworth, 1972). There is, however, no experimental evidence one way or the other.

Probably the rate of loss of water is inversely proportional to tracheal length. But since oxygen moves by diffusion not only through a tracheal path, but also through a short intracellular path before reaching the mitochondrion, the effect of increasing tracheal length on resistance depends upon what fraction of the total resistance is offered by the trachea. Resistance to oxygen diffusion is about 7.8×10^5 times greater in muscle than in air, so that even if the intracellular path is very short (say 0.005 mm) compared to the tracheal path (say 5 mm) the intracellular compartment may still afford almost all the resistance. In this case, doubling (for example) tracheal length would increase total resistance insignificantly. However, it is very difficult to get a good estimate of cellular resistance because of unknown effects of shape, size and other parameters, and further progress with the problem must await further experimental evidence.

There is no doubt of the adaptive advantage of a spiracular closing mechanism. Such a device imposes a very high resistance to the outward movement of water vapour—so that respiratory water loss may be restricted to those periods when the valve must be open to permit ventilation or the inward diffusion of oxygen.

The problem of water conservation would be solved (speaking teleologically) if an arthropod possessed a membrane which was permeable to oxygen but not to water. A layer of air $1\,\mu$ thick imposes resistances to oxygen, carbon dioxide and water vapour of 5.5×10^{-4}, 7.1×10^{-4} and $4.2 \times 10^{-4}\,s\,cm^{-1}$ respectively, but in all known membranes the resistance to water vapour is proportionately less (Waggoner, 1967). Consequently there is no known way in which an arthropod's surface may be waterproofed while remaining permeable to oxygen and carbon dioxide.

In summary, those aspects of the insect tracheal system that concern us most are (1) the fact that the walls of the finer branches are permeable to water, and that their total surface is very large, so that the system is potentially a source of great water loss; (2) ventilation of the larger trunks may increase the potential for water loss, for this would otherwise occur by diffusion only and (3) since contact of the whole system with the outside is reduced to a small number of spiracles, water loss may be reduced to a minimum by closing the spiracles except when the need for oxygen is paramount.

We shall now consider evidence about the functioning of this system.

C. Control of Respiratory Water Loss

I. Tsetse-Flies and Other Insects

Measurements of respiratory water loss have been made in several insects. Those of Bursell (1957a, 1974b) on tsetse-flies are particularly useful because they show not only that respiratory water loss may be controlled by spiracular closure, but also that such closure probably occurs to an extent appropriate to the ambient humidity, as shown in Figure 29. If flies are exposed to different

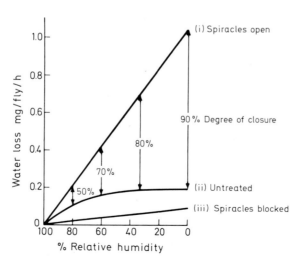

Fig. 29. The rate of water loss from tsetse-flies at different relative humidities, (*i*) in air containing 15% carbon dioxide; (*ii*) in air; (*iii*) in air with spiracles blocked. The deduced degree of effective closures of the spiracles is indicated at four different levels of humidity. (From Bursell, 1970)

humidities while the spiracles are kept open by 15% carbon dioxide, the rate of water loss increases linearly with decreasing relative humidity. If the spiracles are blocked, water loss also increases linearly, but is very much lower, probably representing cuticular loss. If the spiracles are left free to move, however, the rate of loss is intermediate, and no longer linear, but falls off considerably in lower humidities. The extent of this decrease is believed to represent the effects of spiracular regulation, and is shown in Figure 29 as a percentage of total closure. Whether this represents continuous partial closure, however, or total closure for different proportions of the time is uncertain (Bursell, 1957a).

Similar spiracular control is apparent in the cricket *Gryllus* (Jakovlev and Kruger, 1953) and *Drosophila* (Arlian and Eckstrand, 1975). In other insects, including aphids (Lamb, 1968), *Aphodius* (Maelzer, 1961), *Ephestia cautella* (Navarro and Calderon, 1973) and *Glossina* pupae (Bursell, 1958), transpiration is linear with humidity. Examples of these relationships are shown in Figure 30.

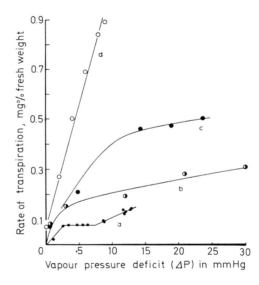

Fig. 30. The rate of transpiration from different insects as a function of vapour pressure gradient (ΔP). *a: Gryllus* at 16°C; *b: Cimex* at 29°C; *c: Glossina* at 25°C; *d: Aphodius* at 13°C. In *a*, *b*, and *c*, the data suggest that transpiration is not linear with ΔP, but falls off at lower ambient humidities, perhaps as a result of spiracular closure. (From Bursell, 1974b).

In *Glossina*, spiracular regulation depends not only on ambient humidity as we have seen, but also on the extent of the flies' water reserves—a relationship shown in Figure 31. Here again there is no difference if the flies are kept in 15% carbon dioxide, when the spiracles remain open, but when the spiracles are free to operate, transpiration into dry air is lower in flies with low water contents. The rate of transpiration from *Glossina* pupae is similarly affected by their water reserves—the latter itself being determined by water loss during the larval stage (Bursell, 1958).

All this evidence is partly circumstantial, since the spiracles were not observed to open differentially in the various conditions—their states were deduced. However, as we shall see later, the correctness of the interpretation is supported by the observed effects of humidity and of blood osmotic pressure on spiracular

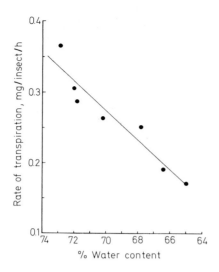

Fig. 31. The relation between water content and rate of transpiration from *Glossina morsitans* in dry air at 25 °C. In tsetse-flies, control of respiratory loss occurs both when ambient humidity is low (Figs. 29, 30) and when the body water reserves are depleted (*above*). (From Bursell, 1974 b)

state in mosquitoes (Krafsur, 1971a, b), dragonflies (Miller, 1964a) and other insects (Chap. 4.D.).

A point of general importance in relation to work on respiratory water loss is well made by Bursell (1974 b), namely that loss depends in part on activity, as Ramsay (1935) originally found for cockroaches, and activity itself may depend on humidity (Perttunen and Hayrinen, 1970; and other refs in Bursell, 1974 b). Thus a tendency for the spiracles to close in response to dry air may be partially or totally countered by the effects of a greater metabolic rate, which keeps them open. The advantage of working with inactive stages, such as pupae, is clear. For active insects, reported relationships between transpiration and humidity should be treated with caution unless activity is allowed for.

II. Locusts

In living *Locusta*, Loveridge (1968 a, b) found that cuticular water loss is less than half the total loss, and that within the limited range permitted by his experimental situation activity does not affect total water loss (and hence spiracular loss). As in tsetse-flies, spiracular water loss is proportionately lower in low humidities (Fig. 32) to an extent that saves up to 5.0 mg water per locust h^{-1} at 30 °C in dry air. For an insect which weighs 1,600 mg this is a substantial saving. However, it seems that spiracular control is only of limited importance in locusts, ventilatory movements being of much greater significance in relation to water loss, as shown in Table 8. In 20% carbon dioxide the spiracles fluctuate between open and closed, abdominal ventilatory movements are large and rapid, and water loss is 10.1 mg g^{-1} h^{-1}—about double that in air. In 50% carbon dioxide the spiracles are held open, ventilatory movements still occur and water loss is more than double that in air. Finally, in 80% carbon dioxide, the spiracles

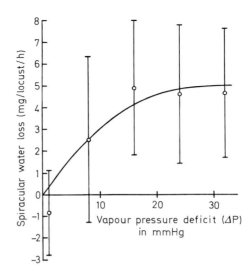

Fig. 32. The relation between vapour pressure gradient (ΔP) and spiracular water loss in *Locusta* at 30°C. (From Loveridge, 1968a)

Table 8. The effect of carbon dioxide and ventilation on spiracular water loss in *Locusta migratoria*. (Data from Loveridge, 1968a)

Concentration of CO_2	Condition of spiracles	Ventilation	Water loss in $mg\,g^{-1}h^{-1}$
Air	Closed	Absent	5.4
20%	Fluctuating	High	10.1
50%	Open	High	11.9
80%	Open	Absent	5.8

are wide open but ventilatory movements have ceased, and water loss is down to normal. If the temperature is raised from 30°C to 42°C ventilation becomes violent and water loss rises dramatically.

If locusts are fasted in dry air instead of 96% R.H. for 24 h their subsequent rate of water loss is reduced, and once again the ventilatory movements are also reduced (Table 9). The assumption behind this conclusion is that ventilatory movements are reduced in dry air below "normal" in moist air. However, it seems open to question whether the relatively high rate in moist air is not

Table 9. The effect of body water content on ventilation and water loss in *Locusta migratoria*. (Data from Loveridge, 1968a)

Pre-treatment during fasting for 24 h	Abdominal movements min^{-1}	Water loss in $mg\,g^{-1}h^{-1}$
Dry air	20	3.2
96% R.H.	34	5.3

so much a reflexion of normal oxygen demand (for if so the "pre-dry" insects would suffer from anoxia) but of a need to eliminate excess water, after having been in very moist air for 24 h.

At all events, ventilation certainly plays a preponderant role in the water conservation mechanisms of locusts, and it is interesting that such control may be triggered both by ambient humidity and by water reserves of the insect.

Normally, in resting locusts, air flow within the large tracheal trunks is unidirectional (Miller, 1960a, b, c), the first two spiracles being used for inspiration, the tenth for expiration; and Buck (1962) has suggested that this might conserve water. The water content of air in the tracheae is unknown. Beament's (1964) observations in which he found that a very small drop in temperature (less than 0.2°C) is sufficient to cause the condensation of droplets inside a trachea of a cockroach, strongly suggest that the air there is practically saturated, but Hamilton (1964) has evidence that in the main trunks bursts of ventilatory activity tend to flush out moist air. In any case, it is not clear how unidirectional, rather than tidal flow, would reduce water loss, unless air is expired at a lower temperature than that at which it becomes charged with water vapour (Heinrich, 1974), and the matter deserves further experimentation.

The amount of water lost from the respiratory system is certainly considerable, particularly during ventilation, and experimental results agree fairly well with prediction, as the following calculation shows. If we assume that all expired air is saturated, then at 30°C, the water lost by any insect in dry air will be that contained by air saturated at that temperature, namely, 0.03 mg ml^{-1}. Weis-Fogh (1967) found that in resting *Schistocerca* air is expired at $30 \text{ ml g}^{-1}\text{h}^{-1}$, which represents a loss of 0.9 mg water $\text{g}^{-1}\text{h}^{-1}$. An active locust, however, expires about $320 \text{ ml g}^{-1}\text{h}^{-1}$, or 9.6 mg water $\text{g}^{-1}\text{h}^{-1}$. Loveridge's experimental measurements of water loss in *Locusta* gave 2.9 and $11.9 \text{ mg g}^{-1}\text{h}^{-1}$ respectively in resting and active insects.

D. Spiracular Control Mechanisms in Dragon-Flies and Other Insects

The muscles that control the spiracular valves of many insects are known to respond to carbon dioxide. This is true for dragon-flies (Miller, 1962) and locusts (Hoyle, 1960) amongst others, in both of which carbon dioxide acts directly on the closer muscles and causes them to relax. We have seen in the previous section that spiracles play a part in water conservation in some insects, closing more fully, or more frequently, when the insect is short of water. The question now arises as to the causal link between water status and spiracle condition. Miller's work (1964 a, b) with dragon-flies goes a long way towards supplying an answer for that insect at least.

If the dragon-fly *Ictinogomphus ferox* is allowed to become fully hydrated by drinking water up to 10% of its total wet weight, then in an atmosphere of 3% oxygen in nitrogen spiracle 2 remains nearly 100% open, whereas in a dehydrated dragon-fly the same spiracle is almost closed (Fig. 33). Further, the spiracle in dehydrated insects closes immediately, after even a short flight, but it may remain open in hydrated ones. Finally, in 2% carbon dioxide, the spiracle opens completely and continuously unless the insect has previously lost water, when it remains closed.

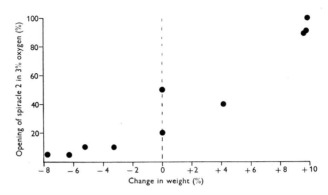

Fig. 33. If dragon flies (*Ictinogomphus ferox*) are allowed to gain water by drinking, or are dehydrated, and then flown, the degree of spiracular opening during subsequent exposure of 3% oxygen in nitrogen is determined by the extent of the insect's water reserves. (From Miller, 1964a)

The above evidence confirms that in dragon-flies, as in locusts and tsetse-flies, spiracular closure is affected not only by respiratory requirements, but by the state of hydration of the insect. By measuring the frequency of impulses in motor nerves to the spiracle closer muscles, Miller established that in the presence of carbon dioxide the frequency of nerve impulses necessary to cause contraction of the closer muscle is raised, and further that in dehydrated insects, the frequency of impulses to the closer muscle is higher than in hydrated ones. In other words, during water shortage a higher concentration of carbon dioxide locally is necessary to open the valves, and in this way water is conserved. Lack of oxygen has the opposite effect, decreasing the frequency of motor impulses, thus tending to prevent spiracle closure. The effects of dehydration can be mimicked by irrigating the mesothoracic ganglion (from which the motor nerve concerned originates) with a saline solution, although it seems that the response is caused not by an overall high osmotic pressure, but rather by high concentrations of one or another of the solutes.

A tentative model showing the various factors involved in spiracle control, and the pathways of their effects according to Miller (1964b) is shown in Figure 34.

Effects similar to the above have been observed by Krafsur (1971a, b) in *Aedes* mosquitos, where exposure of 5 min or less to 5% R.H. reduces the duration and frequency of spiracular opening in response to 5% carbon dioxide. In this insect the effect is greater in fed than in unfed individuals, the duration of opening (but not the frequency) declines with age, and potassium in the

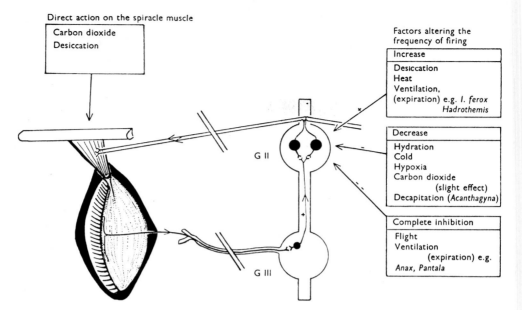

Fig. 34. The valve of thoracic spiracle 2 in dragon-flies is closed by contraction of a muscle. Locally high carbon dioxide concentration causes the muscle to relax, but the concentration necessary to effect this relaxation itself depends on several other factors, including the temperature and the water content of the whole insect. These effects are exerted by appropriate changes in the frequency of nerve impulses in the relevant motor nerve. The above diagram (based on data from several species) summarizes some of these relationships and indicates the pathways that may be involved. *G II, G III*: 2nd and 3rd thoracic ganglia. (From Miller, 1964b)

diet increases spiracular control. Hoyle (1961) also found that in locusts, high blood potassium is necessary for the local control of spiracles by carbon dioxide.

Further work in this field will certainly be rewarding. At present it seems that active ventilation, when it occurs, leads to much water loss. Spiracular water loss in the resting insect, even with spiracles open, is rather low, and spiracular control in this case acts in the manner of a fine adjustment. However, spiracular control when linked with intermittent carbon dioxide release may be effective for water conservation, as the following section suggests.

E. Intermittent Carbon Dioxide Release

The early work of Punt (1950) and Punt *et al.* (1957) showed that carbon dioxide release may be intermittent in several insects. Further analysis by Buck and Keister (1955) and Schneiderman and Williams (1953, 1955) showed that in pupae of *Agapema* and *Hyalophora*, oxygen uptake is continuous even though carbon dioxide release is indeed intermittent. The functional anatomy of the spiracular mechanism in *Hyalophora* was described by Beckel (1958), and Buck

(1958) proposed a model involving the continuous inward movement of air through the spiracles in response to a partial intratracheal vacuum caused by oxygen uptake, to explain the observations. This model has been modified and amplified by the valuable and percipient work of Schneiderman and his school (Schneiderman, 1953, 1960; Levy and Schneiderman, 1966a, b, c; Schneiderman and Schechter, 1966; Brockway and Schneiderman, 1967; Burkett and Schneiderman, 1974a, b) to provide an admirably full and well-documented account of the process in the diapausing pupae of *Hyalophora*—an account that may well be satisfactory for other insects as well (Fig. 35).

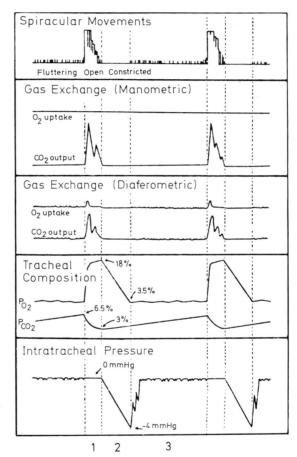

Fig. 35. Summary of the burst cycle for carbon dioxide release in a diapausing pupa of *Hyalophora*. For further discussion see the text. (From Levy and Schneiderman, 1966c)

In this insect a complete cycle occupies several hours. At the beginning of the first phase (1 in Fig. 35) a high P_{CO_2} causes the spiracles to open fully and they remain so for some thirty minutes while some $350 \mu l$ of carbon dioxide in a 5 g pupa are released, most of which must come from bicarbonate stores since the total tracheal volume is about $400 \mu l$. At the end of this phase, P_{CO_2} has fallen from 6.4% to 3.0%, and the spiracles close. In the second phase, P_{O_2} falls to 3.5% and the intratracheal pressure falls to -4 mm Hg. Presumably a small flow of air must pass through the closed spiracles, since oxygen uptake continues although there is no discharge of carbon dioxide or nitrogen. During the third, and longest phase, the spiracles open briefly for seconds at a time, internal pressure rises in a stepwise manner, leading to a period of microcycles when the internal pressure cycles between -0.03 and -0.15 mm Hg. There is some nitrogen loss, but P_{CO_2} gradually builds up until it triggers the return of phase 1, when carbon dioxide and water vapour are lost rapidly (Kanwisher, 1966).

For this process to work, carbonic anhydrase is necessary to permit the rapid release of carbon dioxide from solution in the form of bicarbonate. This enzyme is in fact known to occur in insect tissues (Buck and Friedman, 1958; Edwards and Patton, 1967), although it has only once been demonstrated in insect blood (Byzoua, 1974). Control of the spiracular movements (which is of course critical for the whole process) is exercised by the direct effect of P_{CO_2} on the spiracles themselves, and by the effect of hypoxia on the ganglia involved. Burkett and Schneiderman (1968, 1974 a, b) have found that neuromuscular control of the spiracles continues down to $-5\,°C$, but ceases below $-10\,°C$ when the spiracles freeze, probably in the closed position (Burkett and Schneiderman, 1974 b).

The advantages of discontinuous carbon dioxide release may lie in water conservation according to Buck and Keister (1955), Buck (1958), and Kanwisher (1966). During phases 2 and 3 the slow inward movement of air perhaps prohibits the outward diffusion of water (as well as carbon dioxide). If so then the time available for water loss is restricted to phase 1, and this may result in reduction, although it is not necessarily true that continuous though very partial opening would cause extra loss. Buck's (1958) calculations for a hypothetical situation are consistent with such conservation but conclusive experimental evidence is so far lacking. Sláma (1960) did indeed find that pupae whose normal environment is in high humidities do not show cyclical carbon dioxide release; but as he points out, these are saw-fly not lepidopteran pupae, and the evidence is thus only circumstantial. Further, as Schneiderman and Williams (1955) found, low P_{O_2} tends to reduce or abolish the cyclical nature of carbon dioxide release in their pupae, and Sláma's saw-fly pupae live in moist soil where P_{O_2} is generally rather low.

Other insects in which cyclical carbon dioxide release has been observed include *Triatoma, Rhodnius, Carabus, Popillia* pupae, *Agriotes* pupae, and others (Punt, 1950, 1956; Punt *et al.*, 1957), but often with rather short periods. In *Triatoma* a cycle occurs every few minutes, and such a short period raises the question whether the process is comparable with the typical several-hour cycle of *Hyalophora* pupae.

In *Periplaneta* (Wilkins, 1960), and in *Schistocerca* (Hamilton, 1964), each cycle lasts about 2.5 min and corresponds with a ventilation cycle. Here again water conservation has been suspected, but the evidence is not compelling. In locusts, pauses between ventilatory movements become longer in drier air (Loveridge, 1968 a), and Ahearn (1970 a) suggests that unidirectional air flow and discontinuous ventilation may conserve water in the desert tenebrionid *Eleodes*. However, as he implies, the evidence is equivocal.

Kanwisher (1966) suggests that because *Hyalophora* pupae use fat as a fuel during cyclic carbon dioxide release they could even show a slight increase in water content. However, this seems unlikely; for the same energy output, oxidation of carbohydrates yields more water than fats do (Chap. 8.C.).

F. Respiratory and Cuticular Water Loss Compared

No general statement applicable to all arthropods at all times regarding respiratory versus cuticular water loss is possible because cuticular permeability is so variable between species, and also because respiratory water loss varies enormously according to degree of activity. Loss through both avenues is, of course, strongly affected by ambient humidity.

In tsetse-flies, data obtained by Bursell (1957a, 1959b), when set out as in Table 10, show that cuticular loss represents 75% of the total in resting

Table 10. Water loss in mg per fly h^{-1} from tsetse-flies in dry air at 25°C. (Data from Bursell, 1957a, 1959b)

	Flies at rest	Flies active 30% of the time
Spiracular	0.03	0.12
Cuticular	0.09	0.09
Total	0.12	0.21
Cuticular as % of total	75	43

flies, but only 43% when the flies are active for 30% of the time. Similarly, in flying *Schistocerca*, about 65% of the water loss is spiracular, while in the resting insect very little is lost via that route. In flying *Locusta*, cuticular loss is only about one-fourth of total loss (Loveridge, 1968a). In the tenebrionid beetle *Eleodes armata* where secretion of defensive, quinone-containing fluids may account for up to half of the total water loss, Ahearn (1970a) found that of the transpiratory loss, spiracular is less than 3% at 24°C (less than 2% of total loss), but increases progressively to 36% at 40°C (about 20% of total loss) as a result of increased metabolic activity. In the same species, Bohm and Hadley (in preparation) found that complete deprivation of food and water in simulated summer conditions in the laboratory leads to a four-fold decrease in surface area-specific rate of water loss. Whether this is due to a change in cuticle permeability, to spiracular control, or to some other mechanism is not yet clear. Kanwisher (1966) found that water loss increases by as much as twenty five times when the spiracles of *Hyalophora* are fully opened during carbon dioxide bursts, but this was only for short periods.

The relationship between total oxygen uptake and water loss has not been fully investigated. One would expect water loss to increase with oxygen uptake if the spiracles have to be kept open to a greater extent or for longer periods, but if enhanced oxygen diffusion inwards is caused by a lowering of the P_{O_2} at the tracheolar surface (thereby increasing the gradient along the tracheal tube), this by itself should not affect water loss. What evidence there is suggests that the relationship is not a simple one: in tsetse-flies, for example, the metabolic rate at 30% R.H. and 25°C increases by a factor of 22 during flight, but the water loss increases only six times (Bursell, 1959b). This implies that for each mg fat oxidised, 4.5 mg water are lost during flight, compared with 1.2 mg mg^{-1} in flies at rest. The ability to reduce P_{O_2} in the tracheal system from 18% to 3.5% shown by pupal *Hyalophora* (Levy and Schneiderman, 1966c) (Fig. 35) is remarkable and could be of value for water conservation. For resting man the figures are only from 19.7% (in inspired air) to 15.5% (in expired air) (Dejours, 1975).

This is a technically difficult area for research, but nevertheless one well worth further exploration.

G. Respiratory Water Loss in Arthropods Other Than Insects

Organs associated with gas exchange in land arthropods include the book lungs of spiders, gills of land crabs, pleopods of isopods, and the poorly developed tracheae of some spiders and millipedes. In most cases where the question has been investigated, water loss from these surfaces has been found to be high.

I. Crustaceans

There is little evidence about the loss of water from the gills of crabs. These are reduced in number and size in terrestrial forms, as might be expected from the fact that the availability of oxygen in air is far greater than that in water, in terms of volume per unit volume (refs in Edney, 1960a; Bliss, 1968). But the gills must be kept moist, and the gill chambers of *Uca* spp. are irrigated for this purpose.

Terrestrial isopods provide an interesting series from littoral forms such as *Ligia* spp. to much more xeric animals such as *Armadillidium* spp. *Venezillo arizonicus* and *Hemilepistus* spp., and the respiratory surfaces are suitably constructed (Edney, 1954, 1968a). In the more hygric species, the pleopods are simple plate-like surfaces on the ventral pleon, while the more mesic isopods have short, tuft-like invaginations of tubules in certain pleopods. These "pseudo-tracheae" as they are called, serve to increase the surface for gas exchange, and if their orifices are closable, perhaps by apposition to adjacent pleopods (they lack valves or sphincters), water loss could also be controlled (Fig. 36). However, in several species oxygen uptake occurs through the general body

80

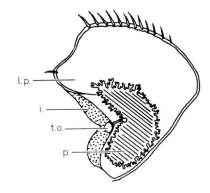

Fig. 36. Exopodite of the first pleopod of an isopod (*Porcellio scaber*) to show the extent of the pseudotracheae. *i*: interior border of the pleopod; *l.p.*: lamina of the pleopod; *p*: pseudotracheal area; *t.o.*: tracheal opening (shown by the presence of an air bubble). (From Edney, 1960a, after Verhoeff)

surface as well as the pleopods (Edney and Spencer, 1955), but this requires a moist surface, and then the total water loss is greater (Edney, 1951b; Table 11). The proportion of the total water loss that is respiratory varies from 25% to 42% in different species, but these values represent loss from the whole pleopodal area of the body rather than from the respiratory surfaces alone.

Table 11. Oxygen uptake and water loss in isopods

Species	Habitat	% of normal O_2 uptake through integument if the pleopods are blocked[a]	Total water loss as % original wt h^{-1}[b]	Proportion that is pleopodal[b]	Area-specific rate of water loss (arbitrary units)[b]	
					Dorsal surface	Gill area
Ligia oceanica	littoral	52	10.6	0.25	11.5	58
Oniscus asellus	hygric	52	—	—	—	—
Porcellio scaber	mesic	34	6.9	0.42	8.4	97
Armadillidium vulgare	mesic	26	4.5	0.32	6.5	83

[a] Data from Edney and Spencer (1955).
[b] Data from Edney (1951b).

II. Myriapods

Some myriapods have a well developed tracheal system with occlusible spiracles (Blower, 1955; Lewis, 1963; Crawford, 1972). According to many authors the tracheal system is the major site of water loss (Kevan, 1962; Toye, 1966; O'Neil, 1969), but Cloudsley-Thompson (1950) found the reverse in *Paradesmus*, where exposure to carbon dioxide causes no increase in water loss (further refs in Stewart and Woodring, 1973). In a thorough study of two contrasted species, *Pachydesmus crassicutis* from a hygric habitat and *Orthoporus toxicolens* which is more mesic, Stewart and Woodring (1973) concluded that both species lack spiracular closing valves, but in *Orthoporus*, the spiracles may be closed by diplosegmental overlap and coxal appression. At 24 °C in 27% R.H., *Orthoporus* loses water equal to about 2.5% of its weight per day, of which about half is cuticular; while in *Pachydesmus*, total water loss is 26.7% per day, including only 3% as the cuticular component—a remarkably high loss from the respiratory surfaces. *Paradesmus gracilis* loses about half its weight in 6 h at 20% R.H. and 25 °C, so that it must have either a much more permeable cuticle than that of *Pachydesmus*, or a very high respiratory water loss (or both). Cloudsley-Thompson (1950) did not exclude respiratory loss.

Crawford (1972) also found that *Orthoporus* loses water less rapidly than other millipedes, and that coiling reduces evaporative loss. Increased activity, brought about by higher temperatures from 35° to 40 °C, results in increased water loss, but cuticle permeability is apparently unaffected by this temperature change (except in very large specimens).

Turning now to centipedes, the common belief that all lithobiomorphs have spiracular closing devices while geophilomorphs do not (refs in Lewis, 1963) has recently been questioned by Curry (1974) who found no such mechanism in species of either group. Nevertheless, Lewis (1963) found that *Lithobius variegatus* has a higher rate of water loss than does the geophilomorph, *Strigamia*. He also found a correlation between spiracular cup structure and rate of water loss in four species of geophilomorphs, but a causal relationship was not demonstrated since he did not measure differences in cuticular permeability, and this may have been at least partly responsible. Curry (1974) suggests that the hair-like processes (trichomes) that line the atrial cavities of all species may serve to hold the spiracles open when body movements might tend to block them. But other functions, such as particle filtration, have also been proposed (Kaufman, 1962). The partitioning of water loss between respiratory and cuticular pathways in centipedes has not been investigated, except for some preliminary work by Curry (1974) which suggests that cuticular loss is at least as great as spiracular.

III. Arachnids and Others

Among the Arachnida are found the smallest (mites) and some of the largest (scorpions) of terrestrial arthropods. Special respiratory organs are absent in some mites, and book lungs are the predominant form in spiders and scorpions. The sun spiders (solpugids) are aberrant—not only do they have a segmented opisthosome ("abdomen") but their tracheal system is as fully developed as that of insects.

In the spider mite *Bryobia*, water content is regulated between 66% and 68% of wet weight in all humidities (Winston and Nelson, 1965), but mites exposed to 10% carbon dioxide equilibrate at a lower level. Water loss is about three times higher in dead than in living mites. (It is at least somewhat higher after death in nearly every arthropod species that has been studied.) Mites treated with carbon dioxide lose water at an intermediate rate. Presumably, respiratory water loss is controlled by spiracles in *Bryobia*, and may be rather higher than cuticular loss when the spiracles are fully open, but the proportion of total loss that is respiratory in resting mites is unknown.

82

Spiracular control was also observed in the spider mite *Tetranychus telarius* by McEnroe (1961). In several mites the situation is complicated by the fact that there may be a net absorption of water vapour in high humidities, and this is discussed further below.

In ticks, Lees (1946a) found that water loss takes place almost entirely through the cuticle. *Ixodes ricinus*, for example (which has occlusible spiracles according to Falke, 1931), when exposed to 5% carbon dioxide, shows no increase in water loss. On the contrary, blocking the spiracles actually leads to enhanced water loss, perhaps as a result of physiological injury caused by excessive dehydration or by disruption of the cuticular wax layer. Respiratory loss is certainly very low compared with that through the cuticle. In *Ornithodorus moubata* also, Browning (1954) found that unfed nymphs suffer very little increase in water loss when exposed to 5% carbon dioxide—possibly because active water vapour-absorbing areas are present in the spiracles and tracheae—but in engorged ticks and in unfed ones in higher carbon dioxide concentrations, total water loss is high, perhaps because the water absorption mechanism breaks down.

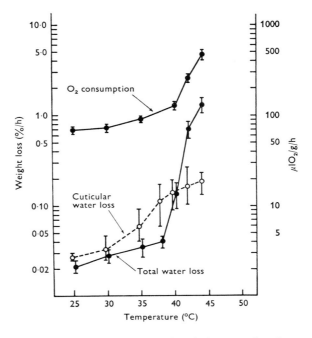

Fig. 37. The effect of temperature on oxygen uptake and on total and cuticular water loss from the scorpion *Hadrurus arizonensis* in dry air. Vertical lines represent 95% confidence limits. Cuticular loss appears to be higher than total loss because the former was measured in dead scorpions, the latter in living ones. A marked increase in oxygen consumption above 35 °C indicates increased activity which may be responsible, at least in part, for the steep increase in transpiration. (From Hadley, 1970b)

In engorged ticks, spiracular loss appears to be greater than cuticular—if 5% CO_2 does in fact cause the spiracles to open. In *Hyalomma*, and in *Ornithodorus* there is a good spiracular control according to Hefnawy (1970), the mechanism being sensitive to carbon dioxide in unfed, but not in engorged females. In these ticks, as in tsetse-flies, low ambient humidities lead to stricter spiracular control.

Kaufman and Philips (1973a) did not distinguish between cuticular and spiracular loss in *Dermacentor andersoni*; however, they found that together these two avenues account for less than 3% of the total water eliminated by an engorged tick, the remainder being excreted by the salivary glands as described below (Chap. 6.C.III.).

Mites, like ticks, are believed to have an epicuticular water barrier (Gibbs and Morrison, 1959), and in some of those that have tracheae, the openings are occlusible by movements of a mandibular plate. In *Tetranychus telarius* which feeds copiously on watery plant fluids (probably cell contents), diffusion from the tracheae accounts for a large part of total water loss in low humidities (McEnroe, 1961). Sealing the tracheal openings causes a reduction in water loss to one fifth. The remarkable mechanisms whereby these mites short-circuit and excrete the excess water in their food is referred to below (Chap. 6.C.III.).

As regards scorpions, Hadley (1970b) found that in the desert form *Hadrurus arizonicus*, total water loss increases greatly after death, and he attributed this to failure of an active cuticular water control mechanism. Cuticular loss exceeds respiratory loss (from the book lungs via the spiracles) at temperatures up to 38 °C, above which respiratory loss increases greatly, as does the total metabolic rate. Cuticular loss was measured in dead scorpions with blocked spiracles, and is actually greater than the total water loss as measured in live scorpions with spiracles open, at all temperatures below 40°C, so that the separate contributions of respiratory and cuticular loss are difficult to assess (Fig. 37). Overall water loss is low in desert scorpions (Chap. 3.C.). In *Hadrurus* it amounts to only 0.028% wet wt h^{-1} at 30°C in dry air.

Crawford and Wooten (1973), working with the more mesic scorpion *Diplocentrotus spitzeri*, found circumstantial evidence for restriction of water loss by spiracle closure at temperatures between 30 and 40°C. They also found an increase, both in total water loss and in oxygen uptake at higher temperatures (as did Hadley with *Hadrurus*). They suggest that respiratory water loss is very small in all but young scorpions at very high temperatures. But again, cuticular loss was measured in dead animals, total loss in living ones, so that the allocation between the two avenues is not clear.

In araneid spiders, there does appear to be a general correlation between cuticle permeability and habitat according to Davies and Edney (1952), Nementz (1954), Vollmer and MacMahon (1974), and other authors quoted by Vollmer and MacMahon (1974), but very little is known about respiratory water loss in these animals.

Davies and Edney (1952) found that in *Lycosa amentata*, 10% carbon dioxide causes total transpiration to increase from 16% to 23% of body weight in 24 h and they attributed this to opening of the spiracles. In dead spiders, transpiration increases from 16% to 19%/day, but carbon dioxide has no further effect.

With book lungs blocked, living spiders (which survive well for 15 min exposures) show a decrease in rate of transpiration by about a third to a fifth at temperatures up to 40°C, above which evaporation is much more rapid. Consequently, respiratory water loss represents about a third or less of total transpiration in this species.

Seymour and Vinegar (1973) measured water loss in the tarantula *Aphonopelma* sp (Table 12) but they did not distinguish between respiratory and cuticular loss. In another tarantula, the common American *Dugesiella hentzi*, Stewart and Martin (1970) found that 15% carbon dioxide causes the rate of water loss to double (from about 2.5% to about 4.4% of original weight in 24 h), presumably as a result of open spiracles.

Table 12. Water loss from the tarantula *Aphonopelma* sp. Numbers are means of four animals except those asterisked which are means to two animals. (Data from Seymour and Vinegar, 1973)

	10°C	30°C
Initial wt (g)	5.99	5.81
Time to death (days)	15.50*	11.25
Rate of loss as % initial wt day^{-1}		
a) living	0.34	1.76
b) dead	0.47*	2.50
% of initial wt lost at death	5.53	18.32

Finally, in the uropygid (whip-scorpion) *Mastigoproctus giganteus*, Ahearn (1970b) measured oxygen uptake and cuticular and total water loss at various temperatures. A transition point is apparent and this is referred to above (Chap. 3.B.III.). Cuticular loss was measured in smaller, dead animals after spiracle blockage, so that interpretation is again difficult. Ahearn suggests that respiratory water loss is less than cuticular at all temperatures, but the evidence is inconclusive. Crawford and Cloudsley-Thompson (1971) did not attempt to differentiate between respiratory and cuticular loss in *Mastigoproctus*.

H. Conclusions

It is only in small arthropods that special respiratory organs are absent and gas exchange occurs through the integument. Examples include collembolans, some proturans, some mites and some ticks. In other perhaps primitive apterygotes, even when tracheae are present, these are without closing mechanisms (Hassan, 1944); but most of these arthropods live in moist habitats. The rather primitive gill-like pleopods of certain isopods *(Ligia, Oniscus, Philoscia)* seem to serve very well for mesic habitats, especially when pseudotracheae are present

(Porcellio, Armadillidium). For the rest, there is no clear ecological association of tracheae or of book lungs with particularly dry habitats, for both are found among the most hygric and most xeric of land arthropods. The main difference between these two systems is independent of water affairs, namely, the need to transport oxygen by blood in arthropods that have book lungs—and this may account for the fact that spiders and scorpions, while capable of short periods of very high activity, are incapable of the sustained level of energy release found for example in flying insects (Seymour and Vinegar, 1973). However, recent work by Stewart and Martin (1974) on the tarantula *Dugesiella hentzi* shows that these animals can develop high blood pressures, normally from 8 to 22 mm Hg, but rising on occasion to 120 mm Hg, and a well developed vascular system conveys blood to the book lungs.

As regards the contribution of respiratory to total water loss, there is room for a lot more work. At present it seems as though the ability to close entry to the respiratory surfaces by spiracles permits the control and regulation of respiratory loss, but it also seems clear that ventilation of the system, when this occurs either tidally or unidirectionally, leads to a very much greater rate of loss. There is no easy way (speaking teleologically) to avoid water loss and permit oxygen uptake by interposing a differentially permeable membrane (Waggoner, 1967), but in theory at least the device of intermittent release of carbon dioxide could result in water conservation. However, both here and in the relationship of oxygen uptake to water loss in general, the evidence is too scanty, or too indecisive, to permit good comparative generalizations. The problems are in some cases technically difficult. For example, efficient sealing of spiracles to eliminate respiratory water loss inevitably leads to physiological injury in long experiments—and this is a general dilemma. But carefully designed experimentation can hardly fail to shed further light on some of the more obscure areas of this important research field.

Chapter 5

Water Loss by Evaporative Cooling

A. Introduction

It is not our purpose to go fully into questions of body temperature determination, interesting though these matters are. For reviews, see Cloudsley-Thompson (1970); Heinrich (1974, 1975). However, one aspect of temperature regulation is directly relevant to water balance, namely evaporative cooling, and this will be discussed.

Owing to the relatively small size, and consequently high surface/volume ratio of arthropods, long term evaporative cooling is out of the question. Schmidt-Nielsen (1964) pointed out that for any animal to maintain a body temperature by evaporative cooling at a particular level under a given heat load, it is necessary to evaporate water at a rate which is roughly proportional to surface area. This is so because each of the factors concerned: radiation, conduction, convexion, metabolism and evaporation, seem to depend more on surface area than on volume. But of course a given amount of evaporation per unit of surface represents very different proportions of total body water, according to size. Schmidt-Nielsen (1964) calculated that for a standard desert situation, where a man weighing 70×10^3 g, would have to lose 1.47% of his weight each hour to keep cool, a rabbit, weighing 2×10^3 g would need to lose 4.77% h^{-1}. If we extend these calculations to large and small arthropods (Fig. 38) it turns out that a goliath beetle (weight 15 g) would have to lose 25% of its wt h^{-1}; a tsetse-fly (20 mg) 235% h^{-1}, and a flea (0.4 mg) 890% h^{-1}.

These numbers are, of course, imaginary and calculated for a high heat load. Mellanby (1932a) calculated that for a cockroach weighing 1 g to maintain a body temperature 5°C below ambient at 45°C it would be necessary to lose about 40 cal h^{-1}, which is equivalent to the evaporation of 70 mg water, or 7% of its wt h^{-1}. In any case, such rates are intolerable for long periods, even for large arthropods. But there is evidence that evaporative cooling does occur for short term emergencies, and this we shall consider, restricting the enquiry largely to (1) the significance of evaporation in relation to other avenues of heat exchange, and (2) the amount of water lost in the process. The heat of evaporation of water is 10.73 kcal mol^{-1} (595 cal g^{-1}) at 0°C, and 10.24 kcal mol^{-1} (568 cal g^{-1}) at 50°C.

The following discussion is divided into two sections, one dealing with laboratory experiments, the other with field work. In fact it is not always possible, let alone desirable, to make such a distinction, and no important theoretical principle is intended. In general, allocations where appropriate have been made according to the main thrust of the topic concerned.

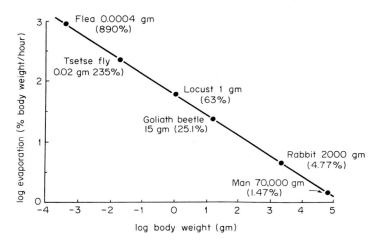

Fig. 38. To maintain a given body temperature depression below ambient, water must evaporate from an animal at a particular rate per unit surface area. In desert conditions such a rate is tolerable for some hours by a large animal because it represents only a small fraction of total water reserves; but a small animal would have to lose several times its own weight of water per hour. Figures in parentheses show the amount of water, expressed as a percentage of total weight, to be evaporated each hour by animals whose weights are also shown. (From Edney, 1960b; data in part from Schmidt-Nielsen, 1964)

B. Evaporative Cooling in Laboratory Experiments

Early work on evaporative cooling, reviewed by Gunn (1942) and by Edney (1957), was largely confined to laboratory studies, and showed that insects of the size of cockroaches can, during short exposures, withstand higher ambient temperatures in low humidities than in saturated air (Gunn and Notley, 1936). Small insects, however, show no such ability—indeed for exposures of 24 h they survive only at lower temperatures in dry air, because they then die from dehydration (Mellanby, 1932a). Small arthropods, such as lice and fleas, are very unlikely to show the effects of evaporative cooling even in laboratory conditions for two reasons: firstly, they are probably well water-proofed because their surface is so large relative to their volume, and secondly, convexion is bound to keep their body temperature rather close to ambient.

The effects of evaporation, body size and period of exposure, are well demonstrated by terrestrial isopods. If these animals are exposed for 15 min to different humidities, the highest temperature they can withstand varies between species but is higher in 50% R.H. than in saturated air. In dry air, upper lethals are slightly lower again (owing to excessive dehydration) except, interestingly enough, for *Ligia*, which is far larger than the rest and which, consequently, can withstand a high rate of transpiration per unit surface for a longer time (Fig. 39) (Edney, 1951b).

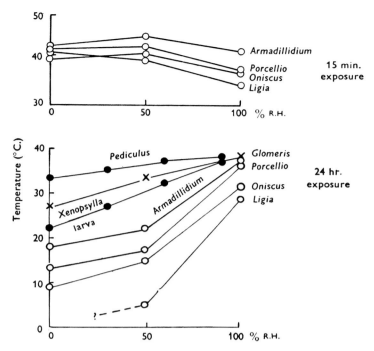

Fig. 39. The effect of humidity and period of exposure on the highest tolerable temperature for various arthropods. *Glomeris* is a millipede, *Pediculus* a body louse, *Xenopsylla* a flea and the remainder are isopods. (From Edney, 1957; data for *Xenopsylla* and *Pediculus* from Mellanby, 1932a; for discussion see the text)

For 24 h exposures the picture is very different—the upper lethals are a good deal lower at 50% R.H. than in saturated air, because too much water is lost if the temperature is higher owing to increased ΔP (vapour pressure gradient) and in dry air the upper lethals are lower still. Again *Ligia* is aberrant, for this animal has a very permeable integument, and it loses so much water in dry air that it fails to survive for 24 h in dry air at any temperature. Measurements of body temperatures showed that these are depressed below ambient in the experimental conditions concerned.

A similar effect has been observed in honey bees by Free and Spencer-Booth (1962). For exposures to high temperatures lasting 1 h, survival is better at lower humidities, and weight loss is then greater owing to enhanced evaporation. For longer exposures, however, lethal temperatures are lower, and survival increases with increasing humidity because desiccation is limiting.

C. Evaporative Cooling in the Field

In field conditions, evaporation from well water-proofed arthropods would not, on general grounds, be expected to play a large part in determining body temperature, and experiments tend to confirm this (Parry, 1951). In *Schistocerca*, for example, measurements by Stower and Griffiths (1966) showed that evaporation (both cuticular and tracheal) is responsible for 0.051 cal per insect min^{-1} (equivalent to 0.0048 cal cm^{-2} min^{-1}) or less than 5% of total heat loss, the remainder being convexional (Table 13).

Table 13. Heat balance for nymphs of the desert locust, *Schistocerca gregaria* in air at 24°C and 65% R.H., moving at 50 cm s^{-1} parallel with the insects' long axis, while the insects are oriented at right angles to the sun's rays. (Data from Stower and Griffiths, 1966)

	cal per hopper min^{-1}
Net radiation load	+0.9874
Metabolism	+0.0900
Total input	+1.0774
Convexion (external)	−1.0206
Convexion (internal through tracheae)	−0.0004
Evaporation (cuticular)	−0.0484
Evaporation (tracheal)	−0.0030
Total output	−1.0724

Similarly, when Hadley (1970a) measured the components of heat exchange in the desert beetle *Eleodes*, he found that a typical heat balance sheet would be (in cal cm^{-2} min^{-1}): radiation (+0.141), metabolism (+0.003), conduction (?), convexion (−0.134), evaporation (−0.008). Clearly metabolism and evaporation are minor components in these circumstances. A similar situation was found by Church (1960) for flying locusts, and by Edney (1971b) for desert tenebrionid beetles (*Onymacris* spp.), but in large flying moths the situation may be different (see below). Further examples and discussion will be found in Cloudsley-Thompson (1970).

Significant evaporative cooling in field conditions has been recorded, but only in larger animals with high permeabilities, or for short periods in others. Fiddler crabs (*Uca* spp.) which weigh about 2 to 5 g, provide good examples of the first alternative. These are decapods that live largely in the littoral zone, feeding at the surface during low tides, where they may be exposed to dry air and direct sunshine. Body temperatures of these animals are about 2 to 3°C higher in saturated air than in air at 80% R.H. (Edney, 1961), and temperature depressions of up to 6°C occur in dry air.

When *Uca* spp. are exposed to direct sunshine in field conditions, the body temperature of living crabs is about 6–7°C lower than those of dead, dry crabs exposed side by side, as a result of evaporative cooling (Edney, 1961). Wilkins and Fingerman essentially confirmed the above results in *Uca pugilator*, and additionally showed how in that crab, surface colouration (which is physiologically variable) affects the radiation input component of the heat balance equation. They also found (as did Edney for *Uca annulipes*, *Uca chlorophthalmus* and *Uca marionis*) that behaviour is an important aspect of thermal regulation (Fig.

40) (Edney, 1961). Indeed behavioural thermoregulation is found to be of central significance in most arthropods where the matter has been studied (the fact that behaviour is not emphasised strongly in the present book is a result of the decision to omit thermoregulation from the enquiry except insofar as water balance is directly involved).

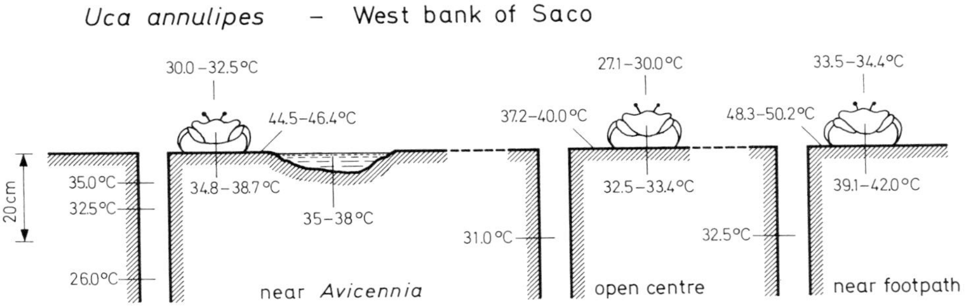

Fig. 40. Body temperatures of fiddler crabs *(Uca annulipes)* in their natural habitat. The crabs' temperatures are usually lower than those of the sunlit ground but even so they may be near the lethal level. Holes and pools form retreats for the crabs if their temperatures become too high or if they lose too much water. The scale shown refers to the holes only. More crabs are present near the mangrove trees (*Avicennia* sp.) than in the middle of an open area, even though temperatures are lower in the latter. Perhaps the need for protection is responsible. (From Edney, 1961)

Rates of transpiration in *Uca* spp. are correspondingly high. Edney (1961) found rates of loss between 2% and 5% of the animals' wet wt h^{-1} in different species, and this is equivalent to about 80–200 $\mu g\ cm^{-2}h^{-1}mm\ Hg^{-1}$, so that permeabilities were $235 \times 10^{-4} - 586 \times 10^{-4}\ cm\ s^{-1}$ (resistances from 42.6 to 17.1 $s\ cm^{-1}$); rather greater than that of most terrestrial isopods.

Experiments in which living and dead, dry *Ligia* (Isopoda) were exposed side by side to sunshine showed that body temperature is depressed by more than 5°C as a result of evaporative cooling (Fig. 41). In these conditions, a typical heat exchange equation (in cal $cm^{-2}min^{-1}$) is: radiation ($+0.287$), metabolism ($+0.0014$), convexion (-0.140), evaporation (-0.142); confirming the significance of evaporation in these circumstances (Edney, 1953). The amount of water lost would have been 2.4 mg $cm^{-2}h^{-1}$ or about 14.0 mg h^{-1}, which is tolerable for an animal that weighs about 1,000 mg.

In natural situations too, *Ligia* makes use of such evaporative cooling. These animals occur below shingle on beaches (among other places), where they may on occasion be subjected to dangerously high temperatures owing to insolation of the stones above them, and high humidities, because of moist sand and vegetation below. In these circumstances they emerge into full sunshine and move about on exposed rock surfaces, where measurements show that they are cooler than those below the stones as a result of convexion and evaporation, the latter playing a major role (Fig. 42) (Edney, 1953). After a time they return to cover and replace lost water by absorption from the moist surfaces below (Spencer and Edney, 1954).

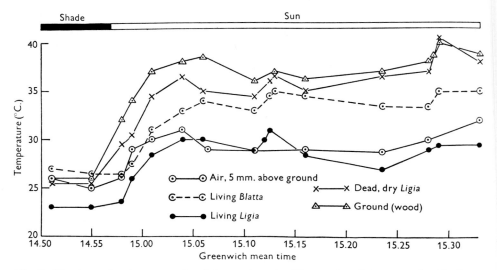

Fig. 41. The internal temperature of a living cockroach *(Blatta orientalis)* compared with that of a living and of a dead, dry littoral isopod *(Ligia)*, during insolation on a wooden surface. R.H. 39–45%; wind speed about 500 cm s^{-1}. (From Edney, 1953)

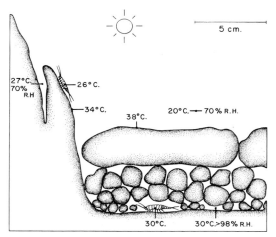

Fig. 42. Section of the base of a red sandstone cliff and shingle inhabited by *Ligia oceanica*, to show the microclimatic conditions and body temperatures of the animals. Owing to insolation of the stones and to the very high R.H. under them, the isopods' body temperature becomes uncomfortably high (Fig. 39 shows that 37°C is lethal). When the animals emerge from their cryptozoic niche, reduction of body temperature occurs by evaporation of water, and by convexion. (From Edney, 1953)

Even in the relatively xeric isopod *Hemilepistus*, evaporative cooling for short periods is effective while the animals seek shelter in their holes if suddenly exposed to sunshine, and this may lead to a body temperature depression of some 3°C from 37°C (Edney, 1960b) (Fig. 43).

92

Among insects, examples of evaporative cooling in real life are less common. Adams and Heath (1964) report that a sphingid moth, *Pholus achemon*, extrudes a drop of liquid from the proboscis if its body temperature rises above 40°C. The liquid is cooled by evaporation after which it is re-imbibed. No measurements of water loss were made in this instance.

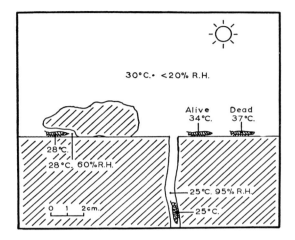

Fig. 43. A set of observations in the habitat of *Hemilepistus reaumuri* (an isopod) in the Algerian Desert near Biskra. Transpiration is sufficiently rapid to reduce the body temperature of a living animal by some 3°C below that of a dead, dry, control specimen. These animals dig vertical holes in which the conditions are relatively equable. (From Edney, 1960b)

Larvae of the saw-fly *Perga dorsalis* show an interesting form of evaporative cooling (Seymour, 1974). These larvae are gregarious, and when, as a result of insolation, their body temperature rises to 37°C or above, each larva raises its abdomen, secretes a watery fluid from the rectum, and distributes this over itself and its neighbours. Evaporation then maintains the body temperature below 42°C (the critical thermal maximum) even when the ambient temperature is 48°C. Since these larvae tolerate a water loss equal to 17% of their weight (such loss being readily replaced by feeding), evaporative cooling may be continued for as long as one hour at ambient temperatures up to 45°C (Seymour, 1974).

When the large sphingid moth *Hyles lineata* flies in air at about 23°C, its necessarily high metabolic rate leads to the production of about 3.14 cal min^{-1} by a 1 g insect. The moth's thoracic temperature is controlled at 40°C, and most of the excess heat is lost by convexion from the body surfaces. But Casey (1976), who has measured these processes, calculates (on the basis of certain likely assumptions) that for a moth flying in dry air, evaporative heat loss from the thoracic tracheae (which are strongly ventilated) is equivalent to about 18% of the total heat production. If the air is moist, then of course the rate of water loss will be less. In perfectly dry air the loss is about 0.8 mg min^{-1}, and this would seem to pose a problem for migrating insects, where even the high metabolic rate (about 58.5 ml O$_2$ g^{-1}h^{-1}) means the production of only 0.5 mg H$_2$O min^{-1}. If the moths are foraging, and feeding on nectar, then of course all the water lost is readily replaced.

Tsetse-flies (*Glossina* spp.) are known to withstand somewhat higher temperatures in dry than in moist air (Buxton and Lewis, 1934; Jack, 1939). The difference amounts to about 2 °C (Bursell, unpublished), and an obvious explanation might be evaporative cooling (Buxton, 1955). To find out, Edney and Barrass (1962) examined spiracle movements of *G. morsitans* in controlled temperatures and humidities. They found that slowly rising ambient temperatures are followed closely by the fly's body temperature until the former reaches about 39 °C, at which point the spiracles (hitherto closed) begin to flutter, and then to open widely, while the fly's temperature remains constant until a temperature depression of nearly 2 °C is reached (Fig. 44). The effect is reversible and repeatable in one fly, but cannot be achieved in saturated air, or even in dry air if the spiracles are blocked.

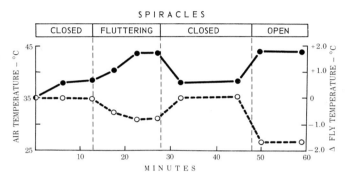

Fig. 44. The effect of spiracular opening on the internal temperature of a tsetse-fly. Closed circles, air temperature; open circles, difference between air and fly temperature. At about 39 °C the spiracles flutter or open and at the same time the body temperature is depressed. Not shown in this figure is the fact that temperature depression is inhibited either by saturated air or by blocked spiracles. (Redrawn from Edney and Barrass, 1962)

It seems unlikely, at first sight, that such a small temperature difference could be of survival value. Most insects, if confronted with uncomfortably high temperatures, simply move away. However, observations in the field suggest that in some circumstances the process might be valuable.

These observations were made on a hot day in the Zambesi Valley in Rhodesia. Flies, observed to feed at a bait ox, were marked, released and observed again while subsequently resting on branches of trees and bushes. Air temperatures in the resting spots were as high as 38 °C and the ambient humidity was 25 to 35% R.H. Shade temperatures on the ox surface were usually between 36 °C and 37.5 °C, and as high as 40–41.5 °C when the skin was exposed to direct sunshine. One skin temperature of 47 °C in direct sunshine was recorded.

Since the flies are bound to feed (and in the process to absorb warm blood), they may occasionally be exposed for short periods to conditions in which an ability to drop the body temperature by evaporative cooling, even by a small amount, may be critical for survival. Assuming that evaporation continues for 2.5 min while the fly feeds, then allowing for an ambient humidity of 25% R.H. and a high temperature of 38 °C, Bursell's (1957a) figure of about 1.0 mg per insect h^{-1} at 25 °C and 25% R.H. for flies with open spiracles, works out at 0.09 mg water for the feeding period, and this would involve a heat loss of 0.05 cal—the amount of heat required to lower the temperature of 50 mg water by 1 °C. Thus the additional amount of water lost is negligible, particularly for a fed fly which will have water in excess.

One *G. morsitans* fly weighs about 23 mg after emergence—hence its surface may be about 1.0 cm^2; so that the rate of heat lost by evaporation would be 0.02 cal cm^{-2} min^{-1}. This is about one seventh of the rate found for *Ligia* when effectively cooling to about 5 °C below ambient, and is 4.2 and 2.8 times as great as the comparable figures for *Schistocerca* and *Eleodes* respectively, where evaporative cooling is insignificant.

Temperature regulation in colonies of social insects, particularly in the brood areas, has been known for a long time. Warming is referred to by Krogh (1916). Cooling by evaporation of water is known to occur in colonies of *Apis mellifera*, and of several wasps, including *Polistes* spp. and *Vespa* spp. (Steiner, 1930; Himmer, 1932). In bees, water is collected and transported to the hive by special workers and either deposited in small drops in or on the brood cells, to be evaporated by fanning, or the water is manipulated by the bees' mouthparts and caused to evaporate in this way (Lindauer, 1954). According to Steiner (1930) a queen *Polistes gallica* can transport 3 ml water h^{-1}, and this, by evaporation, consumes over 1,800 cal.

Bees drink very little at low temperatures, but at ambient temperatures of 35° and 40 °C they drink from 5.8 to 11.4 μl and from 10.1 to 21.7 μl per bee respectively, according to crowding and hence to temperature elevation above ambient (Free and Spencer-Booth, 1958). Further, the frequency of drinking correlates well with external temperatures above 18 °C (Lindauer, 1954).

In spite of the above useful data there is still a lack of quantitative information about water balance in relation to cooling in social insects. A comprehensive study of the water and heat budgets involved would be rewarding.

D. Conclusions

Evaporative cooling is probably not an important component of thermal regulation in arthropods for reasons that have been considered above. But when it does occur, the process has important repercussions for water balance, and for this reason it is confined either to brief periods or to arthropods that have relatively large water supplies. Perhaps the well marked ability of land isopods to use evaporative water cooling is a reflexion of their mode of evolution via the littoral zone (Pearse, 1950), where very large, rapid temperature fluctuations occur while water is abundant. Arthropods that became terrestrial by an underground pathway were not subject to such selection pressures.

Chapter 6

Excretion and Osmoregulation

A. Introduction

In arthropods as in most other animals, excretion of waste or potentially toxic materials is carried out by organs that are also involved in osmotic and ionic regulation, and it is, therefore, convenient to consider these processes together. However, we shall treat nitrogen elimination separately because this is a general problem for all arthropods and one which stands somewhat apart from other aspects of water balance.

Osmotic and ionic regulation are closely related fields in which research is active, and good reviews of the whole or of parts of the area are available. Browne (1964) provides a useful review of the earlier literature; Maddrell (1971) deals with excretory and related mechanisms in insects; Berridge (1970) includes all arthropods in a review of osmoregulation (in a wide sense); Bursell (1967, 1970) deals with several aspects of the problem largely in relation to insects, but including many generally applicable concepts; and Stobbart and Shaw (1974) cover most aspects of water balance and excretion in insects, including a useful review of these matters in aquatic forms. Riegel (1971) deals with excretion in all arthropods, and Cochran (1975) has written a broad review of all matters pertaining to the elimination of excess material in insects, stressing the biochemical aspects. The present chapter owes a great deal to these excellent reviews.

B. Elimination of Nitrogenous Waste

I. Nitrogenous End Products

The elimination of nitrogen is a biochemical necessity resulting from the fact that more protein is usually consumed than is necessary for maintenance, and that the proportions of different amino acids consumed are likely to be different from those required. The process is relevant to water conservation only because the primary end product of nitrogen metabolism, either from deamination or from nucleic acid degradation, is the highly toxic and highly soluble compound ammonia. Thus, a readily available, copious supply of water is generally necessary if ammonia is to be the nitrogenous end product—as it is in most aquatic invertebrates (Prosser, 1973), including several freshwater insects (Staddon, 1964 and earlier papers). Because of this, in terrestrial animals ammonia is usually

96

converted to other, non-toxic nitrogenous compounds including amino acids, urea, uric acid and its primary degradation products allantoin and allantoic acid, and guanine. Uric acid predominates in insects, guanine in arachnids.

The nature of the nitrogenous end product is probably governed by several interacting factors. Figure 45 shows the structure of the common molecules involved

Fig. 45. (a) Pathway for the degradation of uric acid to ammonia (from Prosser, 1973); (b) the properties of the main compounds used in nitrogen excretion

b		H/N	C/N	Solubility g 100 ml^{-1} at 25°C	Toxicity
	Uric acid	1	1.25	0.0006	−
	Urea	2	0.5	119.0	−
	Ammonia	3	0	52	+

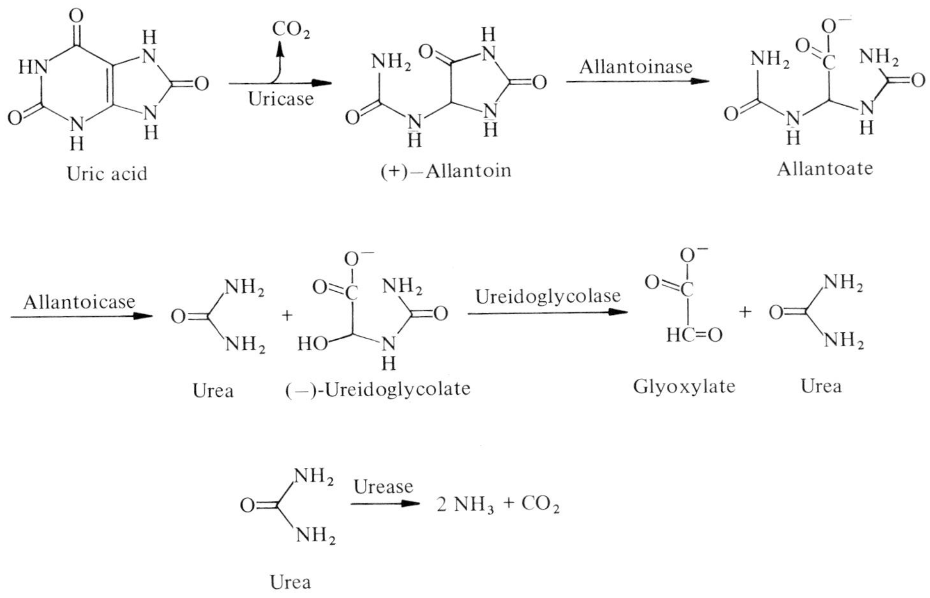

and their relevant properties and relationships. Ammonia (NH_3) has no carbon but its use incurs the loss of three hydrogen atoms for every nitrogen atom eliminated (H/N=3). (Hydrogen atoms represent potential water and their loss is inimical to water conservation.) Urea, $(NH_2)_2 \cdot CO$, is non-toxic, very soluble, and has a H/N ratio of 2. Uric acid, $C_5H_4N_4O_3$, has the lowest H/N ratio of 1, is nearly insoluble and non-toxic, but involves the highest loss of carbon with which energy conservation is linked. Thus, carbon conservation would favour ammonia, toxic considerations would favour urea or uric acid, but water

97

conservation strongly favours uric acid, both on account of its low H/N ratio, and also because being insoluble, uric acid can be concentrated and excreted almost dry without exerting significant osmotic pressure.

As a result, most insects are predominantly uricotelic, although significant quantities of ammonia (Table 14), urea and amino acids are found. (For references

Table 14. The percentage distribution of nitrogen among different excretory products in adult terrestrial insects. (From Bursell, 1970)

Order	Species	% of total nitrogen			
		Uric acid + primary degradation products	Amino acids	Urea	Ammonia
Orthoptera	*Melanoplus bivittatus*	55	11	4	29
Heteroptera	*Dysdercus fasciatus*	61	15	12	—
	Rhodnius prolixus	97	trace	3	—
Diptera	*Aedes aegypti*	66	7	15	12
	Glossina morsitans	82	15	trace	—

see Razet, 1966; Bursell, 1967, and Schoffeniels and Gilles, 1970.) Ammonia has been reported in cockroaches (Mullins and Cochran, 1972, 1973) and in *Sarcophaga* larvae (Prusch, 1972). In cockroaches uric acid is deposited in the fat body (Kilby, 1963) and ammonia forms the largest component of excreted nitrogen.

The purine, guanine, is known to take the place of uric acid in spiders, scorpions and uropygids (Horne, 1969), phalangids, acarines, and ixodid ticks (Hamdy, 1972), (earlier refs in Edney, 1957). Rather surprisingly *Peripatus*, which lives in damp localities, is also uricotelic (Manton and Heatley, 1937), as are some centipedes and millipedes (Wang and Wu, 1948; Horne, 1969) (Chap. 6.C.IV.).

Some species of terrestrial crabs (e. g. *Ocypode*, *Cardisoma*) are ammonotelic only when copious water is available; others (e. g. *Carcinus*), are probably consistently uricotelic (refs in Bliss, 1968; Bliss and Mantel, 1968).

It may turn out that ammonotely in insects is more widespread than was originally thought. Certainly the work of Mullins (1974) and Mullins and Cochran (1974) is throwing new light on this in *Periplaneta*. In this cockroach, uric acid is not excreted but is stored and ammonia is eliminated in the faeces (although the precise form is uncertain), as it is in blow-flies (Prusch, 1972). By feeding cockroaches different proportions of nitrogen, Mullins found not only that the rate of ammonia and of total nitrogen elimination increases with high nitrogen diets (as might be expected) but also that the rate of voluntary water intake by drinking also increases in a similar manner (Fig. 46). Some ammonia is excreted even on a nitrogen-free diet, and then uric acid is mobilized from the fat body. Injection of additional sodium, but not of potassium or ammonium, results in a reduction of ammonium excretion, and Mullins suggests

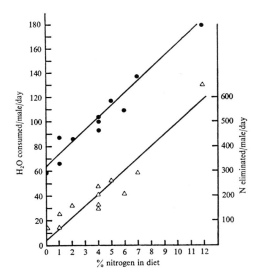

Fig. 46. The effect of nitrogen content of the diet on rate of nitrogen elimination *(triangles)* and on rate of water uptake *(solid circles)* in *Periplaneta*. (From Mullins, 1974)

that ammonotely may in some way be involved in ion conservation, particularly sodium. Mullins and Cochran (1974) also found that the level of whole body urate or uric acid storage varies directly with the level of dietary nitrogen, and that potassium is stored in the fat body together with urates. They advance the attractive hypothesis that the function of uric acid or urates stored in the fat body is to act, together with sodium, potassium and ammonium, as an ion sink, where ions may be sequestered or mobilized according to need. This would have clear advantages for osmotic regulation of the blood (Chap. 6.C.V.2.), and certainly deserves further investigation.

Uricotely, while saving water, is not an unmitigated advantage, especially in arthropods such as ticks and tsetse-flies that feed on high protein diets. Owing to the fact that the synthesis of uric acid requires considerable energy and the loss of considerable carbon, no less than 50% of the dry weight of a tsetse-fly's meal has to be excreted in order that all the surplus nitrogen can be eliminated as uric acid (Bursell, 1964, 1970). If tsetse-flies were ammonotelic, the energy saving would be such that 86% of their food would be metabolically useful.

The mode of formation of uric acid in the excretory system is referred to in Chapter 6.C.V.6; little is known about the formation of guanine.

II. Nitrogen Excretion in Isopods

If the generalisation that terrestrial animals are either ureotelic or uricotelic is true (Needham, 1938), isopods, at least the more terrestrial ones, should differ from their marine ancestors and use uric acid. However, as Dresel and Moyle (1950) found, isopods are largely ammonotelic. These authors believed that the adaptation to terrestrial life in isopods takes the form of a low protein diet and a general reduction in nitrogen metabolism. However, later work by Sloan (unpublished), Hartenstein (1968), Wieser *et al.* (1969), and by Wieser and Schweitzer (1970) showed that there is no reduction in nitrogen metabolism (compared with marine isopods), that only 10% of total nitrogen excreted is contained in the faeces, and that the remainder is released from the integument in the form of gaseous ammonia. As regards the contribution of nitrogenous substrate to energy production (and thus the amount of ammonia released), in *Porcellio scaber* this depends on the amount of energy used; in *Porcellio pictus* it is affected by the rate of energy use, but in *Oniscus asellus* a constant daily amount of nitrogenous substrate is used (Wieser, 1972a).

Excretory ammonia is not carried in the blood as such (indeed this would mean uncomfortably high levels), but rather as glutamine and glutamate, and released through the integument (perhaps via interpleopodal liquid films) where Wieser (1972b) has demonstrated the presence of copious glutaminase, the activity of which varies seasonally in a manner corresponding with variations in ammonia excretion. Wieser (1972b) and Wieser and Schweitzer (1972) also make the interesting suggestion that high ammonia concentrations in the integument may be useful in creating an alkaline environment favourable for the deposition of calcium carbonate. Certain terrestrial snails also excrete gaseous ammonia (Speeg and Campbell, 1968), and they too, of course, lay down calcium carbonate in their shells.

In marine isopods, and in *Ligia baudiniana* (a littoral species which may represent a half-way stage towards the terrestrial habitat), continuous ammonia release occurs only in the presence of a moist substrate, whereas in truly terrestrial isopods ammonia is released in a humid atmosphere, irrespective of substrate—an adaptation to terrestrial life (Wieser, 1972c).

Ammonotely is an energetically cheap way of excreting nitrogen, but it is expensive in terms of water. If, as seems likely, a moist, relatively permeable cuticle is necessary, this may be the reason why most other terrestrial arthropods, even those that live in moist environments such as damp soil, are uricotelic. Maddrell (pers. comm.) suggests that since insects became terrestrial through a freshwater pathway (the soil would be ecologically equivalent to this) it was energetically advantageous for them to develop a cuticle impermeable to water and ions, and this would forbid ammonotely; whereas isopods, becoming terrestrial through the littoral zone, retained a permeable cuticle and could therefore excrete nitrogen as ammonia.

III. Deposit Excretion

We have seen that the deposit of uric acid in the fat body of cockroaches may be adaptive. Other examples of uric acid storage include the cockroach *Arenivaga* (where it is stored in the wings) (Friauf and Edney, 1969), the wings of butterflies (Harmsen, 1966) and the abdominal integument of *Dysdercus* (Berridge, 1965c). In all these cases it is possible that the deposits may be adaptive (as in the white wings of pierids), but in some, e.g. the huge accumulation of uric acid in the accessory sex gland of cockroaches, which is excreted only during copulation, it is difficult to see the value (Roth and Dateo, 1965). For further discussion see Maddrell (1970) and Chapter 6.B.I. where the possible function of the fat body as an "ion sink" is considered.

Deposit excretion has also been observed in the haemocoele of the land crab *Cardisoma*, according to Gifford (1968), and in the pericardial sacs of *Ocypode* (Rao, 1968) (refs in Bliss and Mantel, 1968).

C. Osmotic and Ionic Regulation

I. Introduction

The first five chapters of this book have dealt with water loss, the tacit assumption being that by and large this has to be prevented or reduced. The present chapter will serve to amend this impression, for there are times when water is present in excess and has to be excreted. This is particularly true for species that feed on blood, or on the juices of plants, where the food may be a dilute watery solution.

There are three interrelated aspects to be considered. The first is regulation of the water content as such, i.e. the proportion of water present in the body. This may be broken down in various ways, but we shall mainly be concerned with regulation of blood volume and tissue water. The second aspect concerns osmoregulation, by whatever means this may be achieved. Most information here relates to the blood and the contents of the alimentary canal. Thirdly, we shall deal with ionic regulation, for in addition to their function simply as generators of osmotic pressure, particular ions are clearly involved in particular processes, and the concentration of each is regulated, with greater or less precision, in different circumstances. It is as well to be aware of these aspects, but because the so-called "excretory" systems of land arthropods are involved in all three aspects of regulation, we shall not attempt a separate treatment for each component process.

II. Crustaceans

In marine and freshwater decapods, osmotic and ionic regulation is effected by antennary or maxillary glands, both of which are segmental organs, and by gills (refs in Robertson, 1960). In semi-terrestrial forms such as *Cardisoma guanhumi* and *Gecarcinus lateralis*, the antennal glands produce an iso-osmotic urine even when the blood osmotic concentration is high (Gross *et al.*, 1966). These glands may reduce water loss simply by restricting the amount of urine formed (Gross, 1964; Gifford, 1968), but their function is mainly ionic regulation. For example, the antennary glands in *Cardisoma* and *Sesarma* can concentrate magnesium and calcium in the urine (Gross *et al.*, 1966). Similarly the powerful osmoregulatory abilities of *Uca mordax* are not due to the antennal gland (the urine is always slightly hypo-osmotic) but the gland is known to remove sodium from the urine against a gradient and has the appropriate fine structure for this function (Schmidt-Nielsen *et al.*, 1968). Copeland (1968) found that cells in certain areas of the gills of the land crab *Gecarcinus* also have a fine structure appropriate to the transport of fluids, and he believes that these areas do in fact function in salt and water uptake. The gills rest upon the pericardial sacs, which are well known as organs concerned with water balance, for they swell and store water before ecdysis and at other times, although uptake always takes place from a moist surface, never from humid air (Bliss, 1968). Perhaps the sacs function simply as sponges—the gills doing the necessary osmotic work (Bliss, 1968). However, see Chapter 7.B.II. for an alternative proposal by Wolcott (unpubl.). According to Mantel (1968), the foregut of *Gecarcinus* is also involved in salt and water balance.

Most terrestrial crustaceans seem to have evolved across the littoral zone: they are euryhaline and their blood is hypo-osmotic to sea water (refs in Gross, 1955). *Birgus* (the coconut crab) can tolerate between Δ_i 1.3° and Δ_i 2.2°C (Δ sea water is about 1.86°C).

Most terrestrial isopods also are euryhaline, and, as in land crabs, they are hypo-osmotic to sea water. In *Oniscus*, *Armadillidium* and *Porcellio* Δ_i is 1.04°, 1.18°, and 1.30°C, respectively (Parry, 1953). *Ligia oceanica* is exceptional in having Δ_i about 1.86°C. This species regulates in concentrations down to 50% sea water, but in air it loses water rapidly, and like *Oniscus* (Bursell, 1951) shows little ability to osmoregulate but is strongly euryhaline, tolerating Δ_o 3.48° to Δ_o 1.44°C (Parry, 1953). Horowitz (1970), however, found that in *P. scaber*, although water loss is continuous in dry air, Δ_i remains constant for 8 h at 380–400 mmol liter^{-1} before beginning to rise.

The function of the maxillary glands in terrestrial isopods is uncertain. Although they have been generally thought of as excretory organs (refs in Edney, 1954, 1957), Berridge (1968) found that injected phenol red is eliminated not by these organs but by the hepatopancreas. The gut may play an important role in excretion, and perhaps in osmoregulation, as it does in other crustaceans (Mantel, 1968), for isopods have rectal glands (Gupta, 1961) and the faeces may be dried to some extent according to Kuenen (1959). Recently the fine structure of the hindgut cells of *Armadillidium* has been shown to be typical of those associated with water and ion transport (Schmitz and Schulz, 1969).

In very many of these terrestrial crustaceans, osmotic regulation and water balance generally are maintained by behavioural mechanisms such as choosing favorable microclimates, and using freshwater sources (Edney, 1961, 1968a) (refs in Bliss, 1968; Warburg, 1968b; Cloudsley-Thompson, 1973), but these matters fall outside the scope of the present book.

III. Arachnids

Little is known about the osmotic concentration or the chemical composition of the blood of most groups of arachnids. Sutcliffe (1963) using data obtained by Croghan (1959) and Lockwood and Croghan (1959), concludes that the blood of the spider, *Tegenaria* resembles that of crustaceans, the centipede *Lithobius*, and the primitive apterygote *Petrobius*, rather than that of more advanced insects, since almost all the osmotic effect is caused by sodium and chloride ions, potassium being very scarce. The relationship between insects' blood and their phylogeny is referred to below.

The main excretory organs in arachnids are coxal sacs and Malpighian tubules. In argassid ticks the coxal sacs are excretory and osmoregulatory organs (Boné, 1943; Lees, 1946b), and the structure of one such gland is shown in Figure 47. The internal saccule is very thin-walled and probably functions as

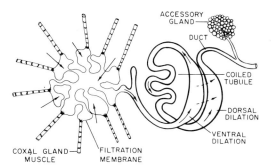

Fig. 47. Coxal gland of *Ornithodorus moubata*. In life the filtration membrane envelopes the coiled tubular region. *Arrows*: direction of fluid flow. (From Berridge, 1970; after Lees, 1946b)

a filtration membrane, for its walls are attached to muscles which, by constriction, can produce a slight reduction in hydrostatic pressure internally. The membrane is structurally similar to the vertebrate glomerulus (Schmidt-Nielsen *et al.*, 1968) and to the crustacean coelomosac (refs in Tyson, 1968).

Morphologically proximal to the filtration chamber are two tubular segments into the first of which the filtration chamber opens. Here the walls are thicker and richly tracheate, and their function is probably reabsorption. According to Lees (1946b) the coxal fluid is formed by filtration, for molecules of the size of haemoglobin, but not casein or albumin, pass through the chamber walls. When the muscles relax, fluid passes down the tubules where salts and other required materials are reabsorbed. The "haemocytes" reported in this fluid by Sidorov (1960) may be the "formed bodies" reported by Riegel (1966) (Chap. 6.C.VII.8.) in the products of many secretory and excretory organs. They are not, therefore, the result of a mass flow of blood into the saccules.

The osmoregulatory function of coxal sacs has been confirmed in *Ornithodorus moubata* by Hecker *et al.* (1969), in *Argas persicus* and *Argas arboreus* by Araman and Said (1972), and in *O. savignyi* by Frayha *et al.* (1974) who measured the concentrations of several organic and inorganic ions in the coxal and other fluids of this tick. The concentrations of inorganic ions are higher in tick blood than in rat plasma or coxal fluid (Table 15), and at the same time the ratio between the concentrations of the various ions in each fluid is somewhat different. Hence the coxal organs are probably both ionic and osmotic regulators, retaining ions while eliminating water, and thereby concentrating the blood meal without causing ionic imbalance.

Table 15. Electrolyte concentrations (meq liter^{-1}) of the coxal fluid and haemolymph of *O. savignyi* and blood plasma of Sprague Dawley rats. (After Frayha *et al.*, 1974)

Electrolyte	Coxal fluid	Haemolymph	Rat plasma
Magnesium	0.09	0.50	0.66
Calcium	1.61	3.01	6.21
Potassium	10.00	8.50	5.70
Sodium	158.00	169.20	151.90
Total cations	169.70	181.21	164.47
Phosphate	0.16	0.45	7.50
Bicarbonate	26.40	22.30	20.90
Chloride	145.00	161.85	129.00
Total anions	171.56	184.60	157.40

Argassid ticks feed rapidly, *Ornithodorus moubata*, for example, which weighs 0.1 g, can ingest about 100 mg blood in 15–30 min. Within the following hour, 45% of the water and 61% of the salts of this meal have been eliminated by the coxal glands, as a result of which the salt concentration of the blood is maintained constant at about 1% NaCl equivalent.

According to Woodring (1973) the structure of the coxal glands of oribatid mites suggests that these organs are also osmoregulatory, and this is supported by studies on the fine structure of the glands in *Phthiracarus* sp. by Dinsdale (1975).

Ixodid ticks do not have coxal glands, and they feed more slowly than argassids. However, very large quantities of water are ingested with the blood meal (*Boophilus* increases its weight from 10 mg to over 200 mg in less than 12 h) and, as in other ticks including *Ixodes* (Lees, 1946a), *Argas* (Tatchell, 1964) and *Amblyomma* (Sauer and Hair, 1972; Frick *et al.*, 1974), the meal is subsequently made two to three times more concentrated by the elimination of water. Gregson (1967) and Balashov (1972) observed that a copious, clear saliva is produced by engorged *Boophilus*, and Tatchell (1967) demonstrated that this is normally injected by the tick's salivary glands into the host at intervals during the process of feeding. The ionic concentrations of the saliva in relation to that of the blood meal and of the tick's blood are such that the meal may be concentrated (i.e. its water content reduced) without unduly disturbing the ionic and osmotic concentrations of the tick's blood. For example, a tick ingested 480 mg blood, but its initial total weight of 10 mg increased only to 250 mg so that it must have lost 240 μl water in the form of saliva (very little by transpiration). The chloride concentrations involved show that there was a correspondence between intake of this ion and the amount present in saliva and in the tick (Table 16). The same is true for sodium, but not for some of the other ions or urea. Now if the salivary secretion mechanism regulates ions as well as overall osmotic pressure, a haemolymph/saliva concentration ratio of more than one will imply retention by the tick of a proportion of the ions concerned, so that the final whole tick/host blood ratio will be greater than one (and vice versa). Table 16 shows that this expectation is indeed

Table 16. The relationships between the ionic and urea content of cattle blood and of the cattle tick *Boophilus microplus*. (From Tatchell, 1969)

Substance	Content (meq × 10⁶)			Ratio whole tick + saliva: host blood	Ratio haemolymph: saliva	Ratio tick: host blood
	480 µl blood	240 µl saliva	250 mg tick			
Cl	42,480	30,960	13,450	1.0	0.9	0.6
Na	54,240	45,120	12,150	1.1	0.7	0.4
K	8,640	2,640	9,175	1.4	1.4	2.0
Ca	1,680	157	2,600	1.6	2.5	3.0
Mg	960	698	1,225	2.0	1.4	2.5
Urea	2,208	2,280	825	1.4	1.8	0.7

borne out for all the inorganic ions; the values for urea being aberrant as they would be if this material is metabolised rather than eliminated. Hsu and Sauer (1974) also found that sodium, potassium, chloride and total O.P. of the blood of *Amblyomma* is regulated by the secretion and elimination of appropriately concentrated saliva. These authors (1975) point out that the salivary glands of ixodids play a similar role to the coxal glands of argassids; but of course the details of both mechanism and control may differ if only as a result of differences in feeding habits between the two groups.

Kaufman and Phillips (1973 a, b, c) went on to examine the situation in another ixodid tick, *Dermacentor andersoni*, where considerable defaecation occurs during feeding. They measured blood volume as well as several other parameters, and as a result were able to propose, and in large part to substantiate, a breakdown of movements of water and inorganic ions during a complete feeding cycle (Table 17). The salivary glands, it turns out once again, are the main organs by which excess water, sodium and chloride are eliminated, and blood volume and ionic and osmotic concentrations are stabilised as a result. The volume of extracellular fluid is maintained at 23% of body weight, although the latter may increase as much as 75 times during feeding. The main pathways of ion and water movements are shown in Figure 48.

Table 17. The routes of excretion for sodium and potassium in the female tick *(Dermacentor andersoni)* during the adult feeding cycle. (Data from Kaufman and Phillips, 1973a)

Total ingested (µeq)	Excreted as % of ingested	% in faeces	% in saliva
		Sodium	
241	89	5	84
537	90	2	88
397	88	4	84
307	88	5	83
		Potassium	
102	65	54	11
228	65	54	11
169	53	42	11
130	56	45	11

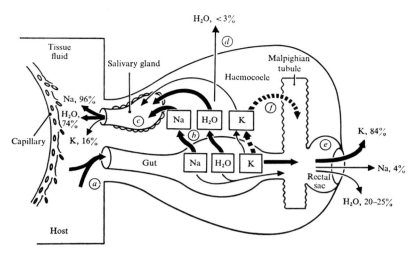

Fig. 48. Summary of ingestion and elimination of ions and water by the female tick (*Dermacentor andersoni*) during a normal adult feeding cycle. *Heavy solid arrows*: major routes; *fine solid arrows*: minor routes; *heavy broken arrow*: possible major route. *Percentage figures*: proportions of the total amount excreted over the complete feeding period. (*a*) Meal derived from a mixture of whole blood and other tissue fluids. (*b*) Na and water, but probably a lesser amount of K is transferred from the gut diverticula to the blood. (*c*) Na (as NaCl), water and some K in excess of the tick's requirements are transferred back to the host in the copious salivary secretions. (*d*) A small quantity of water is evaporated through the integumentary surface. (*e*) Most of the potassium probably passes directly from the gut diverticula to the rectal sac and out through the anus, or alternatively (*f*) potassium enters the blood and then is transferred to the faecal material via the Malpighian tubules. (From Kaufman and Phillips, 1973a)

The isolated salivary glands of this tick will secrete in the absence of an externally applied hydrostatic pressure. Blood from salivating ticks will not trigger secretion in vitro, but adrenaline and several other catecholamines will do so at concentrations of 10^{-6} M (Fig. 49). 5-hydroxy tryptamine stimulates secretion, but only at the high concentrations of 10^{-3} M, and cyclic AMP has virtually no effect at all. (By contrast, the salivary glands of *Calliphora* larvae which are not innervated, are stimulated by 10^{-10} M 5HT, or 10^{-3} M cyclic AMP (Berridge and Patel, 1968; Berridge, 1970).) On these grounds, Kaufman and Phillips (1973b) believe that salivation results from a secretory process rather than filtration and reabsorption, and they tentatively suggest that control is mainly neural, only the immediate transmitter substance being catecholaminergic. Within the gland, chloride seems to be actively secreted with water following, for chloride moves across the acinus wall against a potential difference of 35 mV (Kaufman and Phillips, 1973c).

In *Boophilus microplus* Megaw (1974) found that the injection of several different cholinergic and adrenergic drugs induces salivation, the greatest response being to noradrenaline.

Oral secretion from the lone star tick *Amblyomma americanum* (Barker et al., 1973) and *Haemaphysalis longicornis* (Kitaoka and Morii, 1970) also contains large amounts of chloride, and the glands of the rabbit tick, *Haemaphysalis*

106

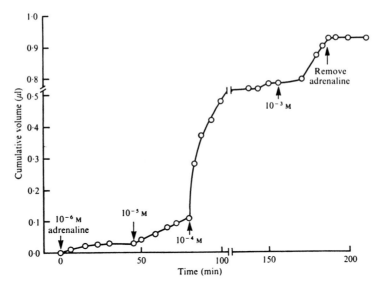

Fig. 49. The effect of adrenalin on the secretion of saliva by the tick *Ornithodorus moubata*. *Arrows*: times at which successively higher concentrations of adrenalin were added to the medium. Concentrations below 10^{-6} have no effect. (From Kaufman and Phillips, 1973b)

leporispalustris and *Dermacentor andersoni*, have been shown by Kirkland (1971) and Meridith and Kaufman (1973), respectively, to possess the type of fine structure usually associated with fluid transport. Sauer *et al.* (1974) confirmed that in vitro chloride uptake by the salivary glands of *Amblyomma* is strongly stimulated by the catecholaminergic drugs adrenaline, noradrenaline, and, rather unexpectedly by cyclic AMP, in finally engorging females, thus largely corroborating the results and suggestions of Kaufman and Phillips.

Finally, Needham and Sauer (1975) found that in *Amblyomma*, as in *Dermacentor*, salivary gland secretion is stimulated by adrenaline and noradrenaline, although cyclic AMP and theophylline are effective only if the gland has already commenced secretion under the influence of adrenaline. This helps to explain the finding by Kaufman and Phillips that cyclic AMP is ineffective in *Dermacentor*. Possible mechanisms involved are discussed by Needham and Sauer.

As a result of this work we now have a good first approach to an understanding of the potentially difficult problem of osmotic and ionic regulation in ticks. As we shall see, the problem is solved differently in insects. Further information about the mechanisms involved, particularly of the fine structure of the salivary acini, will no doubt soon throw more light on this interesting process.

Excretion of potassium via the Malpighian tubules in ixodid ticks has not been ruled out, but a simpler hypothesis is that this ion moves directly to the rectum. Malpighian tubules are indeed present in all ticks, but their function is known to be the formation of guanine, which gradually accumulates in the tubule lumen and rectum during intervals between feeding (Balashov, 1958).

A device of a different kind has been described by McEnroe (1963) in the mite *Tetranychus telarius*, which feeds on tissues from the leaves of plants. In

107

this animal a direct short circuit channel from the oesophagus to the hindgut is established during feeding, through which a large volume of water can be shunted, leaving the contained particulates to be filtered out and digested in the midgut. An advantage of such a system is that no chemical energy is required for the transepithelial transport of inorganic ions, as it is in the usual filter chamber system of insects (Chap. 6.C.VI.2.).

As regards spiders, blood osmolarity in the small number of species studied ranges from about 470 mosmol liter^{-1} for *Lycosa rabida* (a wolf spider) to about 600 mosmol liter^{-1} in *Latrodectus mactans* (a black widow spider); the values for *L. rabida* being significantly lower in females than in males (Pinkerton and Frick, 1973), but there is very little information about regulation. Some preliminary work by Ueda (1974) suggests that osmotic regulation of the blood occurs and that reabsorption of water from the rectum plays a part. In *Nephilia clavata* blood O.P. varies from about 340 to 435 mM NaCl equivalent in humidities from 75% to 10%, while rectal fluid O.P. increases from about 330 to 680 mM NaCl, over the same range of humidities. The number of spiders used was rather small, but the results are interesting. This should prove to be a fertile field for investigation.

For scorpions there was until recently a virtual absence of information about osmotic and ionic regulation. Ahearn and Hadley (1976) point out the advantages of using these large arthropods for such work, and they exemplify this by a preliminary but convincing study of the function of the ileum in *Hadrurus arizonensis*. They used an in vitro preparation of the ileum (which is about 2.0 cm long) to show that in the absence of an osmotic gradient the net flux of water outwards from the ileum is about 2.4 μl cm^{-2} min^{-1}, and consumes energy. If the lumen contains a sodium-free medium, flux is eliminated, but the flux rate is increased by the absence of potassium. Normally fed scorpions have sodium and potassium concentrations of about 123 and 275 mmol liter^{-1} respectively. After fasting for four weeks, sodium rises to 448 while potassium falls to 53 mmol liter^{-1}, and this has the highly adaptive effect of favouring water absorption from the ileum. The precise mechanism and the fine structural basis of this interesting process remains to be studied. In insects (see below) studies on the regulatory function of the ileum have only recently been undertaken, and it is interesting to find that this part of the hind-gut plays so important a role in scorpions.

IV. Myriapods

Centipedes and millipedes have Mapighian tubules similar to those of insects (Füller, 1966); and Wang and Wu (1948) and Horne (1969) showed that those of centipedes probably excrete uric acid. Apart from this, nothing was known of the function of these organs in osmotic and ionic regulation until recently when Farquharson (1974 a, b, c) reported on their structure and function in a millipede, *Glomeris*. She found that the total osmotic pressure of *Glomeris* blood is equivalent to about 87 mmol liter^{-1} of NaCl, and that high sodium and chloride concentrations (58.5 and 53.0 mmol liter^{-1}) are similar to those of *Julus* but lower than those of the chilopod *Lithobius* (Sutcliffe, 1963). Absolute and relative concentrations of ions in the blood of the New Zealand centipede *Cormocephalus rubriceps* are rather similar to those of Sutcliffe's Group 1 of primitive exopterygote insects (215 mmol liter^{-1} sodium and only 4.9 mmol liter^{-1} potassium, for example) (Bedford and Leader, 1975). Amino acids contribute about 10% to the total O.P. but a considerable fraction of anionic solutes is unidentified, and believed to be organic acids other than amino acids.

Binyon and Lewis (1963) reported that three species of geophilomorph centipedes from different ecological situations: *Haplophilus* (terrestrial), *Strigamia* (upper littoral) and *Hydroschendyla* (mid-littoral), all have the same, rather low, total blood osmotic concentration, equivalent to about 45% sea water (about 450 mosm l^{-1}). They suggested (but did not prove) that the enlarged salivary glands of the two littoral species could be the site of salt excretion.

Amino acids of *Glomeris* are also fairly high, at 13% of total O.P., and 13 acids were detected compared with 17 for other diplopod and chilopod bloods by Rajulu (1970). Using in vitro preparations, Farquharson found that the Malpighian tubules of *Glomeris* secrete an iso-osmotic urine (in which they resemble insects), lacking potassium. The rate of urine production is inversely related to blood O.P. (Farquharson, 1974b). The tubule walls appear to be very permeable to larger molecules, but become less so with increasing molecular size (Farquharson, 1974c). Clearly there is room for much fruitful work. It will be of great interest to see what happens to the iso-osmotic urine in the rectum of myriapods, and whether regulatory reabsorption of water and ions takes place as in insects.

V. Insects

There is a large amount of literature on osmotic and ionic regulation in insects [reviewed by Maddrell (1971) and by Stobbart and Shaw (1974)], and the field is currently very active. Recent important advances have been concerned with hormonal control of diuresis, the physiology of ionic regulation, and the interpretation of water and ion movements in terms of fine structure.

The main theme of the following sections will be as follows. First we shall consider the extent to which body water, blood volume and osmotic concentration are regulated, before going on to describe in greater detail the functions of the Malpighian tubules, the hindgut and rectum, at which stage we shall be concerned with ionic as well as total osmotic affairs. Finally, we shall consider fine structure in relation to water and ion movements. During the course of this development, we shall deviate to consider certain special topics including water elimination in blood and sap feeders, and the hormonal control of diuresis.

1. Ionic and Osmotic Concentrations

Information about ionic and osmotic concentrations in the blood of terrestrial arthropods has been summarized by Sutcliffe (1963) and by Stobbart and Shaw (1974) (Table 18) (Fig. 50). The early suggestion of Boné (1944) that Na/K

Table 18. Analyses of the blood of some insects and other arthropods. (After Sutcliffe, 1963)

Group	Species	Stage	Haemolymph $\Delta°C$	Na	K	Cl
Insecta						
Orthoptera	*Locusta migratoria migratorioides*	A	0.841	—	—	98
	Chorthippus parallelus	L	—	72	30	119
Isoptera	*Cryptotermes havilandi*	L	0.730	103	28	82
Dermaptera	*Forficula auricularia*	A	0.736	96	13	90
Homoptera	a jassid	A	0.679	59	21	41
Megaloptera	*Sialis lutaria*	A	0.765	104	—	35
Neuroptera	*Osmylus fulvicephalus*	A	0.805	92	40	62
Hymenoptera	*Neodiprion sertifer*	L	—	3	38	17
Mecoptera	*Panorpa communis*	A	0.690	94	38	34
Chilopoda	*Lithobius* sp.		0.608	157	11	154
Diplopoda	*Iulus scandinavius*		0.457	70	18	61
Isopoda	*Oniscus asellus*		1.093	225	20	256

ratios in the blood of insects are determined by immediate diet has not been confirmed, at least so far as any individual insect is concerned. Rather the ionic composition seems to be species-specific (Boné, 1947; Tobias, 1948); and Sutcliffe (1962) and Duchateau *et al.* (1953), after working with both terrestrial and aquatic species, conclude that blood ion composition is a reflexion of phylogenetic relationships—primitive insects showing high sodium, more advanced ones having high potassium. Sutcliffe (1963) elaborated this proposal, providing greater detail, and assigned insects to several groups according to blood type (Fig. 50). Further data have been assembled by Florkin and Jeuniaux (1974), and there is a good discussion with full documentation of these and related (osmotic and excretory) matters by Jeuniaux (1971) (Table 19).

Amino acids are universally present, and become very important as osmotic effectors in Lepidoptera, Hymenoptera and some Coleoptera (Fig. 50), orders that are believed to be "advanced" evolutionarily. In Lepidoptera and Hymenoptera, and in phasmids to a certain extent, the ubiquitous sodium is replaced by potassium; and magnesium assumes predominant importance in the phasmids.

2. The Extent of Blood Volume and Osmotic Pressure Regulation

Values for total O.P. in normal insects have been referred to above (Table 18). Here we are concerned with the effects of disturbances, largely of volume, on O.P. Mellanby (1939) suggested that one of the functions of insect blood is to act as a water store, from which water lost from the tissues during dehydration may be replenished, and this is probably true. Certainly insects seem to be able to function satisfactorily in conditions where the blood volume is so reduced that it is impossible to obtain good samples for analysis (Hoyle, 1954; Phillips,

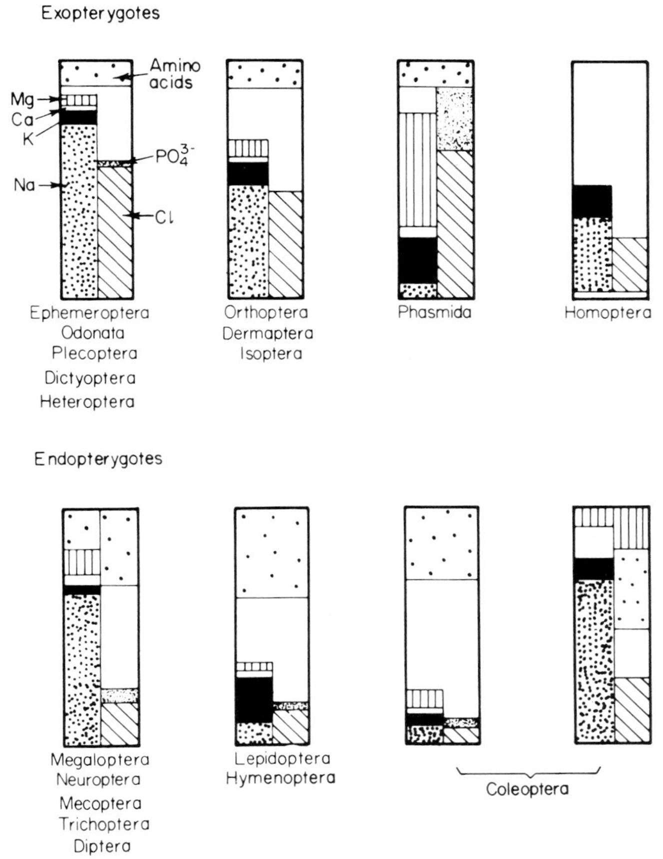

Fig. 50. Osmotic effects of components illustrated as percentages of the total osmolar concentration of blood in pterygote insects. Each block in the figure is visualized as two vertical sections, each section representing 50% of the total osmolar concentration. Left-hand section: the percentage contributions of cations; right-hand section: those of anions. Where possible, amino acids are illustrated in equal proportions in both sections. The large blank area in each block represents the proportion of the total osmolar concentration that must be accounted for by other components of the blood. (From Sutcliffe, 1963)

1964a; Wharton *et al.*, 1965b; Edney, 1968). However, a simple withdrawal of water from the blood would, in the absence of other regulation, lead to a rise in O.P. If the latter is to remain constant, osmotically active material must be proportionately removed or deactivated. Furthermore, there is little point in regulating blood O.P. for its own sake; it is the effect on cells bathed by the blood that matters.

Blood volume is known to change with age and sex. Laviolette and Mestres (1967) showed this for *Galleria*, and Lee (1961) for *Schistocerca*; and volume changes as a result of dehydration have been reported in several insects (early refs in Edney, 1966; Nicolson *et al.*, 1974). Lee (1961) reported a regular fluctuation of blood volume associated with age and the moulting cycle in the locust *Schisto-*

Table 19a. Inorganic cations in the blood of insects. (From Jeuniaux, 1971)

	Species	Cation concentration (meq liter⁻¹)				Sum of cations	Indices (% of the sum)				Reference
		Na	K	Ca	Mg		Na	K	Ca	Mg	
Thysanura	*Petrobius maritimus*	208	5.8	—	—	—	—	—	—	—	Lockwood and Croghan (1959)
Dictyoptera											
adults:	*Periplaneta americana*	157	7.6	4.2	5.4	174.2	90.1	4.3	2.4	3.1	Van Asperen and Van Esch (1956)
adults:	*Carausius morosus*	11	18	7	108	144	7.6	12.5	4.8	75	Ramsay (1955b)
Orthoptera											
larvae:	*Locusta migratoria migratorioides*	60.0	12.0	17.2	24.8	114.0	52.6	10.5	14.9	21.9	Duchâteau et al. (1953)
adults:	*Locusta migratoria migratorioides*	67.4	9.0	15.2	27.0	118.6	56.8	7.6	12.8	22.8	Duchâteau et al. (1953)
Heteroptera											
adults:	*Rhodnius prolixus*	158.0	4.0–6.0	—	—	—	—	—	—	—	Ramsay (1953)
Planipennia (= Neuroptera)											
larvae:	*Myrmeleon formicarius*	143.5	8.7	12.1	31.3	195.6	73.3	4.4	6.1	16.2	Duchâteau et al. (1953)
Trichoptera											
larvae:	*Phryganea* sp.	92.0	6.8	14.4	51.0	164.2	56.0	4.1	8.8	31.1	Duchâteau et al. (1953)
Diptera											
adults:	*Stomoxys calcitrans*	128.0	11.0	—	—	—	—	—	—	—	Duchâteau et al. (1953)
Lepidoptera											
larvae:	*Ephestia kuhniella*	32.6	32.7	41.2	51.1	157.6	20.7	20.8	26.1	32.4	Duchâteau et al. (1953)
	Papilio machaon	13.6	45.3	33.4	59.8	152.1	8.9	29.8	22.0	39.3	Duchâteau et al. (1953)
pupae:	*Deilephila elpenor*	4.7	27.4	41.0	89.3	142.4	2.9	16.9	25.2	55.0	Duchâteau et al. (1953)
adults:	*Bombyx mori*	14.3	36.1	14.5	44.6	109.5	13.0	32.9	13.2	40.7	Bialaszewicz and Landau (1938)
Coleoptera Tenebrionidae											
larvae:	*Tenebrio molitor*	71–75	38.7	11–13	76–83	200–208	36	19	6	38–40	Jeuniaux (1971)
adults:	*Tenebrio molitor*	87.2	30.1	9.9	58.3	185.4	40.7	16.1	5.3	31.4	Jeuniaux (1971)
Scarabaeidae											
larvae:	*Cetonia aurata*	51.3	18.6	22.8	80	172.7	29.7	10.8	13.2	46.3	Duchâteau et al. (1953)
adults:	*Melolontha melolontha*	113	5.8	15.3	41.3	175.4	64.4	3.3	8.7	23.6	Duchâteau et al. (1953)

112

Table 19b. Inorganic anion concentration of the haemolymph and cation-ion balance in some representative species. (From Jeuniaux, 1971)

Species	Stage	Sum of cations (meq liter^{-1})	Anion concentration (meq/liter)			Anion indices (% of the sum of the cations)			Reference
			Cl$^-$	H$_2$PO$_4^-$	HCO$_3^-$	Cl$^-$	H$_2$PO$_4^-$	HCO$_3^-$	
Exopterygotes									
Odonata:									
Aeschna grandis	Larva	169	110	4	15	65	2.3	8.8	Sutcliffe (1962)
Dictyoptera:									
Periplaneta americana	Adult	174.2	144	—	—	82.6	—	—	Van Asperen and Van Esch (1954)
Orthoptera:									
Locusta migratoria	Adult	118.6	97.6	—	—	82.3	—	—	Duchâteau *et al.* (1953)
Dictyoptera:									
Carausius morosus	Adult	197.4 (1)	93 (2)	40 (3)	—	47.1	20.2	—	(1) Duchâteau *et al.* (1953) (2) May (1935) (3) Ramsay (1955a)
Carausius morosus	Adult	154	101	16	—	65.5	10.4	—	Wood (1957)
Endopterygotes									
Megaloptera:									
Sialis lutaria	Larva	167 (1)	31 (1)	5 (2)	15 (1)	18.5	3	6	(1) Shaw (1955) (2) Sutcliffe (1962)
Diptera:									
Gasterophilus intestinalis	Larva	264	14.8	4	14.5	5.6	1.5	5.7	Levenbook (1950)
Lepidoptera:									
Bombyx mori	Larva	150	21	3	—	14	2	—	Buck (1953)
Prodenia eridania	Larva	94.7	34	5.8	—	35.9	6.1	—	Babers (1938)
Samia walkeri	Pupa	128.6	10.4	3.5	—	8	2.7	—	Gese (1950)
Telea polyphemus	Pupa	147.2	19.5	—	—	13.2	—	—	Carrington and Tenney (1959)
Coleoptera:									
Dytiscus marginalis	Adult	231.6 (1)	44 (1)	2.8 (2)	—	19	1.2	—	(1) Sutcliffe (1962) (2) Buck (1953)
Popillia japonica	Larva	84.3	19	4.9	—	22.5	5.8	—	Bishop *et al.* (1925)
Hymenoptera:									
Apis mellifera	Larva	52.7	33	10.3	—	62.6	19.5	—	Bishop *et al.* (1925)

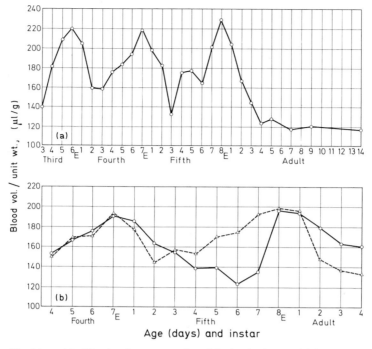

Fig. 51a and b. Blood volume per unit weight in the locust (Schistocerca). (a) Variation with age and instar on a normal (fresh grass) diet. (b) The effect of a dry grass diet *(circles)* and an artificial diet with which water was available *(triangles)*. Water was added to the dry grass diet on the fifth day of the fifth instar, and the effect of this on blood volume is clear. E. *Ecdysis.* (After Lee, 1961)

cerca (Fig. 51a, b), and he also found a low volume in nymphs that had fed on dry grass compared with controls given water in addition (Fig. 51b). In the dry fed locusts loss of blood volume was not accompanied by loss of tissue water, which suggests that water was moved from the blood, but we do not know the effect of this on blood O.P.

In the grasshopper *Chortoicetes*, Djajakusumah and Miles (1966) found a normal blood volume of 112 μl per insect ($= 335 \, \mu l \, g^{-1}$). This fell by 25% during dehydration for 1 day, but the change caused only a small rise in O.P.: from 229 to 251 mM NaCl equivalent (a rise to 305 mM would be expected in the absence of other changes). Restoration of blood volume by drinking was also accomplished without much O.P. change. In this insect free amino acids constitute about 15% of the blood O.P., and in dehydrated insects amino acid concentration fell to 6% and soluble proteins rose, thus reducing the total O.P. However, calculation shows that the change in O.P. so achieved would contribute only one third at the most of the osmotic regulation observed, and the impression remains that inorganic salts must in some way be sequestered during dehydration. Amino acid/protein interchange certainly acts as a major osmoregulatory mechanism in certain aquatic insect larvae (Beadle and Shaw, 1950; Schoffeniels,

114

1960), but in terrestrial forms such a process has been convincingly demonstrated as an important mechanism only in *Dysdercus* (Berridge, 1965b) and the desert tenebrionid *Trachyderma philistina* (Broza, 1975).

In *Dysdercus* Berridge observed a marked drop in blood volume during the middle of an instar, associated with diuresis necessary for the elimination of excess dietary potassium and other inorganic ions. Sodium and chloride ions are retained and their concentration in the blood rises accordingly, but the total blood O.P. remains constant as a result of a decrease in amino acid concentration. In the second half of the instar, the insect drinks water (in culture at least), blood volume is restored and its O.P. is maintained by an increase of amino acids (Fig. 52).

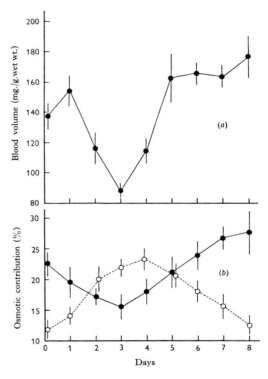

Fig. 52. (a) Changes in blood volume during the fifth instar of the hemipteran *Dysdercus*. (b) Parallel changes in the contribution of chlorides *(open circles)* and amino acids *(solid circles)* to the total blood osmotic pressure. The mid instar drop in blood volume is associated with diuresis necessary for the elimination of excess dietary potassium. Total blood O.P. remains constant as a result of a decrease in amino acid concentration. Later in the instar the insects drink water (if this is available), and amino acids rise to maintain total O.P. (From Berridge, 1965b)

Another example of osmoregulation is that found by Marcuzzi (1955) in *Tenebrio molitor*, whose normal O.P. is equivalent to about 356 mM NaCl. The blood O.P. of these insects varies only between 223 and 365 mM NaCl after exposures of 5–8 days to either moist or dry conditions, respectively (Marcuzzi, 1956), so that some regulation is apparent. However, we do not know how total blood volume or water varies, so that the mechanism of regulation is not clear.

In *Periplaneta*, Munson and Yeager (1949), Wharton *et al.* (1965b) and Yeager and Munson (1950), all suggest that both water and ions may be moved between blood and tissues after dehydration. Pinchon (1970) also found that in *Periplaneta*, blood ions vary in concentration in different degrees of dehydration, and concluded that ions must be sequestered and released somewhere in the body.

Edney (1968) measured changes in water, blood volume and blood O.P. occasioned by dehydration in a desert cockroach, *Arenivaga investigata*, and in *Periplaneta americana*. These experiments showed (Fig. 53) that in *Arenivaga*,

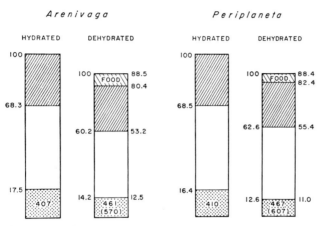

Fig. 53. The effect of dehydration on the water content, blood volume and blood osmotic pressure in *Arenivaga* and *Periplaneta*. *Hatched areas*: dry weight; *blank areas*: tissue water; *stippled areas*: blood water. Numbers in the stippled areas indicate actual osmotic concentrations in mosmol liter^{-1}, with expected values (assuming no osmoregulation) in brackets. *Numbers on left of each column*: percentages of total wet weight at the time of measurement; *numbers on right of "dehydrated" columns*: percentages of pre-dehydration wet weights. (From Edney, 1968b)

during dehydration water is lost from the blood and from the rest of the body in proportional amounts, and that the fall in blood volume does not cause as much rise in blood O.P. as would be expected without some form of O.P. regulation. (The observed rise was from 407 to 461 mosmol liter^{-1} against an expected 570 mosmol liter^{-1}.) The results point to the removal of osmotically active substances from the blood, and very similar results were found in *Periplaneta* (Fig. 53). In *Arenivaga* subsequent rehydration by water vapour absorption (Chap. 9.C.) leads to an increase in total body water, and to a fall in blood O.P. to an extent less than expected without regulation (Edney, 1966) (Fig. 54). In this case blood volume was unknown, so that we cannot say certainly that osmotically active materials were added, although this seems very probable.

Wall (1970) confirmed and amplified the above results. She determined that in *Periplaneta*, blood volume, which falls during dehydration, is more than

116

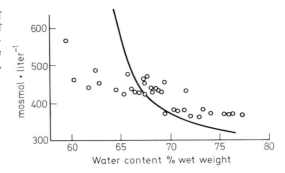

Fig. 54. *Open circles*: osmotic pressure of the blood of *Arenivaga* as a function of the insects' water content. The line represents the expected relationship in the absence of regulation. (From Bursell, 1970, data from Edney, 1966)

replaced during subsequent rehydration by drinking (Table 20), but O.P. varies little during these processes. Wall went on to measure the amount and concentration of Malpighian tubule fluid which unexpectedly turns out to be hyper-osmotic to the blood in dehydrated animals. The tubule continues secretion even after

Table 20. Changes in blood volume in *Periplaneta*. (From Wall, 1970)

Condition of animal	Body wt (mg)	Blood vol (%) $\left(\dfrac{\text{Blood wt}}{\text{Body wt}} \times 100\right)$	Blood vol (μl)
Normal	698 ± 36	18.0 ± 1.1	134 ± 6.3
Dehydrated (8 days)	513 ± 9	9.6 ± 0.8	53 ± 5.3
Rehydrated (24 h)	721 ± 43	19.1 ± 2.1	146 ± 12.2

considerable dehydration, and measurement of the extent of this and of the ionic concentration of the urine, showed that the excess solutes which calculation suggests would have to be removed from the blood to account for O.P. regulation could indeed be secreted by the Malpighian tubules; but analysis of the faecal pellets showed that the NaCl removed by the tubule is not so eliminated. During eight days dehydration 723 μg NaCl were apparently removed from the blood, but only 24 μg appeared in the faeces. Presumably the rest remained in the rectum or elsewhere—at least they were available for redeployment in the blood when this was increased in volume during rehydration. The probable movements of water and ions during a cycle of de- and rehydration is shown in Figure 55.

Nicolson *et al.* (1974) similarly found that in *Carausius*, a fall in blood volume to one seventh as a result of dehydration causes an increase in total O.P. of only 20%, the main blood ions increasing in concentration by like amounts. Laird and Winston (1975) working with *Leucophaea*, also found regulation of blood O.P. in this insect, and Okasha (1973) found little effect of dehydration and rehydration on sodium or potassium concentration in *Thermobia*, although blood volume varies considerably as a result of these processes. Blood volume is strongly affected by dehydration and by drinking in *Trachyderma*, but O.P.

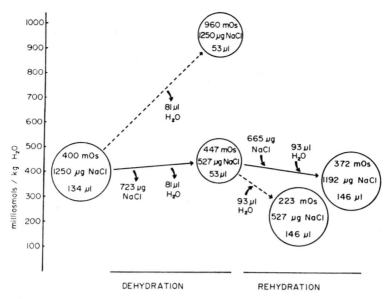

Fig. 55. Summary of changes in the blood of the cockroach *Periplaneta* during dehydration and rehydration. In normal animals *(left circle)* the blood has an O.P. of 400 mosmol liter^{-1}. This consists of 1,250 μg of NaCl, and the total volume is about 134 μl. After nine days of dehydration, the blood volume decreases to an average of 53 μl, a loss of 81 μl of water. If no solutes are removed from the blood, the blood O.P. should increase to 960 mosmol liter^{-1} (upper middle circle). Since blood O.P. does not increase to this extent about 723 μg of NaCl have been removed from the blood. When a dehydrated animal is given water but no food, volume increases to the normal level within 24 h. If no solutes are added to the blood during rehydration, the blood O.P. should decrease to 223 mosmol liter^{-1} (lower right circle). Since blood O.P. in rehydrated animals is actually 372 mosmol liter^{-1} (upper right circle), about 665 μg NaCl have been added. The source of these solutes is not known. (From Wall, 1970)

again seems to be quite well controlled (perhaps by amino acid concentration) (Broza, 1975).

However, such regulation is not always evident. For example, short period fluctuations in blood O.P. and volume associated with feeding have been reported in *Locusta* by Bernays and Chapman (1974a). The effect of feeding is to reduce the blood volume by 13%, but here the O.P. rises in a manner suggesting that water movement unaccompanied by solutes is responsible (Fig. 56). Normal blood volume is restored within an hour or so after feeding has ceased. Larvae of the rock-pool chironomid, *P. tonnoiri*, when undergoing dehydration, lose water from blood and tissues in equal proportions, and again there is little regulation of blood osmotic concentration, for this rises from Δ 0.380° to Δ 0.925 °C during 30 days dehydration in warm, dry conditions (Chap. 2.B.), while the blood volume falls appropriately so that Δ°C = k × 1/blood volume (Jones, 1975).

The results of dehydration on the blood O.P. of several arthropods are shown in Figure 57. There is a clear difference between an isopod *(Oniscus)* and the two cockroaches while the behaviour of the cotton stainer *Dysdercus*

118

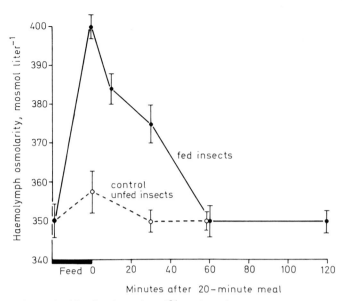

Fig. 56. Feeding in *Locusta* reduces the blood volume by 13%, and as the upper curve shows, this is reflected in a rise in blood O.P., suggesting the absence of O.P. regulation in this case. (From Bernays and Chapman, 1974a)

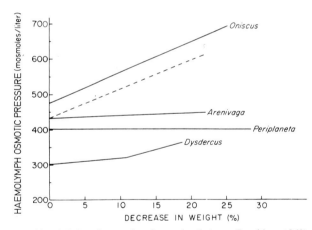

Fig. 57. The effect of water loss on the blood O.P. of several arthropods. *Oniscus* (Berridge, 1968) is a crustacean and shows no regulation. *Arenivaga* (Edney, 1966) is a desert cockroach; *Periplaneta* (Wall, 1968) is a domestic cockroach, and both show good regulation. *Dysdercus* (Berridge, 1965a) is a hemipteran, and shows some degree of regulation. The dashed line shows values for *Arenivaga* in the absence of regulation. (From Berridge, 1970)

seems to be intermediate. There is now sufficient evidence to conclude that in insects at least, blood O.P. is at least partially regulated, and also that the blood serves as a source of water during periods of dehydration; but the emergent question as to the fate of the ions involved is still unanswered.

119

A useful theoretical analysis of the extent of the problems involved, and of possible solutions, has been undertaken by Shaw and Stobbart (1972). Making certain reasonable assumptions, they constructed models to show the effect of dehydration at different rates of cuticular water loss on blood volume, tissue water volume and blood O.P. in fasting *Schistocerca*. The predictions of this model are consistent with the data for *Schistocerca* (Lee, 1961; Fayadh, 1969) (quoted by Shaw and Stobbart, 1972), and for *Periplaneta* (Wall, 1970) and *Arenivaga* (Edney, 1966, 1968). Their models show that if *Schistocerca* is given a diet of moist grass whose ionic content is such that the fluid ingested would be hypo-osmotic to the blood, then the Malpighian tubules and rectum should be able to regulate blood volume and O.P. However, when such insects are presumed to feed on dry food in dry air, and the ingested material forms a hyper-osmotic fluid in the midgut, solute absorption in the midgut would have to be reduced because the Malpighian tubules would not be able to excrete the solutes rapidly enough.

If water absorption against a gradient from the midgut lumen is not possible (and there is no evidence that it is) then perhaps water and solutes could be passed directly to the rectum, where water can be absorbed without solutes down to an osmotic gradient of about 1 osmol liter^{-1} (Phillips, 1964a) (Chap. 6.C.V.6.).

These considerations, as Shaw and Stobbart stress, do involve assumptions and do not take account of the possibility of ion storage in the fat body as envisaged by Mullins and Cochran (1974) (Chap. 6.B.I.), so that the conclusions must be clearly distinguished from experimental evidence. Nevertheless, Shaw and Stobbart's treatment does lead to more precise thinking about physiological processes involved, and as such is a salutary exercise.

3. The Structure and Function of Malpighian Tubules

In one way or another, the whole of the alimentary canal of arthropods is involved in osmotic regulation, and an outline of its structure in insects is shown in Figure 58, but by far the most important (and well known) organs

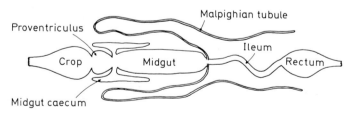

Fig. 58. Diagram to show, for a generalised insect, the relationship of the Malpighian tubules with the rest of the alimentary canal. Note that fluid from the Malpighian tubules passes through the ileum before reaching the rectum. (From Maddrell, 1971)

are the Malpighian tubules and the walls of the hind gut, including the ileum and rectum. Other organs that are certainly or probably involved are referred to below (Chap. 6.C.V.9.); here we shall be concerned first with the Malpighian tubules and thereafter with the rectum.

The Malpighian tubules of insects are probably the best known and best understood of all invertebrate excretory organs. Wigglesworth (1972) reminds us that they were named by Meckel in 1829 after Malpighi, who referred to them as "vasa varicosa" in his work on the silkworm in 1669. Long regarded as biliary organs, their excretory function was first suggested by Cuvier in 1802. The literature has been admirably reviewed by Maddrell (1971).

The tubules occur in a variety of forms. They vary in number from two (in certain coccids) where they are generally very long, to well over a hundred in orthoptera, where they are short, the amount of available surface being approximately constant in relation to the total weight of the insect (Schindler, 1878, quoted by Wigglesworth, 1972). They generally lie freely in the haemocoele, exceptions include the various cryptonephric and filter chamber systems, considered below. Distally they end blindly, and proximally they open into the alimentary canal close to the junction between the mid and hindgut. Their walls are composed of one layer of cells, but often these are of two or more types, either interspersed (*Calliphora*) or separated into regions (*Rhodnius*). In *Musca domestica*, according to Sohal (1974), there are four cell types. Frequently the lower (proximal) part of the tubules, or the rectal walls, are differentiated into regions known as ampullae or rectal glands, where the cells are concerned with the reabsorption of materials rather than with elaboration of a primary fluid. Figure 59a, b shows diagrammatically the general anatomy and the fine structure of the regions of *Rhodnius* tubules, and we shall return to this later.

As a first approximation it can be said that the Malpighian tubules of most insects produce a fluid which is iso-osmotic with the blood, but with a high potassium concentration, and this fluid, as it moves down the tubule or, more usually, when it arrives in the hindgut, is subject to reabsorption of water and/or ions and other substances to an extent necessary to maintain water content, osmotic concentration and ionic proportions in the blood constant, according to the general system shown in Figure 60.

The now classical work of Ramsay on the stick insect *Carausius* (*Dixippus* as it then was) laid a firm foundation for later comparative work on other insects as well, by Maddrell, Berridge, Pilcher, Phillips and others. Most of the information now available could not have been obtained in vivo, and it was Ramsay's (1954) development of an in vitro technique that provided the key (Fig. 61). By this means the performance of a single tubule included in a drop of serum could be monitored. Subsequent modifications by Maddrell (1963), Berridge (1966a) and Pilcher (1970a) showed that tubules of *Calliphora* and *Carausius* function for many hours in an artificial medium as long as this is well aerated and provided with an energy source such as glucose.

Using this isolated tubule technique, Ramsay (1953, 1954, 1956) found that the tubule fluid is nearly always slightly hypo-osmotic to the medium, and that the potassium concentration of the fluid is always higher and the sodium concentration usually lower than that of the medium. In *Carausius*, the rate of secretion is closely related to the potassium concentration, but is not greatly influenced by sodium. Calcium, magnesium and chloride ions are always lower

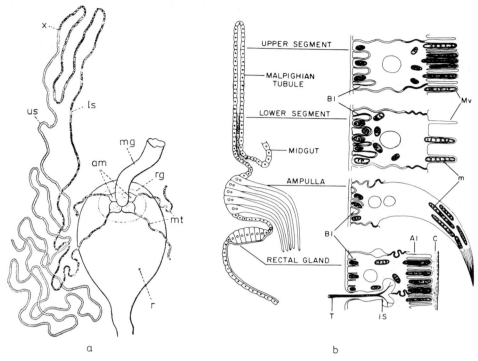

Fig. 59. (a) A general view showing one Malpighian tube in full; *am*: ampulla; *ls*: lower (proximal) segment of Malpighian tube; *mg*: mid-gut; *mt*: Malpighian tube; *r*: rectum; *rg*: rectal gland; *us*: upper (distal) segment of Malpighian tube; *x*: junction of upper and lower segments of tube. (b) The fine structure of a *Rhodnius* Malpighian tubule (diagrammatic). *AI*: apical infold; *BI*: basal infolds; *C*: cuticle; *IS*: intercellular space; *Mv*: microvilli; *T*: trachea. (From Berridge, 1970)

in the tubule than the medium, phosphate higher. On these grounds Ramsay concluded that the active transport of potassium is an essential part of urine formation, and this in general has been found to apply to many insects studied, although in some, including *Glossina* (Gee, 1975a, 1976) and *Rhodnius*, sodium may be as important as potassium or even replace it. Electro-potential differences across the tubule wall (Table 21) mean that potassium would have to move up this gradient, even where the lumen is negative, and the effect of solvent drag is negligible (Stobbart and Shaw, 1974), so there is little doubt about the active transport of potassium. Sodium too is actively transported, although the rate is usually much lower than that of potassium. As regards anion movements; in *Carausius* chloride concentration is lower in the tubule than in the medium, and probably enters passively (Ramsay, 1956). On the other hand, phosphate is more concentrated in the *Calliphora* tubule than outside. The rate of fluid secretion is greatly increased by high chloride in the medium, but the transport of this ion is probably a passive one (Berridge, 1969). However, the situation seems to differ a lot between insects. In *Rhodnius* phosphate does not seem to support secretion, and chloride is strongly, actively transported

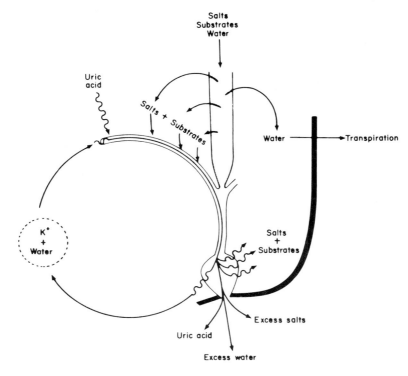

Fig. 60. A diagrammatic summary of the processes involved in ionic and osmotic regulation in insects. (From Bursell, 1970)

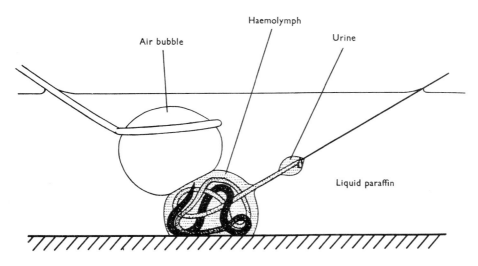

Fig. 61. The experimental arrangement used to investigate secretion in an isolated Malpighian tubule in *Carausius*. (From Ramsay, 1954)

(Maddrell, 1969, 1971). In fact, recent work has shown that many more ions than previously thought are actively transported, including organic anions such as sulphonates and acylamides (Maddrell *et al.*, 1974) (Chap. 6.C.V.5.), and in salt water larvae of *Aedes campestris*, the tubules actively transport magnesium and sulphate ions (Phillips and Maddrell, 1975; Maddrell and Phillips, 1975b). As these authors suggest, it begins to look as if the tubules may excrete most materials if, perhaps as a result of feeding, these concentrate in the blood to a level where they cannot be eliminated by simple diffusion through the tubule wall.

Table 21. Electrical potential across the tubule wall in certain insects[a]

Species	Stage[b]	Fluid	Osmotic pressure mM ($\equiv$ NaCl solution)	Transtubular potential difference[c]	Reference
Carausius morosus	A	Intestinal fluid[d]	171	+21	Ramsay (1953, 1955b)
Locusta migratoria	A	Tubule fluid	—	−16	Ramsay (1953)
Rhodnius prolixus	A	Distal part of tubule	228	−35	Ramsay (1952, 1953)
Pieris brassicae	L	Tubule fluid	—	+28	Ramsay (1953)
Calpodes ethlius	L	Rectal lead (artificial medium)	—	+25	Irvine (1969)
Dytiscus marginalis	A	Tubule fluid	—	+22	Ramsay (1953)
Tenebrio molitor	L	Tubule fluid	—	+45	Ramsay (1953)

[a] From Stobbart and Shaw (1974).
[b] A: adult, L: larva.
[c] Sign refers to lumen.
[d] Fluid from the rectum near the point of entry of the Malpighian tubules.

The rate of secretion in *Rhodnius* is inversely correlated (though perhaps not linearly) with the osmotic concentration of the bathing medium, being greater in lower O.P.s, although the O.P. of both fluid and medium are virtually the same (Maddrell, 1969) (Fig. 62a, b). Rates of secretion and medium O.P. are similarly related in *Dysdercus* (Berridge, 1966b), *Calliphora* and *Carausius*, but a simple osmotic coupling is not necessarily the explanation, for in *Carausius* the luminal O.P. is in fact slightly higher than that in the medium. However, Maddrell (1971) marshalls several compelling arguments for believing that osmotic coupling is the simplest explanation so far. If water moves as a result of ion secretion, this provides a reason for the secretion of potassium and anions, and we can accept that solute secretion is the driving force for movements of fluid. The mechanism proposed, in which osmotic coupling could take part, will be referred to below (Chap. 6.C.VII.2.). (For further discussion of possible models, see Maddrell, 1971.)

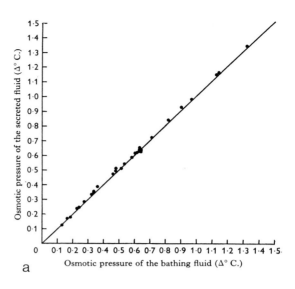

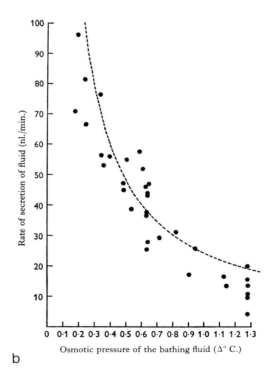

Fig. 62. (a) The osmotic concentration and (b) the rate of secretion of fluid produced by isolated Malpighian tubules of *Rhodnius*, both in relation to the osmotic concentration of the medium. (From Maddrell, 1969)

125

4. Regional Differences Within Malpighian Tubules

In most insects so far studied, certain regions of the Malpighian tubules differ both structurally and functionally from others. Some examples are shown in Table 22. In *Carausius*, where there are three types of tubules, Ramsay (1955a, b)

Table 22. The composition (in mmol liter^{-1}) of fluid collected from various regions of the Malpighian tubules of three insects. (From Stobbart and Shaw, 1974; see also Table 25)

Species	Fluid	Osmotic pressure mM ($\equiv$ NaCl solution)	Na	K	Reference
Carausius morosus	Serum	171	11.0	18	Ramsay (1955b)
Superior tubule cut	Distal piece	171	4.1	147	
into three pieces	Middle piece	162	7.0	129	
	Proximal piece	167	9.7	96	
Rhodnius prolixus					
24–29 h after	Blood	206	174	7	Ramsay (1952)
normal blood meal	Distal part of tubule	228	114	104	
	Proximal part of tubule	199	158	50	
Calpodes ethlius	Medium	—	13	55	Irvine (1969)
	Rectal lead and iliac plexus	—	15	105	
	Yellow region	—	24	41	
	Yellow region and white region	—	60	30	

found the Na/K ratio to vary considerably between types, being higher in the proximal regions of the superior tubules. In the distal region of the inferior tubule, no secretion occurs, but a small potassium reabsorption takes place. In *Calpodes* larvae, where there is a cryptonephric system, the distal regions normally have a high potassium concentration (Irvine, 1969), as in *Carausius* and *Calliphora*, but proximally sodium predominates, and is more concentrated than it is in the medium while potassium is reduced, probably by reabsorption. This is useful to the insect insofar as it returns valuable potassium and water to the blood. In *Rhodnius* the situation is rather different—(see Chap. 6.C.V.6.) secretion occurs in the distal segment of the tubule, but not in the proximal (Wigglesworth, 1931; Maddrell, 1969) where potassium and water are reabsorbed and uric acid is precipitated after the pH has fallen.

In *Bombyx mori*, according to Waku (1974), the Malpighian tubules have two distinct regions. Proximally they lie freely in the haemocoele and excrete calcium oxalate—distally they are cryptonephric, and absorb water from the hind gut. Associated with this functional difference is a fine structural one: the cells in both regions have highly developed basal and apical infoldings, but the mitochondria of the two regions differ in shape. In larvae of the tobacco horn worm, *Manduca sexta*, as in most lepidopterans, there are three morphologically distinct regions in each Malpighian tubule. At some stages in the development of these caterpillars the middle region of the tubule excretes acid (anionic) dyes, while the proximal region deals with basic (cationic) ones (Nijhout, 1975).

It is clear from all the recent work on Malpighian tubule function that osmoregulation is not their role, for the secreted fluid is nearly iso-osmotic with the blood over a wide range of O.P.'s. Is their function then, that of ionic regulation? Again the answer is probably not, because although the fluid is iso-osmotic it is very different ionically from the blood, and hence its production alone would have the effect of altering the blood ionic composition. In order to effect ionic regulation, tubule fluid should contain lower concentrations of any ion that is abnormally low in the blood, and higher concentrations when the blood is too high. But this is not the case except possibly with sodium in *Calpodes* (Irvine, 1969), and the conclusion must be that such regulation is achieved by differential absorption in the rectum—the function of the Malpighian tubules being largely to provide the rectum with a fluid at a *rate* sufficient for homeostasis to be achieved.

5. Malpighian Tubules and Larger Molecules

Early on Ramsay (1958) observed that *Carausius* tubules in vitro will produce considerable quantities of various organic molecules, including alanine, arginine, urea and sucrose (Table 23) if they are present in the medium, and he proposed

Table 23. Concentration of various organic substances in the fluid secreted by isolated Malpighian tubules of *Carausius*. (Data from Ramsay, 1958)[a]

Substance	Concentration in bathing solution (mmol liter^{-1})	U/P ratio[b] (Mean ± S.D.)
Alanine	56	0.25 ± 0.05
Arginine	53	0.35 ± 0.09
Glycine	43	0.17 ± 0.05
Proline	52	0.59 ± 0.04
Valine	59	0.24 ± 0.04
Glucose	53	0.86 ± 0.18
Fructose	53	0.61 ± 0.08
Sucrose	53	0.53 ± 0.10
Urea	56	0.96 ± 0.06

[a] Table from Maddrell (1971).
[b] Ratio of concentrations in secreted fluid and medium.

that this is a reflexion of a basic property of Malpighian tubules: to form a fluid containing samples of any large molecules that happen to be found in the blood, leaving it to the rectum to reabsorb the useful ones. In this way the excretory apparatus could eliminate any unwanted materials without a special transport mechanism for each, so that the only energy-consuming part of the process would be those mechanisms necessary for reabsorption.

There are two aspects to this problem; the elimination of molecules that are potentially toxic, and the retention or reabsorption of those that are useful. As regards the first aspect, Phillips and Dockrill (1968) showed that the rectal cuticle of *Schistocerca*, while being permeable to inorganic and organic molecules up to the size of glucose (M.W. 132), becomes progressively less permeable to larger ones. If such larger molecules are produced by the Malpighian tubules they would not be able to reach, and possibly damage, the rectal epithelium, but instead would pass out with the excreta.

It now appears that potentially toxic molecules may be detoxified by conjugation to produce substances similar in structure to acidic dyes and acylamides (Maddrell *et al.*, 1974), and that these conjugated molecules are then actively transported across the tubule wall. Without such transport, their size would make elimination too slow, but once transported, large size becomes advantageous by preventing back diffusion across the tubule wall or reabsorption from the rectum.

In addition to such active transport there are indications that the Malpighian tubules of several insects are to some extent permeable to quite large molecules from urea (M.W. 52) to insulin (M.W. 5,200) (Maddrell and Gardiner, 1974). On several grounds, including the fact that the ratio between concentration in the secreted fluid and in the medium (the S/M ratio) is not affected by the concentration of M, and is never greater than 1, Maddrell and Gardiner conclude that such large molecules move by passive diffusion, probably between the cells of the tubule (the rate of diffusion is probably too great for any other channel), by what must be leaky lateral junctions. This at present seems to be the most probable conclusion.

As regards the reabsorption of large but potentially useful solutes, it has been assumed for some time that this probably occurs in the rectum, since sugars and amino acids (for example) are present in the tubule fluid of *Carausius* (Ramsay, 1958) and in the hind gut of *Periplaneta* (Wall and Oschman, 1970), but hardly ever in the excreta. (Aphids and other phloem feeding insects are exceptional for obvious reasons.)

In view of the relative impermeability of the rectum, some reabsorption (of disaccharides, for example) probably occurs in the ileum (Maddrell and Gardiner, 1974), but little more can be said on this aspect because the problem has not been studied extensively. In *Schistocerca*, however, Balshin and Phillips (1971) demonstrated the active uptake of glycine and probably of serine and proline, through the rectal wall. They propose a specific pump, in or on the apical plasma membrane. Sodium appears to be involved, since its substitution by potassium abolishes amino acid transport.

These important developments will no doubt be pursued in future research.

6. Reabsorption of Water in the Rectum

The rectum is an organ of central importance for water conservation and ionic regulation in insects. Its structure and function have been actively researched during the last decade, and a very useful account of these matters is contained in a review by Wall and Oschman (1975). In this and the following sections (Chap. 6.C.V.6.—6.C.V.8.) we shall be concerned mainly with the extent and effects of osmotic and ionic regulation by the rectum, leaving a consideration of transport mechanisms at the fine structural level until later (Chap. 6.C.VII.).

Table 24. Composition of Malpighian tubule and rectal fluids in meq liter^{-1} [a,b]

Species	Stage	Fluid	O.P. mM ≡ NaCl	Na	K	Cl	Ca	Mg	PO$_4$	Reference
Carausius morosus	A	Intestinal	171	5	145	65	2	18	51	Ramsay (1953, 1955b)
	A	Rectal	390	18	327					Ramsay (1955b)
Schistocerca[c] *gregaria*	A (water fed)	Intestinal	226	20	139	93				Phillips (1964c)
	A (water fed)	Rectal	433	1	22	5				Phillips (1964c)
Rhodnius prolixus	A	Distal tubule	228	114	104	180				Ramsay (1952, 1953)
	A	Rectal (19–29 h after feeding)	358	161	191					Ramsay (1952)
Calliphora erythrocephala	A	Tubule	221	20	140					Berridge (1968)
	A	Rectal (water fed)	34			6				Phillips (1969)
	A	(no water for 2 days)	501			238				Phillips (1969)

[a] Data from Stobbart and Shaw (1974).

[b] For further data about aquatic insects see Stobbart and Shaw (1974).

[c] For further data on *Schistocerca* see Table 26.

One important rectal function is to precipitate uric acid in an insoluble crystalline form from the soluble anionic form in which it exists in the Malpighian tubules. In *Rhodnius*, this precipitation takes place in the proximal segment of the tubule itself, as mentioned above, but in other insects it occurs in the rectum. The process is encouraged by lowering the pH of the rectal contents, thus causing a high proportion of the urate ions to recombine. But the means whereby this is achieved, and its effectiveness in terms of the proportion of uric acid precipitated, are unknown.

Additionally, the rectum is a region in which the amount and composition of the Malpighian tubule fluid may be changed, and the data in Table 24 illustrate this point by comparing O.P.s and concentrations of various ions in the Malpighian tubule fluid (usually taken from the intestine where the tubules open) and those in the rectal fluid. The differences are striking. In insects which need to conserve water, the rectal fluid is highly concentrated, and sodium, potassium and chloride concentrations are high. However, when there is excess water, as in water-fed *Schistocerca*, *Periplaneta* and *Calliphora*, the ionic concentrations are very low, and the urine becomes hypo-osmotic to the blood (except in *Schistocerca*, where the total O.P. is still fairly high), perhaps as a result of organic solutes (Phillips, 1964 b, c).

Concentration of ions in the rectum could be achieved either by secretion from the rectal walls, or by absorption of water from the lumen. However, the rate of discharge of water from the rectum of *Rhodnius* after the first diuresis has been completed is very low, although the tubules continue to produce fluid (Wigglesworth, 1931), so that reabsorption of water, rather than secretion of ions seems to occur in this insect. Similar absorption is apparent in *Periplaneta* (Wall and Oschman, 1970). Water in these cases must, of course, move against considerable osmotic gradients.

Data indicating the opposite process (secretion of ions) have been found by Prusch in the blow-fly *Sarcophaga* where an electropotential gradient exists across the gut wall equal to the effects of two (proposed) electrogenic pumps, one for anions, the other for cations (Prusch, 1974). As a result Prusch (1973, 1976) believes that ions are secreted into the rectum and a hyper-osmotic fluid is formed in this way, rather than by the withdrawal of water.

If this mechanism is confirmed it will be the first example in which hyper-osmotic conditions result from the secretion of ions into the lumen, but other examples of such secretion are now being reported. They include the mechanism of ionic transport in the desert millipede *Orthoporus*, where it looks as if potassium may be actively transported into the lumen (Moffett, 1975), and that of the salt water mosquito larva, *Aedes taeniorhynchus*, where a high concentration of potassium in a solution otherwise iso-ionic with sea water is secreted by the rectum (Bradley and Phillips, 1975).

The more usual process of water reabsorption has been studied in detail in *Schistocerca* (Phillips, 1964a, b, c) and in *Calliphora* (Phillips, 1969). *Schistocerca* were fed either on tap water, or on hyper-osmotic saline (1 osmol liter^{-1}) for 4 to 6 days. Their blood, Malpighian tubule fluid (derived from the hind gut), and rectal fluid were subsequently analyzed. As Table 25 shows the insects regulate their blood concentration fairly well—total O.P. was $\Delta 0.74\,°C$ in water-

130

Table 25. Osmotic and ionic concentrations in the blood, hind gut (from Malpighian tubules) and rectal fluid of *Schistocerca*. (Data from Phillips, 1964 a, b, c)

		Osmotic concentration		Ion concentrations (mmol liter^{-1})		
		$\equiv$ NaCl meq liter^{-1}	(Δ°C)	Na	K	Cl
Insects fed on	Blood	214	0.74	108	11	115
tap water	Hind-gut	226	0.78	20	139	93
	Rectal fluid	433	1.52	1	22	5
Insects fed on	Blood	275	0.96	158	19	163
hyper-osmotic	Hind-gut	—	—	67	186	192
saline	Rectal fluid	989	3.47	405	241	569

Concentrations of the hyper-osmotic saline in mmol liter^{-1}:

NaCl	KCl	Mg(CH$_3$—COO)$_2$	Ca(NO$_3$)$_2$
300	150	30	30

fed, and $\Delta 0.96$°C in saline-fed insects, and blood ionic concentrations were consistent, i.e. somewhat higher in the salt-fed locusts, but not nearly as high as the concentrations in the diet. Rectal fluids in the two groups, however, were extremely different, those in water-fed insects showing very low ionic concentrations, while the saline-fed group had ionic concentrations even higher than those of the saline. Clearly, absorption of ions in the water-fed locusts, and of water in the other group, had occurred. Interestingly, the rectal fluid of water fed locusts never becomes hypo-osmotic to the blood, although such a response would appear to be adaptive. Apparently *Schistocerca* cannot, or does not, in these circumstances produce a hypo-osmotic urine.

In order to find out more about the mechanisms involved in these regulatory processes, Phillips developed a technique for studying volume changes as well as ionic concentrations in locust recta in vivo. Essentially this consisted in tying off the rectum from the rest of the gut, and sealing in a capillary through the anus, by means of which fluids may be introduced into the rectum, and their volume and concentration changes subsequently measured (Fig. 63).

By these means he found that the volume of fluids perfused into the rectum is changed, either by absorption from, or secretion into, the rectum, the direction and rate of such change depending upon the initial O.P. of the fluid inserted. Even in locusts in good water status, movement of water from the lumen across the rectal wall occurs against substantial osmotic gradients. Figure 64 shows a situation where absorption of water occurred against gradients up to 500 mosmol liter^{-1} (when the O.P. of the rectal contents was about 2.3 times that of the blood). Movement of water in the opposite direction takes place when the gradient is steeper. However, if the insects are first dehydrated, absorption occurs against gradients up to 1,000 mosmol liter^{-1}, thus showing a valuable ability to increase the capacity for absorption under water stress.

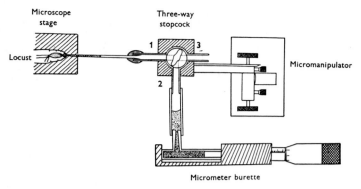

Fig. 63. Diagram of apparatus for measuring water and solute absorption from the ligated locust rectum. *Heavy stippling*: mercury; *light stippling*: aqueous solutions. (From Phillips, 1964a)

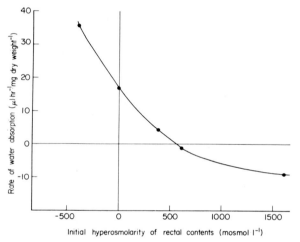

Fig. 64. The rate of water absorption from xylose solutions injected into the lumen of the locust rectum, as a function of the initial osmotic concentration difference across the rectal wall. (From Maddrell, 1971; after Phillips, 1964a)

These observations immediately raise the question as to how such absorption against a gradient can be achieved. Further experiments showed that absorption is not accomplished by hydrostatic pressure or by electro-osmosis, neither does it occur as a result of an active absorption of solutes, for the substitution of trehalose (a sugar which itself is not absorbed) for xylose to raise the rectal O.P. to the required level, has no effect upon the absorption of water, and so far as inorganic ions are concerned, very few indeed move across the rectal wall.

Phillips at this time (1964a) proposed either active transport by a pump specific for water, or water movement following active ion transport within the rectal wall. As we shall see (Chap. 6.C.VII.4.) it has turned out that the second hypothesis is probably very near the truth.

7. Loss of Water with the Faeces

One of the functions of rectal reabsorption of water is to reduce the amount of water lost with the faeces to the extent that may be necessary, and the mechanisms discussed above may profitably be seen in relation to the water status of the insect as a whole. Drying of the faeces in connection with the cryptonephridial system of *Tenebrio* larvae is referred to in Chapter 6.C.VII.7., and water elimination in tsetse-flies in Chapter 6.C.VI.1. Here we shall discuss a well-documented case showing how faecal water content may be used to regulate total water balance in plant feeding insects, (*Locusta, Schistocerca* and *Chortoicetes* (Loveridge, 1974)).

Locusta can reduce the water content of its faeces to 25% or less, and they are then in equilibrium with air at about 85% R.H.—a situation rather similar to that in *Tenebrio* larvae (Ramsay, 1964). Water is withdrawn only to the extent necessary to maintain the normal water content, and the amount of faecal water depends not only on water in the food, but on the water status of the insect at the time, as Table 26 and Figure 65 show.

Table 26. The effect of food water content on weight of food eaten and water content of faeces produced. (Data from Loveridge, 1974)

Pretreatment of locusts	Food	Water content of food %	Dry weight of food eaten $mg\,g^{-1}h^{-1}$	Water content of faeces %
From culture	Grass	85	3.7	77
Fasted at 96% R.H.	Grass	86	3.0	80
Fasted in dry air	Grass	87	4.5	78
From culture	Bran	4.6	1.5	36
Fasted at 96% R.H.	Bran	4.6	2.3	37
Fasted in dry air	Bran	4.6	1.5	36
Fasted at 96% R.H.	Artificial diet	57	1.9	46
Fasted in dry air	Artificial diet	52	1.9	39

When fed on fresh grass that contains about 85% water, locusts take in about 21.0 mg water $g^{-1}h^{-1}$, and produce 9·9 mg faeces whose water content is 77%, thus losing about 7.6 mg $g^{-1}h^{-1}$. In this situation the insects are in positive water balance so far as these two components are concerned. When given dry food containing 5% water, only about 0.08 mg water $g^{-1}h^{-1}$ are taken in, and the faecal water loss is 0.52 mg $g^{-1}h^{-1}$, (the faecal wet weight being about 1·4 mg), so that more water is lost than is gained. In fact, the more food taken, the greater the amount of net water loss, so that it is not surprising to find that in these circumstances locusts severely reduce their food intake, from about 3.7 to about 1.5 mg $g^{-1}h^{-1}$, and this has some mitigating effect. Nevertheless, unless they are given water to drink (up to 0.5 M NaCl is satisfactory), or fresh food, the insects continue to lose weight and eventually die, although during the dehydration and fasting their water content (percent of total weight) remains fairly constant because the dry weight losses keep pace with the water losses, and metabolic water helps to a slight extent (Chap. 8.D.). (For further discussion see Chap. 11.)

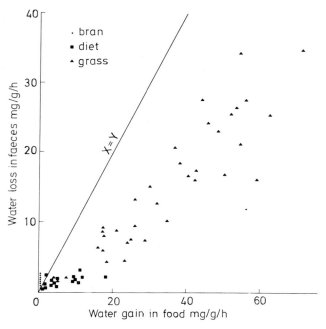

Fig. 65. The relationship between gain of water with the food and loss in the faeces. *Locusta* fed on grass (▲), artificial diet (■) and bran (●): the line $X = Y$ represents a situation in which loss balances gain. (From Loveridge, 1974)

In spite of their pronounced ability to dry out the faeces, it seems that locusts are less well adapted for water conservation than *Tenebrio* larvae, for these, as we have seen, reduce their faecal water content to about 14–17% (Ramsay, 1964) [Schultz's (1930) value of 7–9% is suspect, (see Chap. 7.A.)] and survive indefinitely at 30–40% R.H. on a diet of bran which has a water content of about 8%.

This analysis of the situation in *Locusta* (and Loveridge found *Schistocerca* and *Chortoicetes* to behave essentially in the same way) is a salutory reminder that, although most studies of water balance mechanisms have concentrated on cuticular and respiratory water loss and esoteric processes such as the absorption of water vapour, in a normal rather generalised insect by far the most important components of water balance are intake with the food and loss with the faeces. These two together can effectively regulate the body water content within a fairly wide range of conditions.

134

8. Reabsorption of Ions in the Rectum

The absorption of inorganic ions from the locust rectum as part of the homeostatic process, has also been studied. Phillips (1964a) perfused the rectum with a Ringer solution of similar ionic composition to the blood, but rendered hyperosmotic by the addition of trehalose, so that there was a net flux of water into the rectal lumen. In such conditions, ions are absorbed from the rectum against a concentration gradient, certainly by active transport of chloride and probably of sodium and potassium.

In water-fed locusts, the rate of sodium and potassium absorption is almost proportional to their rectal concentration (although the uptake of potassium is always several times faster than that of sodium at any concentration); but it is interesting to find that in saline-fed locusts the uptake mechanism appears to become saturated at rather low luminal concentrations (about 100 mosmol

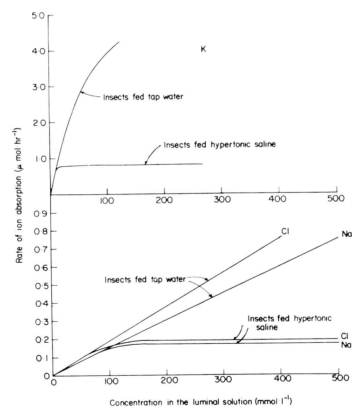

Fig. 66. Rate of absorption of ions from recta of locusts previously fed either tap water or hypertonic saline. In each case the uptake is non-saturable in insects fed tap water but is saturated at relatively low concentration in insects fed hypertonic saline. Chloride ions are taken up at the same rate as the cation in the luminal solution. Thus with a potassium chloride solution in the lumen both potassium and chloride ions are taken up at the higher rates characteristic for potassium ion. (From Maddrell, 1971; after Phillips, 1964)

135

liter^{-1} for sodium and 15 mosmol liter^{-1} for potassium), above which the rate remains substantially constant (Fig. 66). In other words, locusts with higher blood ion concentrations reduce the rate of absorption of further ions from the rectal contents and this is clearly a regulatory process. At the same time, chloride is absorbed at rates consistent with those of the major cations being absorbed. In *Calliphora* too, Phillips (1969) observed essentially the same properties, namely the ability to absorb water and ions against gradients. *Calliphora*, unlike *Schistocerca*, produces a strongly hypo-osmotic urine when given excess water. However, isolated recta in water-fed flies do absorb water until the luminal content is 40% more concentrated than the blood, so that the mechanism responsible for hypo-osmotic urine productions is unclear.

Similar effects were found by Sauer *et al.* (1970) for the isolated rectum of *Periplaneta*, where the rate of movement of water out of the rectum is related to lumen O.P., but is not dependent on the simultaneous absorption of solutes.

9. Regulation by Other Organs

Before closing this section we shall refer briefly to what little is known of the osmoregulatory function of other organs, mainly parts of the alimentary canal or associated glands.

The structure and function of salivary glands as excretory organs in ticks has been referred to above (Chap. 6.C.III.), and those of the blowfly are considered below. These organs are known to be excretory in ticks, but there is no suggestion that this is their function in *Calliphora*. However, if solute concentrations in the saliva of an insect are different from those in its blood, and if the saliva is not completely re-ingested, then salivation must have some osmotic or ionic effect. The saliva of *Calliphora* (Oschman and Berridge, 1970), and of the termite *Hodotermes* (Hewitt *et al.*, 1971) is dilute compared with the blood, but the extent to which water and solutes are lost in this way is unknown. Further references to the physiology of insect saliva are given by Oschman and Berridge (1970).

Several other organs or regions are known to excrete water and ions, and we shall refer to these briefly. The midgut cells of saturniid larvae actively pump potassium from the blood to the gut lumen (Harvey and Nedergard, 1964), and thus prevent the relatively enormous potassium concentration of the larval food (about 153 meq kg^{-1}) from raising blood potassium higher than the normal 27 mmol liter^{-1}. The sodium concentration of the leafy diet is fairly close to that of the blood (4 meq kg^{-1} compared with 6 mmol liter^{-1}) and poses no problem (Nedergard and Harvey, 1968). Analysis by means of microelectrodes shows that the mechanism of this process is in all probability an electrogenic pump situated in or on the apical plasma membrane of the midgut epithelial cells (Wood *et al.*, 1969). In this connection O'Riordan (1969) found an electropotential difference of 12 mV across the isolated midgut in *Periplaneta*, the lumen being negative. She proposed a linked Na-K pump system, whose effect is to transport sodium from the lumen to the blood. However this may be, it is another piece of evidence linking the midgut with ionic regulation, particularly with the absorption of necessary ions.

136

The midgut of *Manduca sexta* larvae is also strongly involved in the excretion of acid dyes by means of what looks like active transport, since movement of the excreted materials is against a gradient, is specific for certain substances, and requires energy, being inhibited by the ATP antagonist DPN (Nijhout, 1975).

The labial glands of certain adult saturniids including *Hyalophora cecropia* and *Antherea pernyi*, secrete a copious iso-osmotic fluid, rich in $KHCO_3^+$ (Kafatos, 1968). The primary purpose of this secretion in some moths is to act as a solvent for a cocoon-ase, but when the secretion continues after emergence, or does not contain an enzyme, its function may be blood volume regulation, for a low blood volume is probably advantageous for a large flying insect. Adult *Calliphora* also excrete a copious fluid before flight (Cottrell, 1962).

Most of the work on reabsorption has been concerned with the rectum, but when an ileum is present, as it is in most insects, Maddrell (1971) points out that there is reason to believe that this is also the site of some reabsorption. In *Schistocerca*, for example, Phillips (1964a) found that amaranth in the hind gut is concentrated before it reaches the rectum, and Ramsay (1955a) found that fluid is absorbed in the ileum of *Carausius*, though such absorption does not alter the composition of the remainder. In *Periplaneta*, Wall (1970) found that while the tubal fluid is slightly hyper-osmotic to the blood, the rectal fluid is hypo-osmotic, perhaps as a result of resorption of ions in the ileum. It may be that the ileum generally functions to reduce the volume of fluid entering the rectum, if this is not done by the proximal region of the tubule itself.

VI. Rapid Elimination of Excess Water

We have seen how ticks deal with the problem of temporary excesses of water (Chap. 6.C.III.). Certain insects, faced with the same problem deal with it either by means of a normal Malpighian tubule–hindgut system (blood sucking species), or by modification of such a system (sap sucking species).

1. Water Elimination by Tsetse-Flies

Tsetse-flies (*Glossina* spp.) feed at intervals of a few days on the blood of vertebrates, at which time large quantities of food, equal to the total weight of the fly, are sucked into the crop. Although the osmotic pressures of host and fly bloods are not very different (about 300 mosmol liter^{-1}), the amount of water taken in with a full meal is much more than a fly needs, and in any case would be a severe hindrance in flight. Consequently, a copious watery urine is excreted during the first hour or so after feeding. Subsequently the fly produces watery faeces, but later in each hunger cycle the faecal material is much drier. Bursell (1960) has described these processes quantitatively in *G. morsitans*, and they show that the process is highly adaptive. During the first 3 h (the "primary excretion" period of Jack (1939)), up to 50% of the wet weight of the blood meal is eliminated; but as Figure 67 shows, the proportion of water excreted (expressed as a percentage of the blood meal) is clearly inversely related to the water content of the fly before feeding. Thus sufficient water (but no more) is retained to replenish the fly's water content if this has been depleted, and in this way the water content of the blood meal may be reduced from 79% to 55% during the first 3 h after feeding. Furthermore, if only a partial meal is taken, then a greater proportion of the water content is retained, and the blood meal becomes dehydrated to a lower level. During the subsequent faecal production, which is necessary for the elimination of excess nitrogen and other

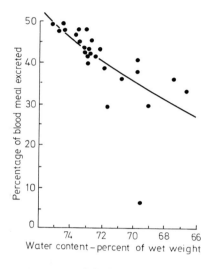

Fig. 67. A lower initial water content leads to a lower proportion of the water in a blood meal of a tsetse-fly being excreted. (From Bursell, 1960)

waste materials in the meal (Chap. 6.B.I.), faeces contain about 75% water when the flies are kept in high humidity, but only 35% in flies kept in dry air—an adaptive process that results in saving no less than 30% of the total water content of a fly in such conditions. As Bursell points out, the abilities of the rectal glands to reabsorb water are utilised to the full only when the fly is short of water; at other times, this activity is closed down.

Since the early work of Lester and Lloyd (1928) rapid diuresis in *Glossina* has usually been ascribed to intense Malpighian tubule activity. However, some interesting experiments with *Glossina austeni* by Tobe (1974) suggest that this may not be the whole story. By labelling the flies' food with tritium as HTO and with ^{14}C as dextran, and taking samples of urine and blood at intervals after the completion of a meal, Tobe found that tritium is much more concentrated in the urine than in the blood, and that ^{14}C sometimes appears in the urine but never in the blood (Table 27). He suggests that these results could be accounted for by a direct passage of water along the gut to the rectum, perhaps after clotting of the blood in the crop and expression of the serum, or perhaps involving the peritrophic membrane as a dialysis bag. Alternatively, or perhaps additionally, if water is absorbed from the crop, its journey through the haemocoele could be restricted to a small compartment thereof, perhaps by means of a short circuit from crop to Malpighian tubules (foreshadowing the filter chamber

Table 27. Radioactivity in dpm $\mu l^{-1} \times 10^{-3}$ of labelled food and subsequent urine and blood in *Glossina austeni*. (From Tobe, 1974)

Time after feeding	Food		Urine		Blood	
	^3H	^{14}C	^3H	^{14}C	^3H	^{14}C
0	125–145	1.5	—	—	—	—
5 min	—	—	122.6	0–1.5	75.5	0
15 min	—	—	119.9	0	81.7	0
30 min	—	—	117.3	0	—	—

system discussed below, in Chapter 6.C.VI.2). The conclusion to which these data seem to point, namely that some of the water in a tsetse-fly's meal does not mix with the blood but moves more or less directly to the rectum, is contrary to the conclusion derived from a good deal of earlier careful work. Thus the confirmation or alternative explanation of Tobe's results will be awaited with interest.

In the blood sucking hemipteran *Rhodnius* also, water is present in excess after a meal. An adult takes up to 180 mg at a single meal and plasticization of the abdominal cuticle occurs (Bennet-Clark, 1962; Reynolds, 1975) to permit the necessary expansion, the process perhaps being under the control of a hormone (Maddrell, 1966) released from abdominal nerve endings (Reynolds, 1974b) (Chap. 6.D.). Two hours later, 76% of the total water ingested has been eliminated as a watery urine which also contains excess sodium, potassium and chloride.

The structure of the Malpighian tubules of *Rhodnius* has been diagrammatically shown in Figure 59. Fluid formed by the upper segment of the tubule is slightly hyper-osmotic, and hypo-osmoticity results from the reabsorption of salts in the lower segment (Maddrell and Phillips, 1975). The relationship between fine structure and function is considered below (Chap. 6.C.VII.2.). In *Rhodnius* and presumably in other blood suckers also, the onset of diuresis is triggered by a diuretic hormone, which is probably not the plasticization hormone (see Chap. 6.D.).

2. Water Elimination by Sap Feeders

Many hemipterous insects and tetranychid mites are plant suckers, some feeding predominantly on xylem sap (cicadoids and cercopoids), others on phloem sap (cicadellids, coccids and aphids) or on cell contents (mites). (For references on feeding habits see Marshall and Cheung, 1974, 1975.) Xylem fluid is a very dilute solution, strongly hypo-osmotic to insect blood. It contains 99.8–99.9% water (Cheung and Marshall, 1973a), small quantities of inorganic ions, particularly potassium and chloride, and variable quantities of amino acids and simple sugars. Phloem sap on the other hand contains relatively large quantities of sugars and amino acids but only small amounts of inorganic salts (Lundegärdh, 1966).

For insects that feed on these fluids, particularly xylem fluid, it is necessary to eliminate excess water and to concentrate ions and useful organic molecules, and the alimentary canal is often modified for rapid short circuiting of water, largely by-passing the midgut, so that the remaining useful materials can be absorbed in the midgut. Several different anatomical and functional arrangements exist, many of them forming so-called "filter chambers." For a review of the structure and evolution of filter chambers see Goodchild (1966), and references in Marshall and Cheung (1974).

On the whole, filter chambers are most complex in the cicadoids and cercopoids (the xylem feeders), and are simpler in the phloem feeders. The structure of a filter chamber is typically (in a cicadoid or cercopoid) as follows. The anterior part of the midgut is dilated, and to its walls are closely applied the proximal regions of the Malpighian tubules and long coiled portions of the posterior midgut and anterior hindgut, the whole being enclosed in an intestinal sheath (Fig. 68). In other groups, e. g. the coccids, part of the rectum may also be included in the complex, while in the fulgorids, a true filter chamber is absent, but an intestinal sheath covers the whole of the long, coiled midgut (Goodchild, 1963).

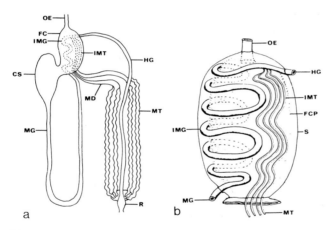

Fig. 68. (a) Diagram of the general organization of the cicadoid and cercopoid gut. (b) Diagram of the filter chamber in cicadoids. *CS*: conical segment; *FC, FCP*: filter chamber; *HG*: ileum; *IMG*: internal mid-gut; *MT*: Malpighian tubules; *OE*: oesophagus; *R*: rectum; *S*: sheath of filter chamber. (From Marshall and Cheung, 1974)

Since the early work of Licent (1912), the function of these complexes has been the object of speculation (refs in Marshall and Cheung, 1974), particularly as regards the mechanism whereby water is filtered off from the dilute plant sap. Goodchild (1963, 1966) has pointed out that there are three likely processes: hydrostatic pressure, active transport and passive osmosis. Gouranton (1968 a, b) advances cogent reasons, based on morphological and histological structure, in favour of such osmosis in a cicadellid, but it was not until Cheung and Marshall (1973 a, b) worked intensively on cicadas that much physiological information was available. These authors, by means of techniques similar to those used for Malpighian tubule study, measured rates of flow, osmotic pressures, and ionic concentrations in various parts of the system.

By weighing before and after feeding, and measuring the urine produced, they found that a cicada completes a full meal in just over an hour, and filtration commences very rapidly after the first ingestion of sap. The rate of flow in the ileum is experimentally increased by ten times immediately after injection of 20 μl xylem sap into the oesophagus. During filtration, O.P. measurements were 14.2, 13.1 and 156 mmol liter^{-1} NaCl equivalent in the sap, urine and blood, respectively, and the potassium concentration was higher in the blood (28.4) than in either sap (7.0) or urine (7.1) mmol liter^{-1}. Potassium is high in the posterior midgut and Malpighian tubules, and only becomes progressively reduced as the fluid flows down the ileum.

In the light of these and other findings, Cheung and Marshall (1973a, b) proposed the following as a possible mechanism. Xylem fluid imbibed into the filter chamber proper (anterior midgut) loses water by osmosis to the proximal Malpighian tubules and posterior midgut whose contents are hyper-osmotic as a result of potassium chloride transport into the distal tubule and into the middle midgut. It also appears that the concentrations of calcium, magnesium, manganese and iron are high in the middle midgut, and associated with the presence there of metal-containing sphaerites (Gouranton, 1968b; Cheung and

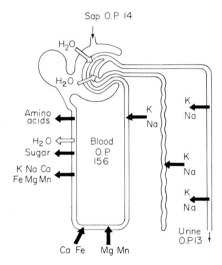

Fig. 69. Postulated mechanism for absorbing water from the anterior mid-gut (filter chamber) and excreting it without passage through the digestive part of the mid-gut. Solutes (Na, K) are actively absorbed from the hind-gut and the anterior part of the mid-gut and transported into the Malpighian tubules and posterior mid-gut, forming a region of high O.P. where these organs are enclosed in the filter chamber. Water is withdrawn from the dilute sap in the filter chamber, and concentrated nutrients (sugars, amino acids) may be more effectively absorbed from the mid-gut. *Solid arrows:* active transport of solutes; *open arrows:* osmotic movement of water. Osmotic concentrations in equivalent mmol liter^{-1} NaCl. (Redrawn from Cheung and Marshall, 1973b)

Marshall, 1973b). This is perhaps a form of ionic storage secretion, but the precise function is obscure. These ions, together with sodium and potassium are probably absorbed into the blood from the anterior tubular midgut, and water is also probably absorbed to some extent both there and in the conical chamber (which follows the filter chamber proper). The overall picture proposed is shown in Figure 69.

Cheung and Marshall then went on to describe the fine structure of the various regions concerned, including the midgut (Cheung and Marshall, 1973b), the hindgut (Marshall and Cheung, 1973) and the Malpighian tubules and filter chamber (Marshall and Cheung, 1974). Let it suffice for the present to say that, according to currently accepted theory about the relationship between fine structures and water and ion transport (Chap. 6.C.VII.3.) the cicada alimentary canal is entirely consistent with the proposed functions of the various parts.

Gouranton (1968a, b) has described the fine structure of the gut of *Cicadella viridis*, and this is consistent with water removal by passive osmosis in the filter chamber (as in cercopids and cicadas) but the rectum shows intercellular channels consistent with ion activated water transport against a gradient. Gouranton (1968a) makes the interesting point that in this insect the fine structure of the cells of the posterior midgut and of the Malpighian tubules is very similar, and in both regions movement of fluids takes place from the basal towards the apical surface. Contrasted with this is the fine structure of the anterior midgut, where movement of the fluids is from the apical towards the basal surface. It would be most interesting to know whether this insect feeds on xylem or phloem, for phloem has a higher O.P. than normal insect blood (Kennedy and Fosbrooke, 1972).

According to a recent report by Marshall and Cheung (1975) the ionic composition of the excreta of xylem sap feeders such as cercopids, machaerotids and cicadas, closely resembles that of the sap on which they feed, if due allowance is made for storage excretion. This should be helpful for further work on feeding habits and regulatory mechanisms on the part of the insects, as well as for studies on the composition of plant saps.

Some aphids produce honeydew (a sugary excretion from the anus), and these do not have filter chambers, although such organs are present in other aphids. The mechanisms for water regulation in aphids have not been researched, and would form an interesting field for study. Malpighian tubules are absent from these insects, and although some of their excess water is eliminated as saliva, much goes out with the honeydew via the rectum. The suggestion that the lack of Malpighian tubules is correlated with high internal hydrostatic pressure is probably unjustified, for although their food source is indeed pressurized, and they use this to facilitate feeding, there is no question about the ability of aphids to control the influx when they are replete (Kennedy and Fosbrooke, 1972). In jassids (*Euscenius*), and cercopids (*Triecophora*), Munk (1968) showed by the use of dyes and radioactive food that most of the sugars and water by-pass the midgut and are excreted as honeydew, while amino acids and proteins are thereby concentrated and presumably absorbed in the midgut.

There is still much to be found out concerning water balance in plant sap feeders. More information is needed particularly on the function of storage

142

secretion, and on the sites and mechanisms of absorption of organic and inorganic materials into the blood. Filter chambers are interesting structures that have been relatively neglected. In the future they will probably yield a lot of useful information about an aspect of arthropod water affairs that is not commonly dealt with, at least in terrestrial forms: situations where water is in excess.

VII. Fine Structure and Transepithelial Transport

1. Introduction

In Malpighian tubules a fluid which is virtually iso-osmotic (but by no means iso-ionic) with the blood is produced by transport of water and solutes across the wall from the blood. In many insects water is transported from the rectal lumen, across the epithelium and into the blood, sometimes against steep osmotic gradients (Chap. 6.C.V.6.). In Malpighian tubules, if the wall behaved as a simple semi-permeable membrane, no net flow of water or ions would occur, and in the rectum, water would flow down an osmotic gradient into the lumen when the latter is hyper-osmotic. In both situations, therefore, active transport of one substance or another is necessary, and recent work suggests that the mechanisms and structures involved in each have a good deal in common: active transport of solutes into long narrow spaces is followed, as a result of osmotic coupling, by a flow of water. Subsequent reabsorption of solutes leads to the production of iso- or hypo-osmotic fluids on the transepithelial side of the epidermis or wall as the case may be. The details vary greatly between species and organs. Excellent reviews of recent developments in this field, and of the theoretical background, are those of Oschman et al., (1974) and Gupta (1975).

2. Fine Structure of Malpighian Tubules

We have already referred briefly to the structure of the Malpighian tubules of *Rhodnius* (Fig. 59) and other insects (Chap. 6.C.V.3.). With certain important reservations (Chap. 6.C.VII.3.) the fine structure of a unit cell of a Malpighian tubule in all insects studied so far is basically similar (refs in Berridge and Oschman, 1969), and this is shown diagrammatically in Figure 70. On the blood side, there is a basement membrane, below which the basal plasma membrane of the cell is invaginated to form a number of deep, narrow tortuous channels into the body of the cytoplasm (but not penetrating the plasma membrane).

The nature of the basement membrane may be of great importance and, one would think, would be matched, so far as its permeability is concerned, with the basal plasma membrane; otherwise, if large molecules such as proteins were allowed through the basement membrane they might accumulate and perhaps block the channels lined by basal infolds, or establish undesirably high osmotic pressures there.

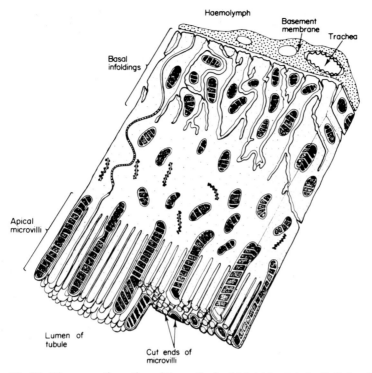

Fig. 70. Diagrammatic section of the wall of a Malpighian tubule. Both basal infolding and apical microvilli form long narrow channels in which standing gradients could develop for overall iso-osmotic urine formation (see text). Mitochondria are closely associated with both infoldings and villi. (From Maddrell, 1971)

The basement membranes of several organs are known to be permeable to proteins (horseradish peroxidase has been used experimentally) (Locke and Collins, 1967, 1968). In *Calpodes*, for example, this protein is readily taken up from the blood by the fat body, but the basement membrane of the Malpighian tubules of that insect is, interestingly, an exception, being impermeable to peroxidase.

The situation in *Calliphora* is similar according to Berridge and Oschman (1969). On the other hand Kessel (1970) reports that in the dragonfly *Libellula*, the Malpighian tubule basement membrane is indeed permeable to HRP which, when injected into the blood, accumulates in the basal channels, and even appears in the cytoplasm, perhaps as a result of micro-pinocytosis. The situation will no doubt be re-examined in these and in other insects, but even if proteins do get through the Malpighian tubule basement membranes there is as yet no demonstration that such molecules interfere with water or ion transport.

In the cytoplasm of a typical Malpighian tubule cell there are mitochondria and a nucleus, the former often being very plentiful in association with the basal infoldings. The apical (luminal) side of the cell, has a large number of

144

often closely packed processes, the micro-villi, some of which again contain elongate mitochondria. (These processes form the "brush border" visible with the light microscope.) Each cell is joined laterally to its neighbour by a septate desmosome, a structure which is usually believed to prevent intercellular trans-epithelial movement of materials. However, in several insects including *Triatoma*, *Calliphora*, *Schistocerca* and *Manduca*, recent work suggests that quite large molecules move from blood to fluid (Maddrell and Gardiner, 1974), so that these junctions are probably leaky.

Finally, the cells are supplied with tracheoles that run in the basal infoldings, and this, together with the large number of mitochondria, indicates that the cells are metabolically very active.

3. A Functional Model

A mechanism to account for the transport of water across epithelia by local osmosis or solute coupled transport was originally proposed by Curran (1960). This "double membrane" theory requires three compartments separated by two membranes (Fig. 71). Solute is actively transported from compartment 1 across membrane α to compartment 2 where it raises the O.P. This is followed by a passive flow of water, increasing the hydrostatic pressure in compartment

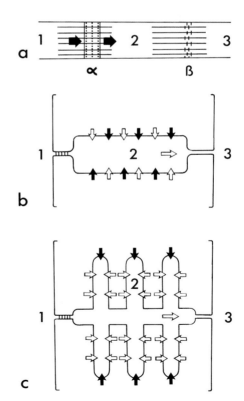

Fig. 71. (a) Curran's double membrane model. (b) an application of Curran's model to transepithelial transport of water; (c) standing gradient osmotic flow model of Diamond and Bossert (1967) involving lateral cell membranes. Curran's model requires three compartments (*1, 2, 3* in the figure) and two membranes (α and β). Solute actively pumped through the semi-permeable membrane (α) causes a rise in osmotic concentration of the solution in *2*, followed by osmotic flow of water from *1* to *2*. The increased hydrostatic pressure in *2* causes a flow of solution through the relatively permeable membrane (β) to compartment *3*. *Solid arrows*: active transport of solutes; *open arrows*: osmotic flow of water. For further discussion see the text. [Partly based on a Fig. by Gupta (in press)]

145

2 and causing the movement of fluid outwards through the relatively permeable membrane β to compartment 3. According to Patlack *et al.* (1963), appropriate physiological parameters and geometry can result in the transported fluid being either hyper-, iso- or hypo-osmotic to the original medium.

This model has since been applied to several real cases (for refs see Gupta, 1975), and membrane α is usually taken to represent the whole of the transporting epithelial cell, the solute pumps occurring on the lateral plasma membranes of the intercellular spaces (Fig. 71b).

A further development of this model by Diamond and Bossert (1967, 1968) emphasizes the importance of having, as compartment 2, long narrow intercellular channels, in which a standing gradient may be generated by the secretion of solute into their blind internal endings, thus leading to the osmotic entry of water and to the mass flow of fluid along them (Fig. 71c). In both models, water is forced to pass from compartments 1 to 2 through cells rather than intercellularly by the existence of tight junctions (appearing as septate desmosomes in invertebrates) between appropriate lateral cell membranes.

Berridge and Oschman (1969) used a standing gradient model to explain the production of urine by *Calliphora* Malpighian tubules, along the following lines. Ions, particularly potassium but probably sodium as well, are actively pumped from fluid in the basal infoldings, across the basal membrane, into the narrow cytoplasmic processes lying between the infoldings (Fig. 72). Because

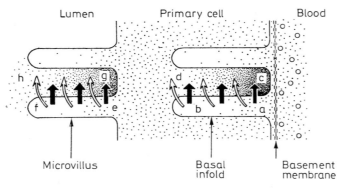

Fig. 72. Standing gradient model for urine formation by Malpighian tubules. (From Berridge and Oschman, 1969.) *Solid arrows*: active transport of solutes; *open arrows*: osmotic flow of water. Solute transport might set up standing gradients in the basal infolds (*a–b*) and intervening cytoplasmic processes (*c–d*), and in the microvilli (*e–f*) and intervening spaces ("micro-crypts") (*g–h*) on the apical side of the cell. Concentration of solutes is indicated by density of stippling. Large' molecules (*circles*) are excluded by the basement membrane

the cytoplasm here is in the form of long narrow processes, transported ions cannot diffuse away quickly, and thus local regions of high O.P. are set up, into which water from the fluid in the basal infoldings follows osmotically. There may also be a thin protein coat on the cytoplasmic side of the infoldings (Gupta pers. comm.) and this would further restrict the dispersion of the transported ions. Again as a result of the shape of the channels, water drawn in

146

osmotically moves down the channels away from the basal region, forming a standing gradient, until it enters the main body of the cytoplasm where the solution is iso-osmotic with the blood. At the apical (luminal) side of the cell, a repeat process occurs, solutes are transported into the micro-crypts, followed osmotically by water. The resulting solution moves along the channels, becoming more dilute as it goes, and again forming a standing gradient, until it emerges into the lumen in an iso-osmotic condition.

This hypothesis, together with that for rectal reabsorption of water against a gradient put forward originally by Berridge and Gupta (1967), is very instructive and certainly acceptable in qualitative terms. There is a need, as Maddrell (1971) has stressed, for mathematical analysis and further experimentation, and Gupta (1975) in a penetrating analysis of the problem, has pointed out several difficulties in accepting the standing gradient model, although he agrees that some form of osmotic coupling almost certainly occurs. In particular Gupta makes the following points; (1) there is so far little experimental evidence that the intercellular channel fluids have the necessary high O.P. called for by the model—the only direct measurements (which were in fact confirmatory, Chap. 6.C.VII.5.) being those of Wall et al. (1970) on the cockroach rectum; (2) cell junctions, assumed to be tight in the model, are known to be leaky in several tissues (e. g. Lane and Treherne, 1972; other refs. in Gupta, 1975), however, as Gupta points out (in correspondence), this objection does not apply to the channels in Malpighian tubules because these are blind at one end and do not involve junctions; (3) solute is required to be transported mainly into the blind ends of intercellular channels if a standing gradient is to be formed, but there is no experimental evidence that transport is so restricted. Meanwhile Sackin and Boulpaep (1976) have developed two models applicable to *Necturus* kidney tubules (and perhaps to other systems), that do involve leaky junctions or the active transport of sodium uniformly along the intercellular channels. According to these models such a system could function to produce an almost iso-osmotic emergent solution.

Aspects of this hypothesis that need further elucidation are:

1. The basal infoldings and micro-villi provide channels that in some cases are very much shorter (5–10 μm) than those considered necessary for the model (50–100 μm) (Diamond and Bossert, 1968). On the other hand, Maddrell points out that the geometry of the situation in the Malpighian tubules is such that Diamond and Bossert's (1968) backward facing and forward facing channels (when the infoldings are directed respectively away from or towards the medium from which absorption is taking place) are both present in the basal region of the cell, one set being represented by the basal infoldings, the other by the cytoplasmic processes between the infoldings, and it is possible that this may permit the development of gradients large enough to offset the comparatively short length of the channels.

2. Differences in fine structure between one region and another will have to be correlated with function. For example the proximal region of the tubules of *Rhodnius* have fewer micro-villi than the distal region; and of course the ampullar cells are different again. (Differences between the proximal and distal regions of silk worm tubules have already been referred to in Chapter 6.C.V.4.)

3. In the Malpighian tubules of some insects *(Calliphora)* but not in others *(Rhodnius)*, the walls in any one region contain more than one type of cell. In *Calliphora*, for instance, "stellate" cells are dispersed among the "primary" secretory cells. Are the former associated with reabsorbtion of sodium from the fluid, as suggested by Berridge and Oschman (1969)? If so, when such cells are absent, does reabsorbtion occur elsewhere in the tubule?

4. Lastly, although the tubule fluid is nearly iso-osmotic, ionically it is quite different from the blood. Somewhere in the cells, machinery responsible for this change must exist, and this remains to be identified. Berridge and Oschman (1969) tentatively suggest that if sodium is selectively absorbed by stellate cells and re-secreted into the blood in exchange for potassium, this would account for the high potassium often found in tubular fluid.

In fact the whole standing gradient hypothesis has been seriously challenged recently by Hill (1975a), on the theoretical grounds that the model, particularly as it relates to the production of an iso-osmotic final secretion, requires much higher osmotic permeabilities on the part of the cell membranes than are either measured or predicted. Instead, Hill (1975b) has proposed a model based on electro-osmosis which does theoretically permit the production of iso- or even hypo-osmotic final products, demands the presence of complex channels and spaces in the lateral membranes (such as we know exist), and avoids problems associated with osmotic permeability and high intercellular osmotic concentrations and hydrostatic pressures. At a simplified level, electro-osmosis may be thought of as a process that occurs when an electropotential gradient across a membrane (however generated) causes the movement of ions to which water molecules are "frictionally" attached. There is no need for either a hydrostatic pressure gradient or a concentration gradient in such a system. Details of the objection, the counter-proposal and Sackin and Boulpaep's alternative, must be sought in the original papers, for they are beyond the scope of the present work.

In sum, recent work on Malpighian tubule function, and on fluid transport in general, has generated interesting and enlightening proposals. Doubtless further experimental work will cause present ideas to be modified, perhaps drastically. Meanwhile one particularly attractive step forward deserves special mention. Several authors have pointed out from time to time that the standing gradient model (for example) will be proved or disproved only by the measurements of ionic concentrations in living cells working normally. This is, of course, technically extremely challenging, but a remarkably near approach to achieving this has recently been reported by Gupta (1975), using electron probe X-ray microanalysis. His preliminary report, which is most encouraging, will not be considered in detail here, for the results are tentative—but they do not conflict with the predictions of osmotically coupled fluid transport by local osmosis. The technique involves deep freezing a tissue in action, bombarding small areas of it with electrons, and analyzing the resulting X-rays which are generated at specific energies by specific elements. This opens the possibility of measuring accurately specific ionic concentrations in spaces as small as the channels between microvilli on a cell membrane.

148

Bearing in mind the caveats regarding interpretation mentioned above, we will now consider how models based on local osmosis have been applied in the interpretation of observed fine structure in various arthropod epithelia.

4. Fine Structure and Function of Rectal Walls

An excellent review of the salient features of rectal fine structure in relation to the reabsorption of water is that of Wall and Oschman (1975). This is a very active research field, and doubtless several of the present hypotheses concerning function will be modified as progress is made.

Unlike the formation of tubular fluid, absorbtion of water from the rectal lumen may occur against considerable gradients, at least 1.0 osmolar in *Schistocerca*, even in the absence of transportable ions in the lumen. The mechanism whereby this is achieved was foreshadowed by Gupta and Berridge (1966a, b) who carried out ultrastructural studies on the rectal papillae of *Calliphora*. Their suggestions were supported by experimental work (Berridge and Gupta, 1967) on the same organ, and have been confirmed by parallel work on other insects, including *Schistocerca* (Irvine, 1966; Irvine and Phillips, 1971). *Aedes* (Hopkins, 1967), *Periplaneta* (Oschman and Wall, 1969; Wall and Oschman, 1970) and *Apis mellifera* (Kuemmel and Zerbst-Boroffka, 1974). Appropriate fine structure has also been described for rectal papillae in the parasitic wasp *Nassonia* by Davis and King (1975). In termites, according to Noirot (1973), rectal papillae are present, the degree of their development and fine structure being related to (1) moisture of the habitat and (2) the extent of evolutionary "advance." Thus in more primitive groups such as *Cryptotermes* living in dry surroundings, there is a strong development of apical and particularly of lateral plasma membranes (as in *Periplaneta*); while in those that live in moister regions such as *Zootermopsis*, or those that obtain water from the soil (as does *Mastotermes*, a primitive Australian genus), differentiation of the papillae is feeble. The "advanced" termites, which obtain all the water they need from the soil, also have less developed papillar cells. Interestingly, in termites there is nothing that compares closely with the highly specialised rectal cells of *Thermobia* (*Lepismodes*), where water vapour is absorbed. Termites (at least the dry wood termite *Incistitermes*) rather surprisingly do not absorb water vapour (Edney, unpubl. experiments).

The fine structure of rectal papillae in *Calliphora*, and of the rectal glands of *Periplaneta*, is basically similar; but the arrangement in *Periplaneta* (Oschman and Wall, 1969) is a little simpler, and we shall consider this first.

In Malpighian tubules, net movement of fluid occurs from the basal to the apical surface of the cells, and there is no cuticle on either surface. In the rectum, however, net flow (during water uptake) is from the apical to the basal sides of the epithelial cells, and the apical side is covered by a cuticle. The general organisation of the area in *Periplaneta* is shown in Figure 73 and further detailed in Figures 74 and 75. Below the cuticle, the apical membrane of each cell is thrown into rather short infoldings, and both apically and basally, the cells are joined at the sides by tight junctions (septate desmosomes). For the greater part of their length, however, the lateral cell membranes are enormously

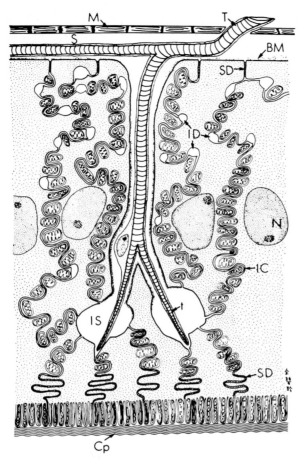

Fig. 73. The general organization of the rectal pads as revealed by electron microscopy. The muscle layer (M) facing the blood is shown at the top and the cuticle (Cp) is shown at the bottom. Tracheae (T) penetrate through the muscle layer and into the subepithelial sinus (S). Tracheae (T) also penetrate into the pads following indentations of the basal surface that are lined with basal membrane (BM). The tracheae branch to send fine tracheoles (t) between the cells and into the large spaces of the intercellular sinus (IS). The lateral membranes of the pad cells (nuclei shown at N) are separated by both narrow intercellular channels (IC) and larger intercellular dilations (ID). These are apparently sealed from the apical and basal surfaces by septate desmosomes (SD). Mitochondria are closely associated with both apical and lateral plasma membranes. (From Oschman and Wall, 1969)

expanded to form a number of membrane stacks, separated from each other by narrow, intercellular channels about 200 Å wide. Mitochondria are closely associated with the membrane stacks. In places the lateral membranes are separated (in other words the intercellular channels are expanded) to form small intercellular dilations. The channels open into larger intercellular sinuses and these in turn lead to a sub-epithelial sinus which lies between the basement

150

Fig. 74. A possible mechanism for the uptake of water from the rectum in *Periplaneta* by solute coupled transport. Cf. Fig. 73. For further description see the text. (From Wall and Oschman, 1969)

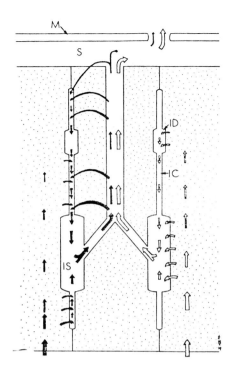

membrane and a thin connective tissue and muscle layer. Tracheae enter the system through openings in the muscle layer and penetrate via the intercellular sinuses deep into the epidermis.

Oschman and Wall (1969) suggest the following mechanism for water absorption. The main driving force is the active secretion of ions (largely potassium and chloride) into the intercellular channels and spaces, creating a standing gradient along which flows water, by passive osmosis, initially from the cytoplasm, but in the steady state from the rectal lumen across the cuticle (which is here probably a molecular filter), and the apical infoldings whose function is not immediately apparent, unless it is to enlarge the surface area, (but see Chap. 6.C.VII.4.).

As a result, fluid flows through the intercellular spaces and sinuses, along the tracheal passages, out to the sub-epithelial sinus (where it is known as "primary absorbate") and thence to the blood through apertures that may be closed when necessary by the action of the muscle layer.

It is essential, of course, for the working of this model that solutes shall continue to be available for pumping into the intercellular spaces to form there solutions of high O.P., and that fluid in the sub-epithelial sinus shall be hypo-osmotic to that in the rectal lumen when the insect is absorbing against a gradient. Oschman and Wall suggest that this could be achieved (1) by recycling, by re-absorption of ions through the walls of the intercellular sinuses from the primary absorbate, or through the basal membrane from the sub-epithelial sinus,

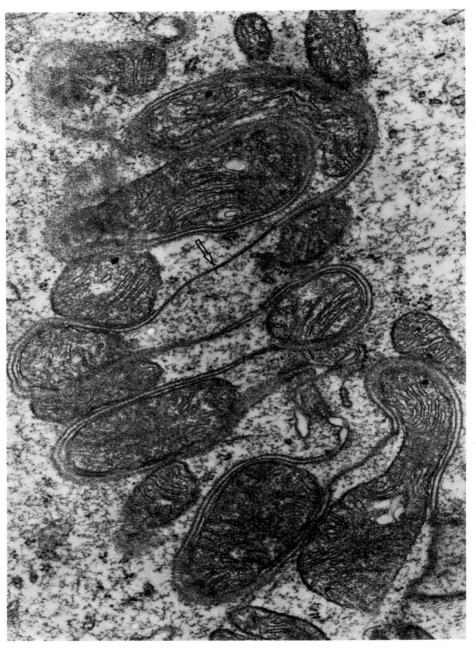

Fig. 75. Interdigitations of adjacent cells in the rectal pads of *Periplaneta*. Mitochondria are closely associated with the lateral membranes. Adjacent to the mitochondria the intercellular channel is about 200 Å wide. In one region the outer leaflets of apposed membranes seem to have fused to form a tight junction *(open arrow)*. The separation between the outer membranes of the mitochondria and the cytoplasmic surface of the lateral plasma membranes is less than 200 Å. × 43,000. According to hypothesis, solutes are actively transported into the narrow intercellular channels, causing water to follow, forming standing gradients along the channels. (From Oschman and Wall, 1969)

or (2), by recruitment of further ions from the blood. They prefer the first alternative on the grounds that the sub-epithelial fluid is in fact hypo-osmotic to the luminal fluid (Wall and Oschman, 1970).

Wall and Oschman (1970) succeeded in sampling the sinus fluid from both anterior and posterior regions of the rectum of *Periplaneta* (a narrow, transverse constriction of the muscle layer partially separates the two compartments) as well as fluid from the blood. These samples on analysis for total O.P. and for potassium and sodium concentrations yielded valuable information on the two points in question (Table 28). They show that in hydrated animals the

Table 28. Osmotic and ionic properties of rectal and sinus fluids in *Periplaneta* in $mosmol\,kg^{-1}$ water.[a] (Data from Wall and Oschman, 1970)

	Normal			Dehydrated			Hydrated		
	Total	K	Na	Total	K	Na	Total	K	Na
Blood	380	20	320	436	35	310	379	20	275
Anterior rectal lumen	409	165	125	572	420	120	275	60	50
Posterior rectal lumen	409	165	125	972	565	115	275	60	50
Anterior sinus	450	115	205	582	105	310	363	80	225
Posterior sinus	460	135	200	620	140	310	310	120	215

[a] The difference between the sum of K and Na osmolalities and total osmolality is due to unidentified substances.

rectal lumen is hypo-osmotic to the blood and the sinus is hyper-osmotic to the lumen (by approximately $100\,mosmol\,kg^{-1}$ in each case). In dehydrated animals, however, the concentration increases posteriorly in the lumen until this is some $540\,mosmol\,kg^{-1}$ hyper-osmotic to the blood and $350–400\,mosmol\,kg^{-1}$ hyper-osmotic to the sinus fluid. Both these situations are, of course, regulatory in effect.

As regards ion balance, in all cases the Na/K ratio in the blood is higher than that in any of the other fluids measured, and this partly reflects the action of the Malpighian tubules. However, in the sinus fluids, sodium concentration is still higher than potassium, which suggests that sodium is the main ion involved in the machinery of the rectal pads. If, as now seems likely, recycling of ions within the pads occurs in *Periplaneta*, reabsorption may occur either in the larger intercellular channels or from the sub-epithelial sinus. But in either case the question arises as to why, if ions are actively pumped inwards, water does not follow, as it is supposed to do in the membrane stack channels. Possibly the limiting cell membranes are impermeable to water where ion absorption occurs, but more probably the explanation lies in the fact that resorption occurs through membranes that are flat, not plicate, so that there is every opportunity for osmotically active particles to diffuse away from the membrane surface, and the high concentrations which characterize the limited spaces in fine channels, and which cause the osmotically linked inflow of water, do not become established.

The mechanism whereby water is encouraged to move in the first place from the rectal lumen into the cell cytoplasm is not certain. Clearly the cytoplasm near the apical membrane must be made hyper-osmotic to the lumen, and this perhaps results from the accumulation of recycled ions. However, the presence of apical membrane infoldings and associated mitochondria suggest that, when they are available, as they usually are, ions from the lumen may be transported into the cytoplasm by pumps situated at the apical membrane infoldings.

We have seen that the rectal pads may produce a primary absorbate that is either hypo- or hyper-osmotic to the luminal fluid in dehydrated or hydrated insects, respectively. The machinery necessary for this change in function is discussed by Wall and Oschman (1970). Possibly hormonally induced changes in the permeability to water of the apical cell membrane may play a part by making water in the cytoplasm either more or less available for dilution of the intercellular channel fluid; but the question is indeed complex and will require much further detailed study before the answer becomes clear.

5. *The Rectal Papillae of* Calliphora

The rectal papillae of *Calliphora* were in fact the first organs whose fine structure was studied in an attempt to understand the mechanism of water transport from the rectum against an osmotic gradient. Amplifying the light microscope studies of Graham-Smith (1934), Gupta and Berridge (1966a, b) by means of admirable electron micrographs, were able to elucidate the structural relationships and the fine structure of these papillae. They found, inter alia, that the lateral walls of the cortical cells (next to the apical surface) are enormously expanded into the stacks of membranes and intercellular channels that we now associate with fluid transport by the standing gradient mechanism. A simplified diagram of the structure is shown in Figure 76.

Each cone shaped papilla contains a central medulla, surmounted by a thick, but unicellular layer of cortical cells, whose apical (luminal) surfaces are of course cuticular. Below the cuticle, the apical plasma membrane is thrown into rather shallow folds from which mitochondria are absent. The basal cell membranes are not folded, but the lateral membranes, secured apically and basally by septate desmosomes, are developed into very large numbers of membrane stacks which open into intercellular channels, from there into sinuses, and finally into a large space or infundibulum which lies between the medullary and cortical regions, and which opens basally by a one way valve into the haemocoele. The whole organ is richly supplied with tracheae via the intercellular sinuses.

Very clearly the structure of this organ is similar in essentials to that of the rectal pads of *Periplaneta* described above, and of *Schistocerca* (Irvine, 1966). The infundibulum of *Calliphora* corresponds with the sub-epithelial sinus of *Periplaneta*, and the valve-guarded entry of this into the haemocoele is functionally and geometrically paralleled by the interruptions of the muscle layers of the cockroach.

It was in *Calliphora* that Berridge and Gupta (1967) first obtained evidence for the mechanism of transepithelial water transport that has since been observed in other insects. They injected fluids of various concentrations into the recta

154

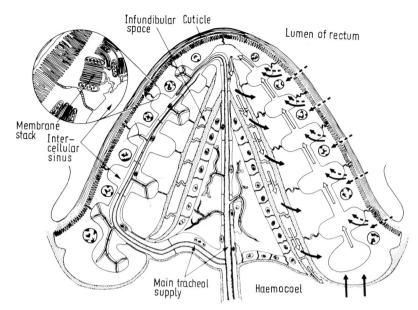

Fig. 76. Diagram of the fine structure of a rectal papilla of *Calliphora* and the proposed mechanism for solute coupled water transport. Solid arrows, active transport of solute; solid dashed arrows, passive movement of water; open arrows, flow of solution in intercellular spaces. (From Maddrell, 1971, after Gupta and Berridge, 1966b)

of starved flies, and observed in many repetitions that the intercellular channels are swollen when fluid is being absorbed, but collapsed when no absorption is to be expected. Finally, they used these data, together with the distribution of mitochondria around the membrane stacks and the geometry of the channels themselves, to propose the mechanism for absorption to which we have already referred. An alternative hypothesis, involving a water specific pump and the active transport of water, has not been ruled out, but the evidence for ion transport is very compelling. Unless the complicated, ubiquitous membrane stacks have some completely unknown function the driving force must be associated with them.

If indeed ion transport and recycling is the mechanism, then the intercellular channel fluid must be hyper-osmotic to the primary absorbate. We have already seen that the sub-epithelial sinus fluid is often hypo-osmotic to the luminal fluid (Wall and Oschman, 1970), and Wall *et al.* (1970) provided further important supporting evidence, when they injected into the recta of intact but dehydrated cockroaches, saline containing a black dye (nigrosine), iso- or hypo-osmotic to the blood, and followed the position of the absorbed fluid visually. By using extremely fine tipped micropipettes, they were able to obtain samples of intercellular channel fluid, and found that its concentration was always higher than that of the injected fluid in the rectum. They were also able to confirm visually, in living *Periplaneta*, the earlier observations of Berridge and Gupta (1967) on *Calliphora*, that intercellular channels become enormously distended when

155

water is being absorbed through the rectal epithelium. (The application of electron microprobe techniques to this problem is referred to in Chap. 6.C.VII.3.)

Assuming ionic transport to be the mechanism, then in *Calliphora*, recycling could very well take place as the primary absorbate passes down the infundibulum over the basal surfaces of the cortical cells. Alternatively the mechanism could be recruitment from the blood via the medulla. The former appears to be more likely.

As Maddrell (1971) comments, it is difficult, though not logically impossible to reconcile all this information with the active transport of water, but final proof is, for the moment, wanting.

6. Fine Structure and Salivary Secretion in *Calliphora*

It will be convenient to take note here of the fine structure and function of salivary glands, for although these organs are not usually thought of as osmoregulatory, in several ticks, as we have seen, they do so function (Chap. 6.C.III.). In *Calliphora* the salivary fluid is dilute and not retained strictly within the buccal cavity, so that water loss in salivary fluid may be a significant component of total loss. Additionally, as Oschman and Berridge (1970) show in an interesting discussion, the fine structure of salivary glands is certainly relevant to general questions regarding the transport of fluids across epithelia—a subject with which we are concerned.

Each of the two salivary glands in adult *Calliphora* consists of a long secretory section that extends far back into the abdomen, and a short translucent reabsorptive region that leads to a small dilation, and thence to a common duct. The secretory region produces a potassium-rich primary saliva, and much of the potassium is absorbed in the short translucent section, leaving a saliva that is hypo-osmotic to the blood (Oschman and Berridge, 1970) (Fig. 77). The main feature of the secretory cells is a system of branching canaliculi whose walls are intricately folded into stacks of leaflets with narrow inter-leaflet channels opening into the canaliculi. Cells in the resorptive region on the other hand, are much thinner. Their luminal (apical) surfaces are again strongly plicate, but there is far less development of canaliculi, and the basal membrane next to the blood has a small number of short, wide infoldings. Reabsorption of potassium may be passive, because the concentration gradient is from lumen to blood, but the secretory process is of course active and mitochondria are associated with the leaflets.

Caterpillars of the moth *Manduca sexta* also produce a hypo-osmotic saliva, (Leslie and Robertson, 1973). Apparently the gland is more complex than that of *Calliphora*. The proximal regions produce a primary saliva which is iso-osmotic with the blood (about 420 mosmol liter^{-1}), and this is modified in the distal regions, probably by reabsorption of ions, to produce a dilute (about 60 mosmol liter^{-1}) saliva. Fine structure of the various regions of this gland is compatible with the above functions, but the precise relationships are not yet clear. Agren (1975) believes that parts of the thoracic salivary glands of the bumble bee *Bombus lapidarius*, are involved in the regulation of ions and water. This conclusion

156

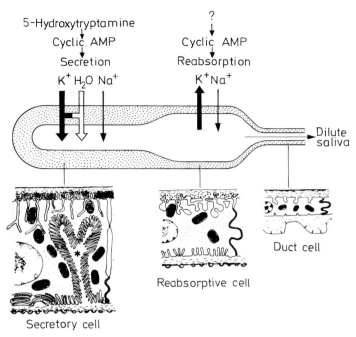

Fig. 77. Summary of the structural and physiological organization of the salivary glands of *Calliphora*. 5-HT, acting via a mechanism involving cyclic AMP, stimulates the secretory cells to produce a potassium-rich primary saliva. This fluid flows into the reabsorptive region where it is modified by potassium reabsorption. Cyclic AMP is an intermediate in the control of the reabsorptive process, although the natural hormone that regulates reabsorption is not known. The result of potassium reabsorption is the formation of a dilute saliva that enters the duct system. The main feature of the cells comprising the secretory portion of the gland is the presence of a branching system of secretory canaliculi (*). The reabsorptive cells have short wide basal infolds, but lack extensive canaliculi. The duct cells do not appear to be specialized for transport, but have a cuticular layer that has periodic thickenings that form ridges around the lumen. (From Oschman and Berridge, 1970)

is so far based on an interpretation of fine structure, and needs experimental confirmation.

Berridge and Patel (1968) and Oschman and Berridge (1970) found that secretion by the distal segment of isolated *Calliphora* glands is fully activated by 5-hydroxytryptamine (5-HT) at a concentration as low as 10^{-9} M, and adenosine-3',5'-monophosphate (cyclic AMP) had a similar effect, although only at 5×10^{-3} M. The role of cyclic AMP and of 5-HT or a similar molecule in other hormonally controlled secretory processes is now becoming better known (although they are not, apparently, involved in the tick salivary gland mechanism (Kaufman and Phillips, 1973 b) (Chap. 6.C.III.).

Interestingly, rapid secretion brought about by AMP or 5 HT does not cause any dilation of the leaflet stacks or canaliculi, although this does occur in the larger intercellular sinuses of the rectal papillae of *Calliphora* and in other such systems (but again not in the fine channels of the membrane stacks). Oschman and Berridge make the point that the function of the system according to the models of Curran (1960), Ito (1961), Diamond and Bossert (1967) and later authors, depends on certain geometrical configurations being maintained, and the diameter of the channels through which the primary fluid flows is of critical importance in this regard. Any enlargement could lead to incomplete equilibration, and an iso-osmotic product would not result. It is therefore of considerable interest to find that in these salivary glands, although the flow rate may be increased 60 times by suitable stimulation, no change of shape is observed in the fine canals or canaliculi. Oschman and Berridge (1970) report the presence of certain strands of material in this and other insects, and these, they suggest, may serve the purpose of maintaining the shape of the channels.

In cells specialised for the reabsorption of solute, it is advantageous for the solute transported to diffuse away rapidly without building up a gradient, and this is reflected in the structure of the resorptive cells in these glands—but whether any part of the cells is also impermeable to water is uncertain.

7. The Cryptonephric System of Tenebrio and Lepidopteran Larvae

In some beetle and saw-fly larvae, and most lepidopterous larvae, the distal ends of the Malpighian tubules are closely applied to the rectum in various conformations; the association being covered by one or more membranes. Such a system is called "cryptonephridial." For a review of the structure and of certain physiological aspects of these systems see Saini (1964). The most complete physiological work so far on such a system is that of Ramsay (1964) and Grimstone et al. (1968) on Tenebrio, and of Ramsay (1976) on Pieris brassicae and Manduca sexta.

Ramsay (1964) confirmed the early suggestion of Wigglesworth, that the rectal complex of Tenebrio is concerned with the removal of water from the faeces, by a series of experiments in which he found that when subjected to drying conditions, Tenebrio larvae produce faecal pellets in equilibrium with air at 90% R.H. ($a_w = 0.9$) or even less, which is equivalent to 136 atmospheres, a solute concentration of 6.1 osmol liter^{-1}, or $\Delta 11.3\,°C$. This is clearly of great value to the insect, because in drying the faeces to this extent it has reabsorbed water from the raw faecal material that arrives at the rectum containing much water and approximately iso-osmotic with the blood. The question is: how does the cryptonephridial system extract water from the faeces down to these low levels?

In Tenebrio there are six pairs of Malpighian tubules, and their convoluted distal ends are applied closely to the rectum. Each tubule bears a number of swellings or "boursouflures," and the whole complex is covered by a perinephric membrane. The latter is not firmly attached to the hindgut anteriorly, but forms a valve-like flap so that the perinephric space enclosed by it is potentially in communication with the haemocoele. Figure 78 shows the general relationships of the system.

The rectal lumen is enclosed by the rectal epithelium, with a cuticle on the luminal side, a thin band of circular muscle and six strands of longitudinal muscle on the basal side. The six convoluted Malpighian tubules lie outside

158

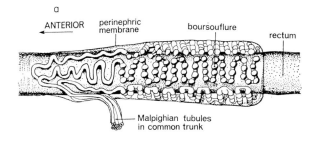

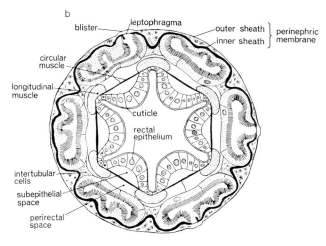

Fig. 78a and b. The cryptonephric system of *Tenebrio* larvae. (a) The whole rectal complex; (b) transverse section of the rectal complex. (From Grimstone *et al.*, 1968)

the rectum and are separated from it by the perirectal space. Finally the perinephric membrane surrounds the Malpighian tubules.

The outer wall of each tubule includes, at frequent intervals, a special "leptophragma cell," whose cytoplasm extends in a very thin layer across a corresponding window or "leptophragma" in the overlying perinephric membrane (Fig. 78) where the latter is represented only by a very thin basement membrane in the form of a blister or "boursouflure" Fig. 79. The cytoplasm of each leptophragma cell extends into the tubule lumen in the form of numerous long, but rather widely separated microvilli, and at these points the tubular lumen is separated from the blood only by a thin outer wall of the leptophragma cell and the basement membrane of the perinephric membrane.

The rectal cells contain many microtubules and mitochondria, and their basal and apical membranes are folded, but there is a complete absence of membrane stacks or marked basal infoldings which are common in the transporting cells of cockroaches and locusts.

The problem of identifying the functions of these various parts is clearly a difficult one. By means of a rare combination of microdissection, electron

159

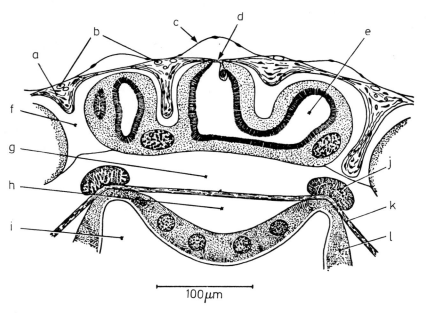

Fig. 79. Detail of the wall of the rectal complex. *a*: perinephric membrane; *b*: tracheae; *c*: blister; *d*: leptophragma; *e*: lumen of perirectal tubule; *f*: peritubular space; *g*: perirectal space; *h*: subepithelial space; *i*: rectal lumen; *j*: longitudinal muscle; *k*: circular muscle; *l*: rectal epithelium. In most sections the tubular lumen underneath the leptophragma is very much narrower than shown here. (From Ramsay, 1964)

micrography and elegant experimental techniques, Ramsay (1964) and Grimstone *et al.* (1968) obtained enough information to propose an acceptable model.

From a physiological point of view the system consists essentially of three separate compartments: (1) the rectal lumen, surrounded by (2) the perirectal space, and (3) the tubule lumen (Fig. 80). The blood (4) may be said to constitute a fourth compartment. In a mealworm in dry air, faecal material that moves into the rectal complex from the ileum is a watery sludge, whose O.P. (about $\varDelta$ 1.4 °C) is similar to that of the blood. As the material moves posteriorly it becomes firmer, and finally, in the posterior region of the complex it is molded into faecal pellets, the spaces between which are air filled and $\varDelta$ is 10 °C or more. Corresponding with these changes the O.P. of the perirectal chamber varies from $\varDelta$ 3 °C anteriorly to $\varDelta$ 9.0 °C posteriorly. Material in the tubule lumen is always somewhat hyper-osmotic to the perirectal fluid, and again increases in concentration posteriorly, although movement of fluid in it is of course anteriorly.

These observations and measurements support the following hypothesis as regards function. Leptophragma cells of the Malpighian tubules actively secrete ions, particularly potassium, followed by chloride, from the blood into the tubule lumen, at the same time preventing an osmotically coupled inward flow of water by means that are not clear. Thus the total O.P. of the tubule lumen is raised above that of the blood and of the perirectal fluid. Evidence supporting

160

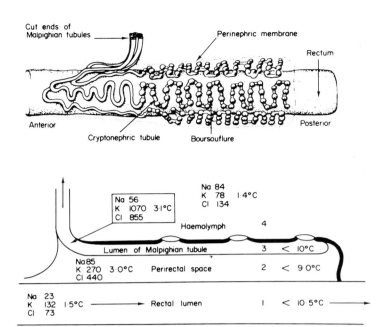

Fig. 80. Diagrammatic longitudinal section through the cryptonephridial system to show the functionally important compartments and the concentrations of ions (in mmol liter^{-1}) and freezing point depressions of fluids in them. (From Maddrell, 1971)

the above is, firstly, that isolated recta are known to be able to take up potassium from a bathing medium and to be permeable to chloride, and secondly, impermeability to water of the perinephric membrane as a whole is suggested by the fact that the addition of distilled water outside the complex has no effect on the rate of secretion of the tubules, while raising the external potassium concentration does increase tubular outflow. It is worth noting that this is the only clear instance so far known, of the formation of a strongly hyper-osmotic tubule fluid in arthropods.

Water then enters the tubule from the perirectal fluid down an osmotic gradient, and the tubule fluid, therefore, moves forwards, becoming more dilute as it goes, until it leaves the rectal complex considerably hypo-osmotic to the posterior rectal lumen, but perhaps iso-osmotic with the anterior perirectal fluid.

At the same time the rectal cells secrete ions and water from the rectal lumen into the perirectal space, thus dehydrating the faeces, although the final stage of the process probably consists in the absorption of water from the vapour phase (Chap. 9.E.) since, as we have seen above, air is present in that part of the rectum, and the faecal pellets are in equilibrium with air at 0.9 a_v (see also Chap. 6.C.V.7., where the water content of locust faeces is considered).

Owing to absorption of water from the rectum, predominantly into the anterior end of the perirectal space, fluid in the latter tends to move posteriorly. As it does so, potassium is secreted from it into the Malpighian tubules; but

161

other osmotically active material, known to be of high molecular weight (10,000 to 12,000) and perhaps a protein, is not so absorbed but accumulates in the posterior region of the perirectal space and helps to elevate the O.P. there.

In summary, a watery, iso-osmotic mixture of gut material and tubule fluid enters the rectal complex via the ileum. What emerges is dry faeces from the anus and a hyper-osmotic fluid, high in potassium and chloride in the distal Malpighian tubules. This fluid contains water that would otherwise have been lost with the faeces, and potassium which is perhaps recycled through the leptophragmata. Presumably the water taken up by the tubules in the rectal complex is reabsorbed from them into the blood in their lower (proximal) regions.

Leptophragmata are absent from the cryptonephridial systems of lepidopteran larvae, so that the mechanism postulated for *Tenebrio* cannot apply. In any case, extraction of water from the faeces of most lepidopteran larvae is unnecessary since they feed on fresh plant material with a high water content, and they are in fact unable to reduce water in the faeces to anything like the level achieved by *Tenebrio*. Ramsay (1976) found that the cryptonephridial systems of *Pieris brassicae* and *Manduca sexta* larvae are deeply involved in salt balance. In fact the ability of *Manduca* to regulate blood sodium and potassium against high concentrations of these ions, introduced by artificial diets or by injection, is far superior to that of either *Schistocerca* or *Carausius*. In lepidopteran larvae, the most probable explanation of the effects observed is that ions enter the rectal complex not by absorption through leptophragmata (since there are none), but by means of a tidal ebb and flow of fluid between the free (proximal) parts of the tubules and the rectal cryptonephric parts of the tubules, coupled with transport of ions from the tubules into the rectal complex. Such a tidal flow would be superimposed upon a net, continuous movement of fluid down the tubules from the rectal complex to the intestine.

As a result of this work, Ramsay is inclined to think that tidal flow may also occur in *Tenebrio*; active transport by the leptophragmata acting as a kind of supercharger. He suggests that cryptonephridial systems in general may have originated in response to a need for salt regulation in insects that have a large and rapid food throughput; in which case cryptonephridial participation in water retention, when this occurs, would be seen as a more recent adaptation of a basic structure and mechanism to a somewhat different function.

Essentially the system provides a gradient with two steps in series against which water has to be absorbed; the presence of the Malpighian tubules, and particularly their ability to recruit potassium from the blood, creates a less extreme gradient than that which the rectal cells would otherwise face.

The work done in this area has been meticulous and remarkably rewarding. Many questions of course still remain, including the rather unexpected fact that the rectal walls, which are supposed to transport water, albeit secondarily, against a steep gradient, possess none of the microstructure usually associated with this process. Maddrell (1971) raises the interesting possibility that the high molecular weight material (perhaps proteinaceous) in the perirectal space, may be directly associated with water vapour absorption from the rectum. The idea is attractive, if only because it would explain why the rectal cells lack the usual microstructure—water being removed by an entirely different mechanism, associated perhaps with the presence of the cryptonephridial system, including a perirectal space. However, direct evidence is so far lacking, and as we shall see below, the thysanuran insect *Thermobia* which also absorbs water vapour per rectum (Noble-Nesbitt, 1970b), has no cryptonephridial system at all.

For a much fuller treatment of these matters than can be given here, see the original papers by Ramsay (1964, 1976), and by Grimstone *et al.* (1968), and the discussion in Maddrell (1971).

The water content of faeces in equilibrium with air at a_w 0.9 is about 14 to 17% of dry weight (Ramsay, 1964). The activity of water in the blood of *Tenebrio* larvae (Δ 1.4°C) is about 0.986, and faeces in equilibrium with this would also be in equilibrium with air at a_v 0.986 (98.6% R.H.). The water content of such faeces is not recorded but at a_v 1.0 (100% R.H.) faecal water content is at least 50% of dry weight. If this value is accepted, then since a larva produces 0.112 mg dry weight of faeces h^{-1}, drying the faeces from 50% to 16% water content would recover 0.038 mg h^{-1} or 0.91 mg day^{-1}, by an insect that weighs perhaps 100 mg. This does not seem to be an enormous saving, but we may have seriously underestimated the water content of moist hind gut material.

8. *The Problem of "Formed Bodies"*

In the excretory systems of a wide range of animals, including frogs, crayfish, brine shrimps and insects, there are many small ($< 1 \mu$m) to larger ($< 20 \mu$m) membrane-bound vesicles, first described by Riegel (1966) in the crayfish antennal gland, and now known as "formed bodies." Riegel (1970, 1971, and earlier papers) has developed an hypothesis concerning the function of these bodies to which we shall now refer.

Riegel suggests that the bodies are formed by pinocytosis within cells, and include lipids and peptides. They are discharged into extracellular spaces within the excretory organ (between the villi of Malpighian tubules, for example) where, owing to their osmotic properties, they swell and eventually burst, their contents having lysed to form small, osmotically active solute particles. As a result of the absorption of water by the intact bodies, concentration of solutes in the ambient fluid increases, and provided that the geometry of the entrance to an exit from the channels is appropriate, this is followed by a movement of water into the spaces occupied by the discharged bodies. In other words, Riegel proposes that formed bodies rather than pumps will turn out to be the mechanism whereby ions are moved across membranes—a concept which, if true, would be of more than ordinary importance for physiology.

It seems that the bodies are not artefacts. However, there are difficulties, pointed out by Maddrell (1971), in accepting the proposal in toto. There is no evidence that the bodies include significant quantities of potassium, sodium or chloride, yet these are known to be secreted in large amounts in Malpighian tubules. There are many fewer bodies per unit volume of primary secretion when *Rhodnius* tubules are secreting fast than when they are less active. The bodies do not occur in the copious salivary fluid of *Calliphora*, although they

163

are present in the Malpighian tubules of that insect. Other points at issue are raised by Maddrell (1971).

None of these difficulties is actually fatal to the proposal. Perhaps the question will be decided only after a rigorous mathematical analysis of its implications. But at present the formed body proposal has to be provided with a good deal of inferred superstructure to make it work, and since we do have a reasonably consistent theory of fluid production by ion secretion, the principle of Occam's razor would suggest caution. Admittedly this leaves us without an explanation for the function of Riegel's formed bodies, unless his original suggestion (Riegel, 1966) is correct, namely that they serve for the transport of recalcitrant materials across membranes.

D. Hormonal Control of Water Balance

It is well known that body water content in vertebrates is under hormonal control, particularly by vasopressin, the so-called anti-diuretic hormone from the neurohypophysis of the pituitary gland. Work has also commenced on the situation in arthropods, but so far mainly insects have been investigated.

Early indications of hormonal effects were obtained by Nuñez (1956), who found that in the beetle *Anisotarsus*, removal of the dorsal surface of the brain leads to water retention, and by Altmann (1956) who showed that injection of corpus allatum extract into honey bees leads to release of water and dehydration while corpus cardiacum extract has the opposite effect. Techniques in those days were less rigorous than they became later, but this early work at least drew attention to important problems. More recently, following the work of Maddrell (1962, 1963) and Berridge (1965, 1966), the field has become very active.

When *Rhodnius* feeds it takes in a volume of blood equal to some ten times its own weight, and during the next few hours eliminates excess water equal to its original weight every 30 min. Clearly this involves prodigious activity on the part of the Malpighian tubules, and Maddrell (1962, 1963, 1964a, b) using a modification of Ramsay's (1954) in vitro technique (Fig. 81) measured an increase in secretion rate of about 1,000 times, to $3.3 \, \mu l \, min^{-1} cm^{-2}$, when blood from a fed *Rhodnius* was added to the medium in which a tubule from an unfed *Rhodnius* was being observed. By a series of ligations followed by microdissection, he traced the source of the diuretic hormone first to the meso-thoracic ganglionic mass, and then to twelve neurosecretory cells at the back of the ganglion.

These cells lie deep within the mass, posing a problem as to how their product could reach the blood. Using electronmicrography Maddrell demonstrated that neurosecretory granules, elaborated by the cell bodies, pass along axons running in abdominal nerves mainly to segments 1 to 3. The surface of these nerves is penetrated in places by the neurosecretory axons, at the so-called "neurohaemal" sites, and in all probability the granules, containing

Fig. 81. Preparation of Malpighian tubules of *Rhodnius* isolated in a drop of blood and producing urine. *MT*: Malpighian tubule; *S*: drop of secretion; *R*: internal surface of a small part of the rectum included in the preparation; *PG*: coiled posterior part of the mid-gut. Photograph kindly supplied by Dr. Maddrell

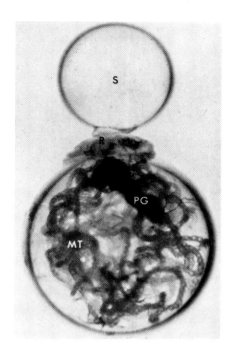

diuretic hormone or its precursor are liberated there (perhaps as a result of depolarisation by action potentials) outside the perineurium and below an acellular sheath through which they can diffuse into the blood close to the target organs—the Malpighian tubules (Fig. 82).

Release of the diuretic hormone in *Rhodnius* is mediated by stretch receptors in the abdominal wall (Maddrell, 1964 b). Collateral branches of the neurosecretory axons run to a part of the neuropile that also receives afferent fibers from the abdominal wall. So perhaps these axon branches are able to convey motor impulses and in this way stretching of the abdominal wall brings about a very rapid response in the form of diuretic hormone release.

Some diuretic activity exists in other parts of the mesothoracic ganglion and in other ganglia, but the activity of the few neurosecretory cells is no less than 50,000 times greater than that of other cells, and theoretically detectable in dilutions of $1:10^8$, so that there can be no doubt of the true source of the hormone.

As regards its mode of action, there is still some uncertainty. Measurement of electric potential across the tubule wall shows that the first effect of the hormone, in vitro, is to make the lumen briefly more negative to the blood (from -13 mV to about -23 mV), then rapidly more positive (to about $+38$ mV), before returning slowly (over a few minutes) to a lower negative state than before diuresis began (-30 mV) (Fig. 83). This level is held while diuresis continues; but as soon as hormone secretion ceases, the electric potential returns to normal and diuresis decreases as the remaining hormone is inactivated. A tentative propo-

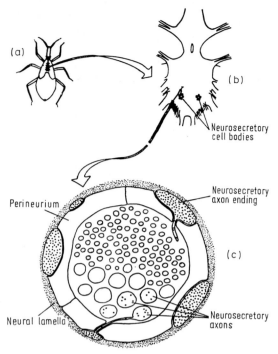

(a)

(b)

Neurosecretory
cell bodies

Perineurium

Neurosecretory
axon ending

(c)

Neural lamella

Neurosecretory
axons

Fig. 82 a–c. Diagram to show where in *Rhodnius* the diuretic hormone is released into the circulating blood. (a) The position and size of the mesothoracic ganglionic mass in relation to the rest of the insect. (b) The position in the mesothoracic ganglionic mass of the neurosecretory cell bodies which synthesize the diuretic hormone and the course of one of their axons along an abdominal nerve. (c) A cross-section of one of the abdominal nerves to show, lying outside the perineurium, the neurosecretory axon endings from which the diuretic hormone is released. The branches to these endings from the protected neurosecretory axons run longitudinally deeper within the nerves

sal by Maddrell as to the ionic events associated with the above electro-potential changes is also shown in Figure 83.

Control of diuresis by hormones has since been found in several insects. In *Dysdercus*, Berridge (1966b) found the highest diuretic activity in cells of the median neurosecretory complex in the brain, extracts from which caused the Malpighian tubules to increase production from the normal 0.78 to $3.1 \ \mu l \times 10^{-3} \ min^{-1}$. Some activity is present in other nervous tissue, particularly the corpora cardiaca and the mesothoracic ganglion (the main seat of production in *Rhodnius*). As a result Berridge believes that when the insect feeds (on plant material) a diuretic hormone is synthesized in the median neurosecretory complex and released into the blood by the corpora cardiaca.

In isolated tubules from *Dysdercus* the fluid is iso-osmotic with the blood, and Berridge (1966b) also found that the O.P. of the fluid is unaffected by an increase in the rate of production under hormonal influence. From this it seems that the hormone's function may be simply to alter the rate at which tubule fluid is sent to the rectum, where its composition may be appropriately modified before elimination.

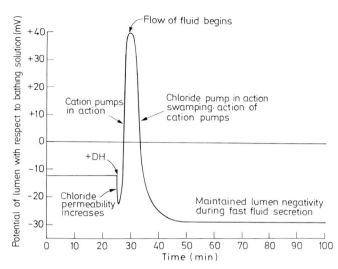

Fig. 83. Proposal by Maddrell linking observed changes in the electropotential gradient in a *Rhodnius* Malpighian tubule with the machinery of urine production. (From Maddrell, 1971)

In the stick insect *Carausius* (formerly *Dixippus*) Pilcher (1970a, b) found a diuretic hormone with high activity in the brain and especially the corpora cardiaca and she proposed a control mechanism similar to that in *Dysdercus*. The hormone in *Carausius* may be inactivated by the tubules themselves, as in *Rhodnius* and *Dysdercus*. In *Glossina*, a diuretic hormone, probably produced in the ganglionic mass, is released by nerve axons very near to the Malpighian tubules (Gee, 1975b). A degradative enzyme is present with the hormone, but the latter may also be inactivated by the tubules themselves.

Pilcher also found that the amount of hormone (and hence the rate of urine production) in *Carausius* varies with the state of hydration—an effect similar to that reported by Wall and Ralph (1962) for the cockroach, *Blaberus*. These are interesting points to which we shall return later.

At one time there was some doubt about the existence of an antidiuretic hormone in insects, but evidence for its presence has been reported by Wall and Ralph (1962) for *Blaberus*, Vietinghoff (1966) for *Carausius*; Wall and Ralph (1964), Wall (1967) and Mills (1967) for *Periplaneta*, and Delphin (1963) for *Schistocerca*. Antidiuresis in these cases is apparently effected by increasing rectal absorption under the influence of a hormone from neurosecretory cells in the ventral nerve cord, particularly the last abdominal ganglion. Nerve axons containing dense granules have in fact been observed within the sheath covering the rectal absorptive cells in *Calliphora* (Gupta and Berridge, 1966b; Oschman and Wall, 1969). But both decreased tubule production and increased rectal reabsorption can apparently be mediated by the same hormone in *Periplaneta* (Wall and Ralph, 1964; Mills and Nielsen, 1967), and in *Schistocerca* (Mordue, 1969).

Hormonal control is not universally present in all arthropods where it has been sought. According to Farquharson (1974b) there is no evidence of hormonal

167

control in the millipede *Glomeris*, neither does cyclic AMP, 5-HT or theophylline stimulate fluid formation by the tubules of that animal. We recall that in the ixodid tick *Dermacentor*, salivary gland control is basically nervous according to Kaufman and Phillips (1973 b), although catecholinergic substances are probably involved as intermediary messengers (Chap. 6.C.III.). For further references and discussion see Berridge (1970) and Pilcher (1970a, b).

As Berridge (1966 b) pointed out, there are formidable technical and interpretational difficulties in this field. Thus almost every tissue, particularly nervous tissue, may have some diuretic activity: the site of production is usually different from the site of release; and if ablation of suspected endocrinal tissue is used as an experimental procedure, several other physiological processes may be affected and the situation thereby obscured.

It is fairly clear that sites of production and release, specific activity and periods of release all differ from one insect to another, but the correlation of these differences with the whole biology of the insects concerned is very desirable. Pilcher (1970a) has made a start along these lines by comparing the situation in *Rhodnius* with that in *Carausius*. The first named feeds on blood at long intervals, the second on moist plant tissue almost continuously, and the adaptive value of the two different control mechanisms becomes clear. In *Rhodnius*, the hormone has a very high specific activity, for when it is needed at all it is needed immediately and in a highly effective form. It is produced only in response to a meal, and liberated, via nerve axons, at neurohaemal sites close to the Malpighian tubules which respond energetically. In *Carausius* on the other hand, the hormone has a much lower specific activity. It is produced continuously by the brain in small quantities and liberated by the corpora cardiaca at a distance from the target organs, whose response is comparatively mild.

More recently Maddrell *et al.* (1969, 1971) have made a study of the pharmacology of the Malpighian tubules in relation to diuretic hormone mechanisms. They found that in both *Rhodnius* and *Carausius* 5-hydroxytryptamine (5-HT) will mimic the activity of the diuretic hormone at very low concentrations down to 10^{-8} (Fig. 84). Furthermore, some of the derivatives of 5-HT, notably those obtained by methylation or acetylation of the terminal amino group do not affect activity, although other changes do, producing molecules that are not only inactive but inhibitory.

However, 5-HT is not the diuretic hormone itself. In the first place, 5-HT stimulates *Carausius* tubules while *Rhodnius* diuretic hormone does not, and other evidence pointing in this direction is discussed by Maddrell *et al.* (1971). Maddrell and Casida (1971) believe that the hormone molecule is, or contains, an indole alkylamine, but its precise chemical nature remains unknown.

The role of cyclic AMP as an intracellular mediator in several hormonal systems is well known (Major and Kilpatrick, 1972), and Maddrell *et al.* (1971) obtained an indication of its participation in the insect diuretic mechanism when they found that in the absence of 5-HT in the medium, large doses of cyclic AMP would stimulate tubule activity. Concentrations of 10^{-4} M were required in *Carausius* for threshold response, and while *Rhodnius* responded to 10^{-5} M, 10^{-3} M was necessary for full response. These rather high concentrations are, of course, probably due to the fact that they are applied to the

168

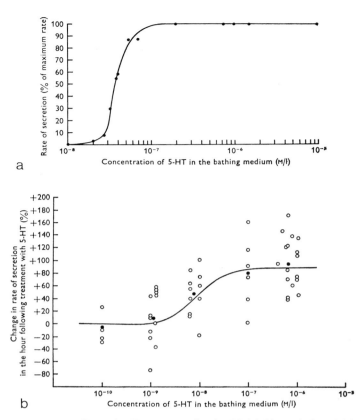

Fig. 84. Dose response curves for the effects of 5-HT on secretion by Malpighian tubules of (a) *Rhodnius* and (b) *Carausius*. (From Maddrell, 1971)

wrong side of the cell wall—intracellular levels are normally of the order of 10^{-8} M to 10^{-6} M (Butcher and Sutherland, 1962). Recently the case for the involvement of cyclic AMP has been strengthened by the discovery that treatment of *Rhodnius* tubules with diuretic stimulants increases intracellular cAMP levels (Aston, 1975).

From this and other evidence, Maddrell *et al.* suggest that the diuretic hormone affects the tubule cell at specific receptor sites, causing an intracellular release of cAMP which then stimulates fluid formation (Fig. 85).

Finally, Maddrell and Gee (1974) have used an elegant technique to identify the sites of release of hormone (the neurohaemal sites) in *Rhodnius* and *Glossina*, and their results support the idea that such release is triggered by depolarization of the axonal membrane. In vertebrates, the release of neurohormones is associated with depolarization by potassium and entry into the axon of calcium. Maddrell and Gee (1974) found a very similar situation in *Rhodnius* and *Glossina*, where the neurohaemal areas are on lateral abdominal nerves. They applied a small drop of a potassium-rich solution to a small part of the insect being tested

169

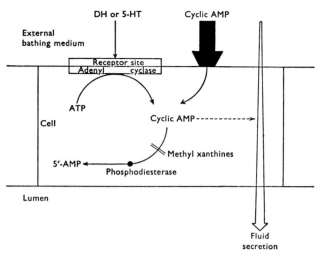

Fig. 85. Possible intracellular events involved in the stimulation of secretion by the Malpighian tubules of *Rhodnius* and *Carausius*. (From Maddrell *et al.*, 1971)

(e. g. the mesothoracic ganglion), and after a suitable interval, measured the effect of this drop on a Malpighian tubule placed in it. If the organ or tissue tested had contained neurohaemal release areas, these would have been depolarized by potassium, hormone would have been released and its presence revealed by diuretic activity of the test drop. In this way they were able to prove (for example) the absence of release sites on the mesothoracic ganglionic mass (although the neurosecretory cells are within the mass) and their presence on the proximal regions of the peripheral abdominal nerves, just where electron micrography had suggested that they would be (Fig. 82).

The importance of hormonal control of diuresis, and through this of water balance, is clear. Two further related points deserve attention. The first concerns the possibility of hormonal control of cuticle permeability reported recently by Treherne and Willmer (1975a, b) and referred to above (Chap. 3.B.IX.). If cockroaches are ligated at the neck, cuticular water loss increases significantly, although this does not occur if the neck nerves are severed without ligation. Brain extract injected into ligated insects restores impermeability, thus strongly suggesting that a hormone, produced in the brain, is responsible. The effect, they find, does not occur in all cell membranes (nerve sheaths are unaffected) and they believe that the effect may be restricted specifically to epidermal cells. This is the first report of a hormonal effect on water balance via the cuticle; if it is confirmed, and if the amount of hormone liberated turns out to be related to the water status of the insect, or to ambient conditions, the system might help to account for several observations of changes in cuticle permeability (see also Chap. 3).

Neuro-hormonal control of certain mechanical properties of the cuticle has already been proposed by Maddrell (1966) and Reynolds (1974a, b, 1975) to account for the plasticization of the abdominal cuticle of *Rhodnius* after a large

170

meal (see also Chap. 3.B.IX.). Maddrell and Reynolds (1972) clearly showed that the diuretic hormone and the proposed plasticization factor are not identical, although they are both released, through their natural pathways (neurosecretory axons in the abdominal wall) when insecticides cause paralysis. As suggested by Bennet-Clark (1961), plasticization of the abdominal cuticle is accompanied by an increase in its water content by more than 20% (Maddrell, 1966; Reynolds, 1975). The water content of the decapitated cockroach cuticle is reduced while its permeability is increased (Treherne and Willmer, 1975b). The implications of these observations are not at present clear, (see Chap. 3.B.IX. for further discussion) but further exploration of the nature and effects of plasticization and cuticle water content in general is likely to yield information bearing on the important question of the control of cuticle permeability.

The second topic concerns an interesting development from the above research field, and concerns the effects of organic insecticides. It has often been observed that the water balance of insects poisoned with these materials is seriously affected (Jochum, 1956; Roberts et al., 1969), and Maddrell and Casida (1971), following up this clue, observed that the effect of several insecticides (they worked particularly with the carbamate "Zectran") is to cause swelling of the rectum by water accumulation in paralyzed insects. Further work showed that the topical application of Zectran is followed by the release of a factor similar to, but not necessarily identical with, the diuretic hormone of *Rhodnius* into the blood, and that this leads to the observed diuresis. The mesothoracic ganglion is again the site of hormone production.

Lethality of these insecticides seems to be related to their effect on the nervous system, particularly insofar as they inhibit anticholinergic activity. There is no direct evidence that their effect on water balance mechanisms is lethal or contributes to lethality, but as Maddrell and Casida point out, these possibilities are worth further investigation, particularly because the answers may be important in the design of new insecticides.

Hormonal control of water balance is clearly an exciting field. As a result of recent work there have been important advances in our understanding of the mechanisms of diuresis in insects. Some links in the chain are stronger than others (for example we do not really know the link between hormonal receptor sites and fluid production, or even whether cyclic AMP is involved) and the means whereby the hormone is destroyed or removed is not clear, but excellent techniques are now available for further work. It would be particularly useful if some of the work on insects were to be followed up on other arthropods.

Chapter 7

Uptake of Liquid Water

The term "preformed water" is sometimes used to describe water taken in with the food or imbibed by drinking, presumably in order to distinguish it from "post-formed" water which results from oxidation of the food. But most of the water content of food can in principle be recovered in the liquid form by expression, and here we shall use the term "liquid water" to include both this and water in the form of droplets and such like. The distinction between these on the one hand and water vapour (which is also absorbed by some arthropods) on the other, is a useful one.

There are three fairly distinct ways in which arthropods gain water: (1) uptake of liquid water, (2) absorption of water from the vapour phase, and (3) production of metabolic or oxidation water. The present chapter is concerned with the uptake of liquid water, which may take place by ingestion with the food, by direct drinking, or by the uptake of capillary water from moist surfaces by mouth, rectum, or by other special organs. Some aspects of this field have been reviewed by Cloudsley-Thompson (1975).

A. Uptake of Water as a Result of Feeding

Food forms the most important source of water for many terrestrial arthropods. We have seen that for those that feed on blood or sap, water at times may be more than sufficient. Predators such as spiders, scorpions and insects, as well as all the many forms that feed on fresh plants, also have a reliable water source provided that their normal food is available. Even desert insects may have a copious supply of water if they are plant feeders, as Casey (unpublished) found for the sphingid larvae *Manduca sexta* and *Hyles lineata*. In others, exemplified by the pests of stored products, and including many insects and mites, the supply of water with the food is limited, and for them water conservation is all important. The water content of fresh plant leaves is often 90% or more, but that of stored grain, bran, hair or feathers is frequently in the region of 15%, and many insects that feed on these materials develop when the water content is even lower: 10% for the beetle *Ptinus*; 6% for *Tribolium* and *Silvanus*; and as low as 1% for the meal moth *Ephestia* (Fraenkel and Blewett, 1943) (refs in Wigglesworth, 1972).

Locusta, when feeding normally (but in culture) on fresh grass which contains about 80% water, are in positive water balance (Loveridge, 1974). If the water content of the food falls to 5%, locusts absorb less water from this source than they lose with the faeces, and they then reduce feeding to a minimum. (See Chaps. 6.C.V.7. and 11.G. for details and further discussion.)

We have seen that *Glossina* retains only that proportion of water in a blood meal that is necessary to replenish its normal content (Bursell, 1960). The bed bug *Cimex* behaves similarly (Mellanby, 1932b), retaining more water after it has been dehydrated and thus remaining independent of humidity provided that it can feed frequently enough.

Apart from this there is very little quantitative information about the water intake of arthropods in natural conditions (or indeed in the laboratory), largely owing to the difficulty of measuring the amount and water content of the food eaten. The development of radioisotope techniques would seem to offer opportunities in this field. Such methods have been used effectively to measure food and energy fluxes in populations of organisms, by Reichle and Van Hook (1970), Van Hook (1971), Reichle *et al.* (1973) and by other groups (refs in Reichle *et al.*, 1971); but so far little use has been made of these methods for water balance studies, apart from the work of Van Hook and Deal (1972) on the cricket *Acheta domestica*. The use of doubly labelled water ($HT^{18}O$) in particular has been developed as a very useful technique for studies of water and energy turnover by vertebrates in the field (e. g., Nagy, 1972; Nagy and Shoemaker, 1975) and we shall undoubtedly see the results of the application of these techniques to field studies on arthropods before long (e. g. Bohm and Hadley, in preparation).

In this connection an important question that has never been fully answered concerns the extent to which arthropods eat food above their energetic requirements for the sake of the contained water. It may well be that some arthropods eat more in low than in high humidities, or eat more moist than dry food when they are short of water, as desert *Onymacris* beetles do (Seeley and Edney, unpubl. experiments); but insects are often more active in dry than in moist air, so that a high metabolic rate, rather than an attempt to obtain more water, may be the cause of enhanced feeding.

Buxton (1924) suggested that desert beetles might maintain their water content by feeding on dead plant material (which has a moisture content of 60% after absorbing water hygroscopically in air at 80% R.H.). Pierre (1958) believes that the process may be important for desert lepismatids, but he offers no data in support, and in fact he found the water content of desert plant material in the Sahara to be as low as 2% after a night in summer.

Broza (unpublished) observed drinking and feeding on moist plant material by three tenebrionid species: *Blaps sulcata*, *Pimelia derassa*, and *Trachyderma philistina*, and in field experiments he found that partially dehydrated *Trachyderma* and *Adesmia dilatata* gained about 14.0% and 7.6% of their weight respectively by eating food containing up to 68% by weight of hygroscopically absorbed water. In this way *Trachyderma* increased its water content by at least 19.5%, *Adesmia* by a little less. There is no doubt that such liquid water intake is all important for the water balance of a desert arthropod. Where drinking occurs, this is clearly done to replenish water reserves, and it seems at least probable (but not certain) that ingestion of moist foods in these conditions may be done for the sake of the food's water content rather than its energy content.

At first sight the work of Schultz (1930) seems to provide conclusive evidence. He found (1) that *Tenebrio* larvae fed on "air dry" bran that contains 10–12% water, ingest much more than is energetically necessary, as shown by the fact that 80% of the ingested bran appears in the faeces in an undigested form, (2) that the water content of the undigested faecal bran is 7–9%, and (3) that larvae, given only completely dry bran, drastically reduce their intake and fail to develop. This seems to show that the excess food is eaten for its water content, but the proportion of undigested food in the faeces is not securely established, and perhaps more importantly, the water content of the faecal material was not determined immediately after production. Murray (1968), however, found that *Tenebrio molitor* larvae raised on bran containing 16.1% water (in equilibrium with 80% R.H.), develop optimally only if allowed to drink. At lower humidities the amount of bran taken is actually reduced, and at 30% R.H. or lower, development is restricted even if drinking water is available. Such larvae reared on whole meal flour develop well even without drinking in all humidities down to 50%. Murray suggests that these results may be explained by the different amounts of metabolic water available from whole meal flour (50%) and from bran (19%). If these figures are even approximately true, they provide a good example of the interaction of humidity and the nature of the diet in determining the need for drinking by developing insect larvae, but the results do not confirm Schultz's conclusions that extra food is taken for the sake of its water content.

Some of the most convincing data so far are those obtained by Sinoir (1966) for *Locusta*. If nymphs of this species are fasted for six hours either at low (45–50%) or high (95–100%) humidity and then for the following eighteen hours offered synthetic food (filter paper laced with sucrose) with different water contents while still being kept either at low or high humidities, then those whose water content is low eat more high water content food than those whose body water is normal (Fig. 86). If we assume that the intake of moist food

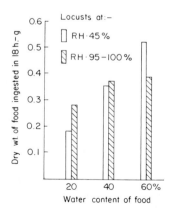

Fig. 86. Locusts kept at 45% R.H. and offered dry food eat less than those kept at 95%, but they eat more than those kept at 95% if both are offered moist food. The absolute amount eaten by both groups in greater if the food has a high water content. (Redrawn from Sinoir, 1966)

by optimally hydrated nymphs is sufficient for energy requirements, then the dehydrated nymphs may be ingesting the extra food for the sake of its water content. However, there are two reasons why this assumption is questionable:

174

firstly the activity and hence the metabolic needs of the nymphs may be higher in dry air (no data on this point were recorded), and secondly, cellulose filter paper is indigestible, so that the energy requirements are probably not fulfilled in either case. It would be interesting to repeat these experiments using normal food and monitoring the insects' metabolic rates.

Figure 86 also shows that when *Locusta* nymphs are offered drier food (water content 20% of wet weight), those in good water status take nearly as much as they do of the moister food (water content 60%), but the dehydrated nymphs take significantly less. Here there is no suggestion that extra food is consumed for the sake of its water content; but perhaps the situation is approaching that described by Loveridge (1974) in which adult *Locusta*, when offered food with only 5% water content, reduce their intake drastically because they lose more water in their faeces than they gain in their food (Chap. 6.C.V.7.).

B. Uptake of Water by Drinking

Many arthropods are known to drink water if it is available (see Table 29). Here we shall consider one or two illustrative cases in some detail.

I. Drinking by Insect Larvae

Mellanby and French (1958) have published a useful review of the extent and biological significance of drinking in larval insects. Among stored products insects, *Tenebrio* larvae after dehydration drink almost their own weight and in this way raise their water content from 65% to 85% (Mellanby and French, 1958). They take this water by mouth, for no gain in weight occurs if water is in contact with the cuticle only, or if the mouth is sealed, or if the larvae are in chill coma while in contact with water. The imbibed water which is taken into the crop, soon leads to a great increase of blood volume in the haemocoele (Mellanby, 1958). Such larvae, given access to water twice a week, become full sized in three months, compared with a year or more for those not allowed to drink. In one month at 25°C, larvae increase in weight from about 35 mg to 145 mg, but only to 55 mg if water is not available. At 35°C larvae without drinking water lose weight and ultimately die. However, much depends upon the nature of the food, as described above (Murray, 1968).

The question arises as to whether such insects would ever come across free water in real life. The answer is uncertain, but as Mellanby and French point out, changes in temperature may lead to condensation, and if this happens, they could then imbibe the drops. Perhaps the dew that forms in deserts on cold nights may also form a source of water for beetles such as *Eleodes* that live there. Certainly advective fog is a source of water in adult desert insects (see Chap. 7.B.II.).

Table 29. A representative list of arthropods known to drink water or absorb it from moist surfaces

Taxon	Remarks	Reference
Onycophora		
Peripatopsis	Drinks from moist surfaces. No eversible vesicles.	Manton and Heatley (1937)
Opisthopatus	Drinks, but also absorbs from moist surfaces by eversible vesicles.	Alexander and Ewer (1955)
Isopoda		
Oniscus	Drinks water drops by mouth. Restores lost weight. Controlled by blood concentration.	Berridge (1970)
Ligia, Oniscus, *Porcellio,* *Armadillidium*	From water surface by mouth and rectum. From moist surface by mouth.	Spencer and Edney (1954)
Myriapoda		
Narceus *Orthoporus*	By eversion of rectal tissue onto moist surface.	O'Neill (1969) Crawford (1972)
Symphyla		
Hanseniella	Eversible vesicles.	Tiegs (1947)
Collembola		
Various collembolans	Eversible vesicles.	Sedlag (1951)
Podura *Onychiurus*	Ventral tube with eversible vesicles.	Noble-Nesbitt (1963) Nutman (1941)
Diplura		
Campodea	Ventral abdominal vesicles, everted by blood pressure.	Drummond (1953)
Thysanura		
Machilis	From moist surfaces by eversible vesicles.	Willem (1924)
Insecta		
Dictyoptera		
Periplaneta	Drinks by mouth.	refs in Leclercq (1946), Buxton (1932a)
	Drinks cyclically during vitellogenesis.	Verrett and Mills (1973, 1975a)
Arenivaga	Drinks by mouth.	Hawke and Farley (1973)

176

Orthoptera		
Gryllus	Drinks by mouth.	refs in Leclercq (1946)
Acridium	Drinks. Cutaneous absorption claimed.	Colosi (1933)
Locusta	Drinks $>100\,\mathrm{mg\,g^{-1}}$ in 30 min from moist cotton when dehydrated. Survives on 0.25 M NaCl	Loveridge (1974, 1975)
Isoptera		
Hodotermes	Newly emerged alates drink and store water in modified salivary sacs.	Hewitt *et al.* (1971), Watson *et al.* (1971)
Hemiptera		
Rhinocoris, Platymeris, *Reduvius*	Drink water drops in laboratory.	Edwards (1962)
Dysdercus	Drinks to replace water needed for elimination of excess inorganic ions in diet.	Berridge (1965)
Coleoptera		
Various adults		refs in Leclercq (1946)
Calandra, Silvanus		refs in Buxton (1932a)
Geotrupes, Trichius		refs in Leclercq (1946)
Clytus	Drink by mouth in culture.	refs in Leclercq (1946)
Oryzaephilus		Mellanby and French (1958)
Psylloides		refs in Mellanby (1958)
Tenebrio larvae	Access to water affects growth rate.	Mellanby (1958)
Agriotes larvae	Drink when dehydrated. Exchange water with soil when abraded.	Evans (1944)
Ptinus	Access to water in culture affects development.	Ewer and Ewer (1942)
Lepidochora *Onymacris*	Desert species. Drink condensation on their surface and on sand from advective fog.	Louw and Hamilton (1972), Louw (1972), Hamilton and Seely (in prep)
Adesmia, Pimelia	Desert species. Drinking reverses strong hygropositive reaction in dehydrated beetles.	Rayah (1970)
Trachyderma	Desert species, drinks rapidly to replace all water lost by dehydration (may be up to 35% original weight).	Broza *et al.* (1976)

Table 29 (continued)

Taxon	Remarks	Reference
Lepidoptera		
Various Rhopalocera	Species from hot tropics.	Buxton (1932a), Norris (1936), Leclercq (1946)
Pierids, papilionids, } lycaenids, hesperids }	Drink from puddles and moist soil.	refs in Leclercq (1946)
Ephestia, Pyrausta	Drinking affects longevity and fecundity of many stored products insects.	Norris (1934)
Noctuid larvae }	Drink by mouth in culture.	refs in Mellanby and French (1958)
Pieris larvae }		refs in Mellanby and French (1958)
Philadoria, Diataraxia	Drink readily if "thirsty" (probably low water content).	Mellanby and French (1958)
Diptera		
Various tabanids and muscids	Drinks by mouth in hot dry conditions.	refs in Buxton (1932a)
Calliphora	Drinks water if sugar soln. not available. No extra water if food is dry.	Leclercq (1946)
Lucilia		Roberts and Kitching (1974)
	Control by blood concentration, especially Cl⁻.	Browne (1968), Browne and Dudzinski (1968)
Lucilia larvae	Drink if dehydrated.	Mellanby (1938)
Various soil larvae	Drink if dehydrated.	Mellanby and French (1958)
Phlebotomus larvae	Drink. Cutaneous absorption claimed.	Theodor (1936)
Phormia	Control by blood volume. Concentration unimportant.	Dethier and Evans (1961)
Stomoxys, Lyperosia	One of few examples of blood feeders drinking water.	Krijgsman (1936)
Syrphid larvae	If dehydrated, absorb by rectal lobes from water surface.	Schneider (1948)
Hymenoptera		
Apis	Prefer water with NaCl or NH₄Cl to pure water. Controlled by amount of water lost. Control by hormones claimed. Drinks own body weight in water in 1 min.	Butler (1940) Altmann (1956) Betts (1930)
Social wasps	Drink for control of nest temperature and humidity.	refs in Leclercq (1946)
Icaria, Odynerus	Solitary wasps. Drink by mouth.	refs in Leclercq (1946)
Polistes	Social wasps. Drinks for distribution to larvae.	refs in Leclercq (1946)

Taxon		Behavior	Reference
Arachnida			
Acarina			
	Echinolaelaps	Drinks when below equilibrium humidity.	Knülle (1967)
	Boophilus	Drinks drops of water by mouth—unusual for ticks.	Wilkinson (1953)
Araneida			
	Tarentula, Lycosa	If dehydrated, drink from damp soil against suction pressures up to 400 mm Hg.	Parry (1954)
	Eurypelma	A desert "tarantula." Drinks feebly.	Cloudsley-Thompson (1967)
	Dugesiella	A tarantula. Drinks readily to restore blood volume.	Stewart and Martin (1970)
Scorpionidea			
	Androctonus	A desert form. Drinks avidly when dehydrated.	Cloudsley-Thompson (1958)
	Pandinus	Drinks but not avidly: 15% weight in 3 h.	Cloudsley-Thompson (1963)
	Leiurus, Buthotus	Drink when dehydrated: 25% in 3 h.	Cloudsley-Thompson (1963)
	Centruroides	Drinks when dehydrated.	Hadley (1971)
	Diplocentrotus	Drinks by sucking from moist soil.	Crawford and Wooten (1973)
	Palamnaeus	Forest scorpion; drinks regularly.	refs in Cloudsley-Thompson (1968)
	Hadrurus	From droplets when dehydrated, but not from moist sand.	Hadley (1970b)
	Centruroides	Drinks from water droplets.	Hadley (1971)
Solpugida			
	Galeodes	Drinks very inefficiently.	Cloudsley-Thompson (1961)
Uropygi			
	Mastigoproctus	Drinks from moist sponge.	Crawford and Cloudsley-Thompson (1971), Ahearn (1970b)

Among phytophagous insects, larvae of the tomato moth *Diataraxia oleracea* drink freely—a 300 mg larva imbibing as much as 40 mg in 1 min (Mellanby and French, 1958). The normal water content of these larvae is about 90% and they drink if it falls much below this. Drinking stops, however, as soon as the normal content is restored.

The food of this and of other phytophagous larvae normally contains so much water that drinking is unnecessary, but the advantage of the ability to do so is manifest when larvae are fed on wilted leaves, for the supplementation of the diet by drinking is obligatory.

As regards soil larvae, *Agriotes*, the wire-worm, is known to become permeable as a result of abrasion (Wigglesworth, 1945). It is then in osmotic equilibrium with soil water at $\simeq 0.35$ M and gains or loses water according to the gradient across the permeable cuticle even when the mouth is blocked (Evans, 1944). However, wire-worms are also known to drink if they become dehydrated (Evans, 1944), and according to Mellanby and French several other larvae including *Lucilia* and the wheat-bulb fly larva *Leptohylemyia*, which do not become abraded, will drink avidly if dehydrated.

Finally, Hodson (1937) believes that many diapausing larvae need cuticular absorption of water to terminate diapause; but Mellanby and French question this, and suggest that drinking is the usual method. Conclusive evidence is lacking.

II. The Extent of Drinking by Arthropods

The rate of drinking and the amount of water drunk, varies greatly among arthropods. Thus the solpugid *Galeodes* scarcely drinks at all (Cloudsley-Thompson, 1961) while honey bees *(Apis mellifera)* drink nearly as much as their own weight of water in 1 min according to Betts (1930). The factors controlling drinking also turn out to be rather various insofar as they have been investigated.

Desert insects are known to drink water deposited as dew or from advective fog. Louw and Hamilton (1972) demonstrated this for the tenebrionid beetle *Lepidochora argentogrisea*; and Seeley and Hamilton (1976) took this further by observing that *Lepidochora* spp. construct small ridges in the sand at right angles to the drifting fog. The ridges concentrate water droplets, after which the beetles burrow back along the same ridges and increase in weight by absorbing water, presumably through the mouth. Hamilton and Seeley (1976) also observed that another tenebrionid, *Onymacris unguicularis*, stands (or "basks" as they say) in a head down, abdomen up position in the path of drifting fog (Fig. 87). Water droplets collect on the insect's body, run down to the head and are taken in through the mouth. By weighing and marking individual beetles before basking, they found that the beetles increase in weight by a mean of 12% as a result of such drinking. Broza (unpublished) observed drinking by *Scanthius aegyptius* (a pyrrhocorid) and *Monomorium subopacum* (an ant) both from stones, and by two beetles, *Carabus impressus* and *Coccinella septempunctata* (predators though they be) from droplets on leaves.

Fig. 87. The desert tenebrionid *Onymacris unguicularis* stilting during a night fog in the Namib desert. Condensation of water droplets can be seen on the body, and near the mouth. Such drinking leads to an increase in weight of about 12% during one night fog (Hamilton and Seeley, 1976). From an unpublished photograph kindly provided by Dr. W. J. Hamilton III

Such drinking by desert forms undoubtedly plays a very important part in water balance, particularly in those species such as *Lepidochora* and certain *Onymacris* species that live in sand dunes where fresh green plants are virtually absent.

Norris (1936), reviewing the incidence of drinking in adult lepidopterans, observes that this is well known in most families of butterflies, and she confirms that aggregations of drinking butterflies are almost entirely composed of males. Downs (1973) further reports that the habit is common in several moth families. Water uptake does not seem to be the main reason for this behaviour, since the latter is confined to one sex, and in any case ample water is available in the nectar on which these insects feed. Downs (1973) has suggested that because evaporation of water concentrates solutes at the surface of mud and puddles, supplementation of the nectar diet to obtain necessary salts may be the reason for "puddling," and that males need extra nutrients because they do more flying. A recent experiment by Arms *et al.* (1974) suggests that sodium ion is the stimulus for the behaviour in *Papilio glaucus*, the need for sodium arising because of its very low concentration in plant tissues and nectar.

181

There is an alternative explanation: that the behaviour is an epideictic display of the kind discussed by Wynne-Edwards (1962), in which an attraction to sodium ions would simply be the means of encouraging aggregation.

If the isopod *Oniscus* is dehydrated it will readily drink pure water until its original weight is restored, but if given a salt solution, it drinks excessively, suggesting that internal receptors responsive to blood osmotic pressure or ionic concentration control uptake (Berridge, 1970). The isopods *Ligia oceanica*, *Oniscus asellus*, *Porcellio scaber* and *Armadillidium vulgare*, when dehydrated, all absorb water if this is available as a free surface, by rectum as well as by mouth (Spencer and Edney, 1954). In addition, all except *Ligia* (which is a littoral animal) maintain their water content in air at 85% (to which they lose water by evaporation from the cuticle) if a moist plaster surface from which they can absorb water by mouth suction is available. Presumably this ability serves them very well in field conditions, but there is no direct evidence on this point.

According to Altmann (1956) honey bees drink more water in low than in high humidities, and this appears to be due to differential loss of water through the tracheal system, for carbon dioxide, which causes the spiracles to open, also leads to increased drinking. Altmann found that removal of the corpora alata increases drinking by bees, while removal of the corpora cardiaca has the reverse effect. Browne (1964) believes that this is not a direct causal relationship, the effects of the hormones on blood concentration and volume forming an intermediate link. However, the evidence is inconclusive.

Among scorpions, Crawford and Wooten (1973) found that second instar larvae of *Diplocentrotus spitzeri*, dehydrated for 24 h, subsequently gain at least enough water from a moist substrate to make up for the loss. Uptake probably occurs by mouth, although the evidence is inconclusive on this point. Adults of this species do not regain lost water in this way, and consequently must rely entirely on their prey for external water. Hadley (1970b, 1971) did not observe drinking in *Hadrurus*, but he reports that another scorpion, *Centruroides*, does drink free water when it is dehydrated, regaining up to 70% of lost weight. Cloudsley-Thompson (1967) found that the desert tarantula, *Eurypelma*, will drink when dehydrated but not rapidly or to the extent necessary to replace all the water lost. Several species of ixodid tick larvae drink from moist paper in the laboratory (Wilkinson, 1953) and in so doing increase their chance of survival in dry conditions (Londt and Whitehead, 1972). Perhaps they drink dew in nature, but this has not been demonstrated.

As regards millipedes, O'Neill (1969) observed that *Narceus* may absorb up to half its body weight in water by mouth from a moist sponge, and Crawford (1972), working with the more xeric species *Orthoporus*, found that while ingestion of succulent food, such as moist carrots, normally enables these animals to maintain their water content in air below 40% R.H., if they become dehydrated they can absorb water by extruding rectal surfaces through the opened anal flaps onto moist soil. This species also may imbibe water through the mouth from a moist sponge.

A similar situation was found by Ahearn (1970b) for the uropygid (whip-scorpion) *Mastigoproctus giganteus*, which, after dehydration, absorbs sufficient water from a moist sponge to restore its normal water content. Crawford and Cloudsley-

Thompson (1971) measured water absorption of 10% of body weight during 24 h in this species. Whether the uptake was by mouth or elsewhere was not determined.

Spiders, represented by two lycosids, *Tarentula barbipes* and *Lycosa radiata*, have been shown by Parry (1954) to suck capillary water from moist surfaces, even against considerable tensions. After dehydration to a water loss of 10% these spiders imbibe water at a constant rate that depends on suction force. *Tarentula* imbibes, albeit slowly, at suctions up to 400 mm Hg; but from lower suctions such as 75 and 150 mm Hg, spiders weighing between 31 and 43 mg, imbibe at rates of 17.4 and 3.3 μg s^{-1} respectively. A 10% loss of weight is, therefore, replaceable in about 2.5 h. Whether such drinking occurs in nature is unknown, but Parry's description of the apparently purposeful searching and sucking behaviour of dehydrated spiders on moist soil certainly suggests that it does. In the laboratory, the tarantula *Dugesiella hentzi* drinks water readily, to an extent necessary to maintain a constant blood volume. Fasted animals drink somewhat more, tending to maintain a constant weight (Stewart and Martin, 1974).

Manton and Heatley (1937) reported drinking by the onycophoran *Peripatopsis*. The process takes place from moist surfaces (e.g., moist filter paper in experimental conditions) by a sucking action of the mouth, the remainder of the cuticle being hydrophobic. Interestingly enough, another onycophoran, *Opisthopatus cinctipes*, also absorbs water by eversible vesicles, which are not present in *Peripatopsis*.

An interesting means of water uptake has recently been found by Walcott (in press) in the ghost crab, *Ocypode quadrata*. After dehydration, these animals absorb water from moist soil against a capillary suction of up to 38 mm Hg. Water is first absorbed by capillarity into tufts of strongly hydrophilic hairs near the bases of the legs. From there the water passes into the gill chamber (where it is presumably absorbed through the gills) by means of mechanical suction. If a fine cannula is sealed into the gill chamber and left open to the outside, absorption ceases, but it begins again if the cannula is closed.

III. The Control of Drinking

Most work on the control of drinking has been done on muscid flies and locusts. Bolwig (1953) obtained some evidence that an increase in blood O.P. causes a heightened response to the presence of water, and Wohlbarsht (1957) found that dehydration causes a similar response. Evans (1961) and Dethier and Evans (1961) analyzed the situation in the blow-fly, *Phormia regina*. The positive drinking response by dehydrated flies is signaled by elevation of the proboscis in response to water on the tarsal or labellar sensory hairs (Evans and Mellon, 1961), and this is reversed by the injection of water into the haemocoele. Injection of solutions of high O.P. does not restore the response in negative flies, but such solutions injected into positive flies does turn them negative. If flies which are somewhat dehydrated but still negative to water are bled, they become positive. Information about blood volume and pressure is conveyed to the brain by the recurrent nerve, and Dethier and Bodenstein (1958) found that section of this nerve causes continuous drinking until the flies become bloated. The combined weight of this evidence points strongly to control of drinking by blood volume, apart from any osmotic or ionic effect, in *Phormia*.

The picture is rather different in the sheep blow-fly, *Lucilia cuprina*. Browne (1968) found that injection of 0.3 M (or stronger) sodium chloride, or equivalent Ringer's solution, causes negative flies to become positive to water, and increases the intake of fluid in already positive flies. Bleeding has little effect, and Browne concludes that in *Lucilia*, blood chloride concentration is predominant in the control of drinking. Removal of the corpora allata or cardiaca has no effect, and he suggests that where such effects were previously found, this might be due to accidental cutting of the recurrent nerve.

Browne and Dudzinski (1968) used ^{14}C injections and measured changes in blood volume, total weight, crop weight, osmotic pressure and sodium and chloride concentration, during water deprivation in *Lucilia*. Again their results showed that the volume and O.P. of the blood has little effect on the response to water; they obtained better correlation between degree of response and blood sodium and chloride concentrations.

This apparent difference in mechanism between *Phormia* and *Lucilia* is somewhat puzzling. However, it is possible that both blood volume and blood concentration play a part in both flies, the former predominantly in *Phormia*, the latter in *Lucilia*. If so, it would be interesting to know whether such a difference is associated with other aspects of the animals' mode of life. Certainly *Phormia* drinks very much less than *Lucilia*; satiety being achieved by the intake of only about 1 μl.

Loveridge (1975) found that adult *Locusta* will drink from moist cotton by expressing water with the mandibles and taking it into the mouth. Locusts that are fasted at high humidity, or fed, drink very little, while those fasted in dry air drink a lot (Table 30), so that dehydration is a stimulus for drinking. Locusts fasted at high humidity and then exsanguinated (without permanent harm) drink copiously. Since locusts fed on dry food also have very little blood, and drink a lot of water, a possible interpretation is that low blood volume in this insect, as in *Phormia*, is a stimulus for drinking.

Table 30. Amounts of water drunk by locusts during 30 min after different pretreatments. The fiducial limits are ± 2 X S.E. 10 insects in each of groups (i)–(v), 9 in group (vi)a, b. (From Loveridge, 1975)

Pretreatment						
(i)	(ii)	(iii)	(iv)	(v)	(vi)	
					(a)	(b)
fed fresh food	starved 24 h 96% R.H.	starved 24 h 0% R.H.	starved 48 h 0% R.H.	exsanguinated	fed dry food 5 days	8 days
Gain 0.24 mg/g	Loss 1.42$\pm$0.27 mg/g	Gain 41.07$\pm$12.07 mg/g	Gain 39.38$\pm$20.38 mg/g	Gain 68.90$\pm$14.64 mg/g	Gain 120.16$\pm$23.42 mg/g	Gain 94.73$\pm$57.48 mg/g

Bernays, Chapman and Middlefell (in preparation) also found that drinking is elicited in *Locusta* after feeding on dry food, or not feeding at all. Dry fed insects lose 23% of their original water content (over half of it from the gut), and their blood volume decreases from 116 to 83 μl in 24 h. The blood O.P., however, rises only from 400 to 435 mosmol liter^{-1}, so that once again osmoregulation is evident. As regards the mechanisms involved, Bernays found that increasing the blood O.P. by injecting solutes does not elicit drinking, and rather unexpectedly, increasing the blood volume does not inhibit drinking in "thirsty" locusts, unless this is done by injecting distilled water rather than iso-osmotic saline. Such treatment would perhaps reduce the blood O.P. slightly.

These results are not easy to interpret. Certainly, as in Loveridge's work there is a clear demonstration that dehydration leads to drinking, but the causal chain involved is not clear.

The factors governing intake of water and sucrose by the Australian locust *Chortoicetes terminifera* have been investigated by Browne et al. (1975a, b; 1976), particular attention being paid to the mechanisms responsible for cessation of drinking. While the whole story is not yet clear, it seems that the amount taken increases with increasing sucrose concentration up to 0.5 M, above which the meal becomes smaller. Adaptation of the receptors on the mouthparts probably plays an important part in regulating meal size, because more sugar solution is imbibed when this is provided in small drops at intervals of 30 s (giving sufficient time for some recovery from adaptation) than when the locusts feed continuously. Interestingly, this is not true when pure water is offered, for the volume taken is then the same in both modes. Water uptake, then, is probably governed by volume or by its effect on O.P.

Further information about the amounts of water drunk in relation to water content and to ambient conditions in these and other insects, and about the nervous and hormonal pathways involved, has been obtained by Bernays and Chapman (1972, 1974c) and others. It is proving to be a highly complex field, the behavioural aspects of which, while they are of fundamental importance for an understanding of the mechanisms involved, are somewhat removed from our present (admittedly rather arbitrarily defined) interests. For a review of this very active field see Browne (1975), Bernays and Chapman (1974c), and Dethier (1974).

C. The Question of Absorption Through the Cuticle

There have been a few reports of water absorption directly through the cuticle, but the evidence is contradictory and often equivocal in any one case. The observation of Pal (1950) of the appearance of water in a large trachea after application of a drop to the outer surface of the cuticle, has been explained by Beament (1964) as being due to a temperature effect (local cooling brings about local condensation from the nearly saturated air in the tracheae). The absorption of small droplets directly through the cuticle in *Periplaneta* as described by Beament (1964) might be explained if the cuticle has a low water activity, in equilibrium with the air above it, so that water would then go into rather than through the cuticle. Loveridge (1968b) suggested this for *Locusta*, Edney (1971) for desert beetles, and Vannier (1974b) for collembolan moults. Spencer and Edney (1954) were unable to detect cutaneous absorption in isopods, and Ahearn's work with *Mastigoproctus*, although not ruling out cutaneous absorption, does not prove it. Colosi (1933) reported the entry of water through the cuticle of the grasshopper *Acridium*. After dehydration to 95% of original weight one group regained water to 97% from a moist surface while another, dehydrated to 91%, regained to 99%. In this latter case it is difficult to ascribe the result to absorption into the cuticle alone (the weight increase is too much), but some absorption through the mouth may also have occurred.

Theodor (1936) observed water absorption by drinking and, he believes, through the cuticle in larvae of the sand-fly, *Phlebotomus*: but the only evidence for cuticular uptake is the fact that dead larvae on moist filter paper gain weight. Vannier (1974a) has demonstrated that cuticles of the collembolan *Tetrodontophora bielensis* absorb and release water according to ambient humidity, but he was dealing with moulted cuticles. Louw and Hamilton (1972) were unable to observe cuticular absorption in the desert tenebrionid *Lepidochora argentogrisea* when droplets were placed on the elytra of dehydrated beetles, but they did establish the drinking of advective fog by this beetle (Chap. 7.B.II.). Loveridge (1975) was unable to find cutaneous water absorption in *Locusta*.

D. Special Organs for Water Absorption

Despite the lack of convincing evidence for general cutaneous absorption there are certain specialized areas or organs through which water, and perhaps salts, may be absorbed. We have already taken note of rectal uptake from moist surfaces in isopods and millipedes (Chap. 7.B.II.) and it remains to consider further examples by this route and by other specialized organs.

A further example of rectal uptake is that of syrphid larvae (Schneider, 1948). If dehydrated larvae have access to a water surface, they make contact with this by extending the rectal lobes through the anus and absorb water rapidly enough to increase their weight by 50% in a few hours.

Other organs involved in uptake usually take the form of eversible vesicles, often associated with the posterior margins of ventral sclerites. Noble-Nesbitt (1963) has described a good example of these in the collembolan, *Podura aquatica*. If these animals are strongly dehydrated and then placed on a water surface, drinking through the mouth begins at once: only later on is the ventral vesicle everted to make contact with the water surface, for such eversion can occur only as the blood pressure becomes great enough. In moderately dehydrated podurids, however, where sufficient blood is still available, the ventral tube is everted immediately and water is taken up through it.

The use of ^{24}Na and ^{42}K in animals with blocked mouths or ventral tubes, showed that sodium is exchanged mainly through the ventral tube, while potassium exchange with the medium is not measureable. Perhaps ventral tube vesicles may confer the ability to absorb water from a moist surface where drinking would not be possible; however, Noble-Nesbitt's experiment did not relate to this point.

Further examples of water absorption by extensible vesicles have been reported in (1) *Onychiurus*, a collembolan, by Nutman (1941), and in other collembolans by Sedlag (1951); (2) *Campodea*, a dipluran, by Drummond (1953) (Fig. 88); (3) *Machilis maritima*, a thysanuran, by Willem (1924); (4) *Hanseniella*, a symphilid, by Tiegs (1947); and (5) *Opisthopatus*, an onycophoran, by Alexander and Ewer (1955). Another onycophoran, *Peripatopsis*, does not possess vesicles and as we have seen (Chap. 7.B.II.), drinks by mouth alone.

It is interesting that the only myriapodous arthropod known to possess water absorbing vesicles is a symphilid, for this group is sometimes held to

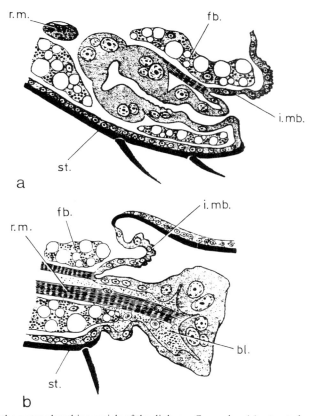

Fig. 88a and b. Sagittal section of the water absorbing vesicle of the dipluran *Campodea*; (a) retracted, (b) everted. This is one of the few examples of organs which are known to absorb water from moist surfaces. (From Drummond, 1953.) *bl*: blood; *fb*: fat body; *imb*: intersegmental membrane; *rm*: large retractor muscle; *st*: sternum

187

be closely related to diplurans, and through them to insects proper (Tiegs, 1947).

When water is taken up through special organs, as in the above examples, the cuticle covering such organs is presumably different from that of the general hydrophobic surface. This has in fact been demonstrated in *Podura* by Noble-Nesbitt (1963), who found that while the general surface is strongly tuberculate, each tubercle being covered by a strongly hydrophobic wax giving a high contact angle with water, the cuticle over the vesicles is clearly demarcated from the rest and is thin, smooth, and hydrophilic, with a contact angle of about 90°. After wetting the contact angle may become even lower.

E. Conclusions

The above discussion makes it all too clear that, with a few exceptions, liquid water absorption by arthropods is a neglected field so far as quantitative experimental work is concerned. When absorption occurs through regions other than the mouth it would be extremely interesting to know something about the fine structure of the organs involved. Data are also needed about the quantitative aspects of water absorbed by drinking or in any other way, and on the relationship of these quantities to internal and external conditions.

Chapter 8

Metabolic Water

A. Introduction

Water that results from the oxidation of organic materials by living organisms is termed "metabolic" or "oxidation" water, and has long been recognized as a component, sometimes an important one, of the total water income of animals (Babcock, 1912). The complete oxidation of a fat such as tripalmitate to carbon dioxide and water yields about 107 g water for every 100 g fat while a similar weight of the carbohydrate glucose yields 56 g, and of protein (taken to urea) about 40 g water (Table 31). Thus an animal while fasting and metabolizing

Table 31. The relationship between oxygen consumed and energy and water produced when various classes of food materials are oxidized. It is assumed that fats and carbohydrates are oxidized to CO_2, and that proteins go as far as urea. Values are representative approximations, since different foods within the same class behave differently. (From Schmidt-Nielsen, 1964; Prosser, 1973)

Food	g water g food^{-1}	$1 O_2$ g food^{-1}	$1 O_2$ g water^{-1}	kcal g food^{-1}	g water kcal^{-1}	R.Q.
Carbohydrates	0.56	0.83	1.49	4.20	0.133	1.0
Fats	1.07	2.02	1.89	9.54	0.112	0.71
Proteins	0.40	0.97	2.44	4.19	0.095	0.79

fat reserves could gain weight were it not for loss of water by transpiration and of other materials by defaecation. It follows that for arthropods living in dry surroundings such as stored grain where there is no free water and the water content of the food is low (Chap. 7.A.), metabolic water may be very important, as was recognized early on by Buxton (1930), Mellanby (1932a, c), Fraenkel and Blewett (1944) and others.

B. The Question of Regulation

Do arthropods, when they are short of water, relieve the situation either by increasing metabolism, or by using fat instead of carbohydrate as a substrate? The early work gave somewhat conflicting answers, and it turns out that the situation is very much more complex than was once thought. Buxton (1930) found that

189

fasted *Tenebrio* larvae maintain a constant water content (percent) even though their total weight decreases, and Mellanby's (1932c) results further suggest that these larvae regulate their water content by varying the proportion of fat used. [At higher humidities *Tenebrio* larvae gain water by absorption of the vapour (Chap. 9.B.) but this is another matter.] Buxton and Lewis (1934) also found that fasting tsetse-flies oxidize additional fat when the air is dry. However, several contrary results were obtained: by Buxton (1932a) for *Rhodnius*, and by Mellanby (1932b, 1934) for *Cimex* and *Tineola*. Fraenkel and Blewett (1944) approached the problem from a different point of view and showed that *Tribolium*, *Ephestia* and *Dermestes* larvae take longer to develop in lower humidities, and although they use more food in the process, they produce smaller pupae. Furthermore, in *Dermestes* reared at 30% R.H., only 30% of the water in the pupae can have been derived from the larval food as liquid water, and for *Ephestia* reared at 1% R.H., the comparable figure is only 7.6%. The remainder must in both cases have been derived from oxidation, so that these data provide good evidence that dry larvae consume more food. Whether they produce more metabolic water depends on the extent of food utilization.

This is not universally true, however, for *Ptinus tectus* larvae eat less food when they (and the food) are dry, unless they have been allowed to drink previously (Ewer and Ewer, 1942) and the same is true for *Locusta* (Bernays and Chapman, 1974b).

When additional food is taken, the fate of the additional energy generated is an interesting question that has not yet been considered. In most animals, once the normal quota of ATP has been synthesized, production of energy is curtailed and metabolism drops, and Berridge (in conversation) suggests that it would be interesting to look for some kind of uncoupling mechanism in *Ephestia* whereby excess energy could be eliminated, perhaps as heat, so that metabolism may continue. For further discussion of the earlier work see Edney (1957), Loveridge (1975) and Loveridge and Bursell (1975).

Mellanby (1942) made the important point that an increase in metabolic rate, or a shift from carbohydrate to fat metabolism, may involve an increase in oxygen demand, and lead to more frequent, or wider, spiracular opening, or to more ventilation. Any or all of these could lead to enhanced spiracular water loss, so that the overall effect on water balance of such changes could be in doubt. Bursell (1975b) referring to Buxton and Lewis's (1934) results, also points out that dehydrated tsetse-flies are more active than others, so that this, rather than an attempt to produce more water, may account for a higher metabolic rate.

These objections are not crucial. Evidence from flying *Glossina* (Bursell, 1959b), *Locusta* (Loveridge, 1967) and *Schistocerca* (Weis-Fogh, 1967) goes to show that increased oxygen uptake does not necessarily imply a proportionate increase in water loss. Such an increase could be achieved, for example, by allowing P_{O_2} at the mitochondrial surface to fall, thus increasing the oxygen concentration gradient across the cell, the tracheole wall, and thus along the trachea, leading to more rapid diffusion inwards of oxygen, without increasing the water vapour pressure gradient, and hence the diffusion of water vapour, in the opposite direction (Chap. 4.B.).

Loveridge and Bursell (1975) also point to the fact that, although fat metabolism does indeed yield more water than an equal weight of carbohydrate, if the fat stored had to be synthesized from other materials in the first place, water would have been used in that process. If so, then the synthesis and subsequent oxidation of these fat reserves would not result in any net effect on the animal's water balance. However, to the extent that the original food contained fat, or that fat stores were synthesized during periods of water abundance (e.g., locusts feeding on moist grass), and oxidized during water shortage, such fat metabolism could be of advantage to the insect. Both in *Leucophaea* (Scheurer and Leuthold, 1969) and *Periplaneta* (Verrett and Mills, 1973, 1975a, b) water is imbibed several days before its massive incorporation into developing eggs. During this intervening period the water must be stored somewhere, and these authors suggest that storage may occur in the fat body and in the integument (Verrett and Mills, 1975a; and see Chap. 2.C.). The liquid water content of the fat body certainly varies appropriately, but a high concentration of ATP in the fat body at the time of maximum water release (Lüscher and Wyss-Huber, 1964) offers some support for another interpretation: that imbibed water is stored by fat synthesis, to be released by oxidation at the appropriate time. There is, however, no proof at present that this does occur.

C. Respiratory Quotient and Metabolic Water in *Locusta*

A further important point has been raised by Schmidt-Nielsen (1964) and elaborated by Loveridge and Bursell (1975). If metabolism is measured, not in terms of oxygen used, or of substrate oxidized, but as energy produced as ATP, then a different picture emerges. As Table 31 shows, 0.133 g water $kcal^{-1}$ are produced by oxidizing carbohydrates, while only 0.112 g water are produced by the same energy output using fat. Experiments with *Locusta* by Loveridge and Bursell (1975) illustrate the above point very well. They found that fasting or dehydration, while having no measurable effect on metabolic rate, significantly reduce the R.Q. from 0.83 to 0.77. Since protein is unlikely to be used as a respiratory substrate (Loveridge, 1967), this indicates a shift towards the use of a higher proportion of fat.

At first sight this appears to confirm the prediction that dehydration would initiate a shift to fat metabolism. In fact the data are equally well explained if a shift to fat results from starvation. But more important from the present point of view is that knowing metabolic rates, R.Q.s (and thus the proportion of fats and carbohydrates used), and the energetic values for the two substrates (38 ATP mol^{-1} or 4.2 kcal g^{-1} for glucose and 411 ATP mol^{-1} or 9.5 kcal g^{-1} for fat), the rates of energy output for fed and fasted insects can be calculated. These turn out to be very similar. In other words, extra fat is indeed oxidized by starved insects, but this is offset by a reduction in carbohydrate oxidized, to an extent that the total energy output remains unchanged. But such a switch also has the effect of reducing the metabolic water output because fats produce less water per calorie than carbohydrates do. Fed locusts with an R.Q. of 0.83 produce 0.35 mg water $g^{-1}h^{-1}$ while fasted or dehydrated locusts whose R.Q. is 0.77 produce only 0.28 mg water $g^{-1}h^{-1}$.

Thus the use of fat rather than carbohydrate as a respiratory substrate is, on an equal energy basis, counterproductive so far as water is concerned. This is a principle of general significance for water balance, and Figure 89

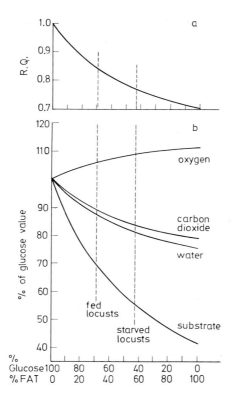

Fig. 89a and b. Certain effects of the use of various proportions of carbohydrates and fats for energy production. (a) The effect on R.Q. (b) The effects on oxygen consumed, carbon dioxide evolved and on metabolic water produced, all per unit ATP formed. Consumption and production are shown as percentages of the respective values if glucose were the sole substrate. (From Loveridge and Bursell, 1975)

shows the theoretically calculated values of the parameters involved (oxygen uptake, carbon dioxide output and water output) for a constant energy production with different proportions of fat and carbohydrate in the respiratory fuel.

D. Metabolic Water in Flying Locusts and Aphids

Metabolic water is a component of water gain in all arthropods, but its significance can be judged only in relation to all the other components of the water budget, and such information is very hard to come by. One instructive case has been described by Weis-Fogh (1967) for the desert locust *Schistocerca*. These insects often fly in swarms over arid country for many hours, and the question of their water balance at such times is particularly interesting because both loss of water by ventilation and gain of water by metabolism are enhanced during flight.

Weis-Fogh's measurements show that during level flight at 3.5 m s^{-1} the insect's metabolic rate is 65 cal g^{-1} h^{-1} and if fat is the sole fuel, 7.28 mg water g^{-1} h^{-1} would be produced (from Table 31). Weis-Fogh arrives at the somewhat higher value of 8.1 mg water g^{-1} h^{-1}. Thoracic ventilation in a resting locust is about 30 ml g^{-1} h^{-1}, but this rises to 320 ml g^{-1} h^{-1} during flight; and whether or not a locust remains in water balance depends largely on the respiratory loss occasioned by such ventilation—a loss which is in turn affected by the temperature of the thorax and the humidity of the air being pumped through it. At 30°C and 60% R.H. a flying locust's thoracic temperature is about 36.5°C, and it loses water at about 8.0 mg g^{-1}h^{-1}, so that if we accept Weis-Fogh's values for metabolic water production, the insect is in positive water balance. In higher radiation loads, the thoracic temperature excess may be increased by 2° or even 4°C (Weis-Fogh, 1967), and this of course leads to greater evaporation.

Taking account of these variables, Weis-Fogh has devised a model showing the effect of various combinations of ambient temperature, ambient humidity and radiation load, upon the water balance of a flying locust, and this is shown in Figure 90. For any given radiation load (curves A, B or C), at points above

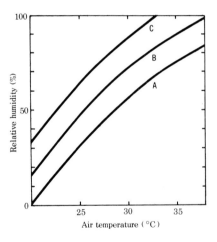

Fig. 90. Each curve represents a set of combinations of ambient temperature and humidity at which water loss by a flying locust is balanced by the production of metabolic water. *A*: No net radiation load, *B* and *C*: thoracic temperature is assumed to be increased by 2° or 4°C respectively as a result of solar radiation. (From Weis-Fogh, 1967)

the line, locusts would gain water, but at points below they would suffer a net loss. Thus, unless the R.H. of the air is 90% or higher, continuous flight in direct sunshine (curve C) will lead to water loss if the ambient temperature is 30°C or above; but in the absence of a net radiation load (curve A) in air at 25°C, a relative humidity of 30% or so is sufficient to maintain water balance.

In arid conditions it is therefore advantageous to fly high where temperatures are lower, and Weis-Fogh summarizes the situation as follows: "A decrease in temperature from 35 °C at ground level to 23 °C at 3 km altitude should permit sustained flight at 35% R.H. even in sunshine. At ground level this amount of water corresponds to only 18% R.H. as in a true desert. Provided that a swarm of locusts is lifted to a height of 2–3 km by means of thermal upcurrents it should be able to cross a large desert without suffering from water shortage when on the wing."

It should perhaps be stressed that the arguments in the preceding paragraphs involve several approximations. For example, cuticular evaporation is not taken into account. But the heuristic value of the model is very great for it helps to explain the relationships between the variables involved, and of course points to gaps in our knowledge where further information is required.

Aphids also fly for several hours continuously, and Cockbain (1961 a, b) has made some useful measurements of metabolism and water content in *Aphis fabae*. During tethered flight for 6 h at 25°–26 °C and 57–82% R.H., an aphid weighing 700 μg loses 9% of its wet weight, mainly by evaporation. During this period it oxidizes about 30 μg fat and in so doing generates 32 μg of water. In fact, as a result of metabolic water production, the percentage water content remains almost constant at 70% during this long period. (See Table 32 for

Table 32. Water content of *Aphis fabae* during a six hour flight at 25°–26 °C and 57–82% R.H. Weights in μg. (Data from Cockbain, 1961 a, b)

	Total wt	Water wt	(% of wet)	Dry wt	(% of wet)
Original	700	480	(69)	220	(31)
Loss of water after 6 h	-63	-63			
	637	417			
Loss of fat after 6 h	-30			-30	
	607			190	
Metabolic water after 6 h	$+32$	$+32$			
Final after 6 h	639	449	(70)	190	(30)

details.) These experiments were carried out with tethered aphids in a laboratory, but Cockbain believes that the results are applicable in real life, and that water is not likely to be a limiting factor for aphid flight. This certainly seems to be true, provided that a particular water content (percent of wet weight) rather than a particular absolute amount of water is the criterion for well-being.

E. Conclusions

Metabolic water is clearly a component of the water income of all organisms. The water so produced enters the general reserves, and whether or not it is necessary for survival depends on the relative sizes of all the other components. For dry-living arthropods there is evidence that metabolic water is an essential component (Chap. 8.B.). In other cases, the relative importance of metabolic water may vary with the nature of the food, or with the intensity of activity. For insects such as locusts, which feed on moist plant material, metabolic water at about 0.6 mg per locust h^{-1} is insignificant compared with liquid water gained by feeding (50.7 mg per locust h^{-1}). However, if these insects are fed on wilted grass, the rate of feeding decreases greatly but the metabolic rate remains constant, so that metabolic water, still at 0.6 mg per locust h^{-1} is greater than water input from the food at 0.2 mg per locust h^{-1} (Loveridge, 1975).

When no other source of water is available, e.g., in pupae, or in insects on long migratory flights, metabolic water is clearly of the greatest importance. During flight, the rate of production increases and at least helps to counterbalance additional evaporative loss.

The interesting question as to whether metabolism is increased, or the substrate switched, during water stress can now be partly answered. The total amount of metabolism during a developmental stage may be increased by ingesting more food and using the excess for water production. (The problem of energy disposal in this circumstance is referred to in Chap. 8.B.). The rate of metabolism may increase in dry air, perhaps as a result of greater activity, but nevertheless producing more water. However, switching the metabolic substrate from sugars to fats is counterproductive for water conservation if the rate of energy production remains constant.

There is little evidence about metabolic water production in field conditions. Metabolic rates have sometimes been measured for the assessment of energy budgets, and some of this work is referred to above (Chap. 5.C.), but these have been relatively short period measurements, and in general they suggest that metabolism should not normally be expected to play an important role (except, of course, during periods of high activity). But the field is virtually unexplored.

Probably the greatest need now is for further careful measurement of metabolic water production in relation to other components of water balance by analytical work in the laboratory and, eventually, by measurements of these variables in field conditions. Short term regulation of water balance by varying the rate of metabolic water production has not been demonstrated. But information about the extent of such production is essential for an understanding of other processes that are regulatory in effect.

Chapter 9

Absorption of Water Vapour[1]

A. Introduction

One of the most intriguing aspects of water balance in land arthropods concerns the mechanisms by which water vapour is absorbed from unsaturated air. The process has been recognized for a long time, but an explanation in physiological terms is still lacking. Relevant reviews are those of Beament (1961, 1964, 1965), Locke (1964), and Noble-Nesbitt (1976).

B. Distribution of the Faculty Among Arthropods

In 1930 Buxton described how larvae of the meal worm *Tenebrio molitor* increase in weight, even though starved, at high humidities, and ascribed this to the overproduction of metabolic water. Mellanby (1932c) found that production of metabolic water is insufficient to account for the increases in weight and water content observed, and suggested that water vapour is condensed and absorbed through the tracheae. Soon thereafter Ludwig (1937) observed uptake in nymphs of the grasshopper *Chortophaga* (above 92% R.H.); Lees (1946a) observed it in various ticks with lower limits from 84% to 94% R.H.; and Edney (1947) found uptake in flea prepupae *(Xenopsylla brasiliensis)* down to 50% R.H.

Since then several species of arthropods have been found to take up water vapour in a variety of circumstances, and it is worth considering these species to see whether any regularities or trends appear. Table 33 which is based on a list prepared by Berridge (1970) with some later additions, shows that all the arthropods reported to take up water vapour are either acarines (ticks or mites) or insects. Furthermore, none of the insects included is winged. This sometimes means that the faculty is restricted to the immature forms, as it

[1] Parts of this chapter are reprinted from a recent review by the present writer (Edney, 1975). I am grateful to Dr. Vernberg, the editor of that publication, and to Intext Educational Press, for their agreement to this procedure.

196

Table 33. A representative list of arthropods that are known to take up water vapour from unsaturated air, and their critical equilibrium humidities (C. E. H.)[a,b]

Species	C.E.H. (as R.H.%)	Reference
Arachnida		
Acarini		
Acarus siro	75	Knülle (1962)
Amblyomma cajennense		
Larvae	80–85	Knülle (1965)
Adults	91	Lees (1946a)
A. maculatum	88–90	Lees (1946a)
A. variegatum	80–85	Rudolph and Knülle (1974)
Boophilus microplus	>70 <95	Hitchcock (1955)
Dermacentor andersoni		
Larvae	80–85	Knülle (1965)
Adults	85	Lees (1946a)
D. reticulatus	86–88	Knülle (1965)
D. variabilis	80–85	Knülle (1965), McEnroe (1972)
Dermatophagoides pteronyssinus	73	Arlian (1975)
Hyalomma anatolicum excavatum	—	Rudolph and Knülle (1974)
H. schulzei	—	Rudolph and Knülle (1974)
Ixodes canisuga	94	Lees (1946a)
I. hexagonus	94	Lees (1946a)
I. ricinus	92	Lees (1946a)
Echinolaelaps echidnina	90	Wharton and Kanungo (1962)
Ornithodorus moubata	85.5	Lees (1946a)
Rhipicephalus appendiculatus	>70	Londt and Whitehead (1972)
R. bursa	—	Rudolph and Knülle (1974)
R. sanguineus	84–90	Lees (1946a)
Insecta		
Thysanura		
Ctenolepisma longicaudata	≃70	Lindsay (1940)
C. terebrans	47.5	Edney (1971a)
Thermobia domestica	45	Beament *et al.* (1964)
Mallophaga		
Goniodes colchici	60	Williams (1971)
Psocoptera		
Liposcellis bostrychophilus	60	Knülle and Spadafora (1969)
L. knulleri	≃70	Knülle and Spadafora (1969)
L. rufus	58	Knülle and Spadafora (1969)
Siphonaptera		
Ceratophyllus gallinae	82	Humphries (1967)
Xenopsylla brasiliensis (Pre-pupae)	50	Edney (1947)
X. cheopis (Larvae)	65	Knülle (1967)
Coleoptera		
Tenebrio molitor (Larvae)	88	Mellanby (1932c)
Orthoptera		
Arenivaga investigata	82.5	Edney (1966)
Chortophaga viridifasciata	92	Ludwig (1937)

[a] After Berridge (1970) with additions.
[b] Water activity in the vapour phase (a_v) is equal to R.H./100.

is in beetles and orthopterans. It does occur in some imagines, but if so, they are wingless, as are psocids, mallophagans, thysanurans, and the neotenic adult females of the desert cockroach *Arenivaga*. However, by no means all wingless insects possess this faculty; caterpillars, for example, so far as they have been studied, do not. The restriction to wingless forms is probably more than a mere coincidence; it may be an indirect reflection of the adaptive significance of the process whereby insects are able to live in environments where free water either is not available or is in short supply. This is true for the following groups: stored-products arthropods (whose original habitat was often birds' nests) such as *Tenebrio* larvae; mites, such as *Acarus* and *Laelaps*; flea larvae and prepupae; firebrats; desert thysanurans; and sand cockroaches. Adult fleas are indeed wingless, but they have no difficulty in obtaining water from the blood of their hosts, while flea pupae may have to wait a long time in dry places before the adults which emerge find a host. Most adult fleas do not take up water vapour; the hen flea *Ceratophyllus gallinae* which apparently does during the first day of adult life (Humphries, 1967) is an exception. Ticks obtain water from the blood of their hosts, and if the tick spends most of its life on the host, the need for water vapour uptake is less immediate. However, many ticks drop off and have to find a new host after each moult.

The adaptive significance of uptake may be seen in the fact that the lowest humidity that permits survival in culture is the same as that which permits uptake both in the mallophagan *Goniodes colchici* (Williams, 1970, 1971) and in the psocid *Liposcelis rufus* (Knülle and Spadafora, 1969). The limit for these species is about 60% R.H. ($0.6\,a_v$).

Arenivaga nymphs and adult females (which are larviform) live in sand dunes. The adult males become winged and can presumably find water more easily; in any event, they lose the capacity to absorb. The nymphs and females feed on dead leaves of desert shrubs. They become active at night just below the surface, presumably searching for food (Edney *et al.*, 1974) (Fig. 91). Even at night the surface is hot—35 °C where the nymphs are feeding—and in the summer, very dry—below 10% R.H. But the insects have only to move downwards, usually 10 cm, at the most 45 cm, to find an area where the R.H. is above 80% (a_v 0.8), and where they can replenish lost water (Fig. 92).

In *Ixodes* too, the faculty is closely linked with behaviour (Lees, 1948a). These sheep ticks crawl to the top of tall grasses and "quest" for hosts, and in so doing they may be exposed to dehydration. But after losing a certain amount of water their behavior changes and they move downwards, returning to an area of high humidity at the base of the grass, where they can absorb water vapour to make up for that lost.

There are, of course, many wingless insects, termites for example (Edney, unpubl.) and adult tenebrionid beetles with fused elytra, that live in dry areas and do not possess the faculty. So the reason for the limitation of the faculty to certain species within the phylum Arthropoda, and particularly within the Insecta, is by no means clear. In general, however, where it does occur, the adaptive significance of uptake is abundantly clear, and we need not labour the point further.

198

Fig. 91. (a) Adult males (winged) and females of the desert cockroach, *Arenivaga investigata*, from sand dunes in the Colorado Desert near Palm Springs, California. (The coin is 2.4 cm in diameter.) (b) By night, adult females and nymphs move about just below the surface, leaving ridges in the sand *(top and upper right)*. Occasionally they emerge and leave tracks shown in the center of the photograph. The smaller sinusoidal ridge is made by the larva of a therevid fly. (From Edney, 1974)

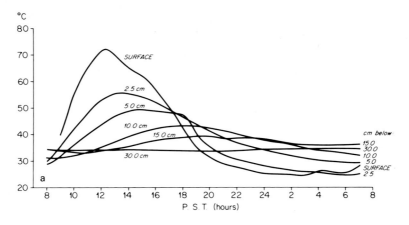

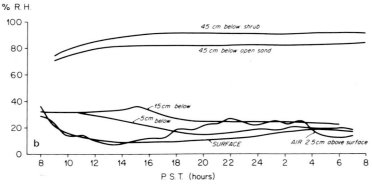

Fig. 92. (a) Temperature. (b) Humidity. Microclimatic data for a hot day in summer in the habitat of *Arenivaga*. At 12.00 h the surface temperature was 72 °C but tolerable conditions occurred at −15 cm or lower. Surface R.H. was below 10%, but at −45 cm, R.H. was 82%, the critical equilibrium humidity for water vapour absorption. (From Edney *et al.*, 1974)

C. Limiting Conditions

It was established early on that for any one species, uptake would occur only if (1) the individual was to some degree dehydrated (there is an important qualification to this to be considered below), and (2) if the relative humidity is above a critical equilibrium humidity (C.E.H.—a term coined by Knülle and Wharton, 1964).

Satisfaction of the second criterion can be seen clearly in flea prepupae (Edney, 1947). Figure 93 shows that these insects gain about 14% of their net weight during the first 3–4 days after ceasing to feed and evacuating the gut. The amount of water gained is not determined by R.H. provided the latter is at 50% or above. After the pupal moult (which usually occurs inside the cocoon), the insects begin to lose weight, and they do so faster in lower humidities,

200

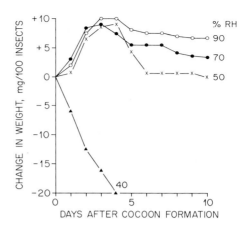

Fig. 93. Change in weight caused by uptake (or loss) of water by prepupae of the flea *Xenopsylla brasiliensis*. The prepupal period lasts for 3 days after cocoon formation. (Redrawn from Edney, 1947)

so that the adults which finally emerge start life with water contents determined by the R.H. during the pupal period. That the changes in weight are indeed due to changes in water content was demonstrated by measuring the water contents of aliquot batches of fleas before, during, and after exposure.

When this experiment was done at different temperatures and over a narrower range of humidites, about half the fleas avoided dehydration at 45% R.H.; nearly all fleas died of dehydration at 40% and nearly all survived and gained water at 50%, irrespective of temperature (12°, 24°, or 35°C) in each case (Table 34).

Table 34. The percentage of prepupae of the flea *Xenopsylla brasiliensis* that avoid desiccation at various combinations of temperature and humidity. The "critical equilibrium humidity" (45–50% R.H.) is unaffected by temperature. (Data from Edney, 1947)

Temperature °C	Relative humidity percent				
	35	40	45	50	55
12	0	0	45	90	95
24	0	0	50	95	100
35	0	0	20	95	95

To clinch the matter of the lower humidity limit, 50 prepupae were enclosed in about 90 ml of air at 70% R.H. in an airtight container with a paper hygrometer. The humidity was quickly reduced to 50% by the prepupae, and held there until after the third day, when the insects pupated and lost the power to absorb water. This result is shown in Figure 94.

In a somewhat similar experiment, Kalmus (1936) put *Tenebrio* larvae in a small enclosed space and found that they generated a R.H. of 90% until they died; and, in a rather different but relevant experiment, Solomon (1966) found that *Acarus siro* maintain R.H.'s of 90%, 87%, or 76% when kept in small closed vials. These values correspond with the humidities at which they have been previously cultured.

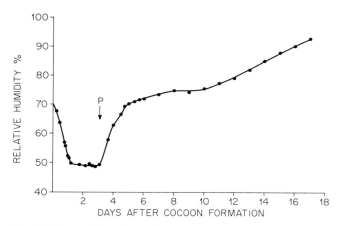

Fig. 94. The variation in relative humidity in a closed vessel containing about 90 ml air and 50 *Xenopsylla* prepupae. The fleas take up water vapour and reduce the humidity to 50%. At *P* the prepupae moulted to become pupae and began to lose water. (Redrawn from Edney, 1947)

In flea prepupae, as we have seen, and in several other arthropods, e.g., the psocid *Liposcelis* (Knülle and Spadafora, 1969), and *Thermobia domestica* (Noble-Nesbitt, 1969), the final water content is uninfluenced by humidity during rehydration. This is not always so, however, as Solomon (1966) found for the flour mite *Acarus siro*, which has body water contents of 70% at 95% R.H. and 66% at 75% R.H., and as Knülle (1967) found for *Xenopsylla cheopis* larvae, where the equilibrium body water content increases or decreases as the larvae are transferred to higher or lower humidities, provided these are above the C.E.H. Mc Enroe (1971, 1972) also found that at very high humidities (a$_v$ 0.98) the equilibrium weight of *Dermacentor variabilis* is affected quite strongly by temperature, the highest weight being achieved at 20 °C, the temperature at which the rate of uptake is greatest. These are important exceptions to the general statement made above that all arthropods which exhibit water vapour uptake gain water at all humidities above the C.E.H. Knülle also found that as soon as the prepupae of fleas are formed they take up more water, irrespective of the humidity to which they have previously equilibrated. This would account for the fact that Edney (1947) found that prepupae absorb much water even though the larvae from which they develop have not been dehydrated.

An example of the combined effects of temperature and R.H. on vapour uptake is shown by the desert cockroach *Arenivaga investigata* (Edney, 1966). Figure 95 shows that after dehydration to about 80% of their original weight, both sexes of nymphs and adult females (but not males) regain water at 95% R.H. *Periplaneta* or *Blatta* nymphs, however, do not. Figure 96 shows further that nymphs take up water at 82.5% or above regardless of temperature, although

202

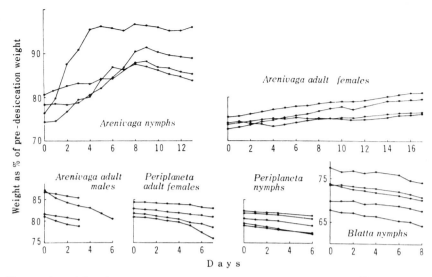

Fig. 95. Changes in weight of *Arenivaga investigata* and other cockroaches in 95% R.H. after previous dehydration to between 75% and 85% of their original wet weight. *Arenivaga* nymphs and adult females gained weight by water uptake, the remainder lost weight. (From Edney, 1966)

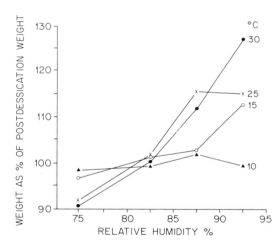

Fig. 96. The effect of temperature and humidity on direction and rate of water exchange in *Arenivaga investigata* nymphs. The critical equilibrium humidity is about 82.5% R.H. irrespective of temperature and, therefore, of vapour pressure deficit. (From Edney, 1966)

the rate is faster at higher temperatures. Moulting inhibits uptake, and so does mild abrasion of the cuticle, although the damage is repaired in a day or so.

In *Arenivaga* the gain in weight at high humidities is indeed due to water uptake because the amount of the body water increases by about 15.7% of original weight, while metabolically produced water during the same time would have been only about 3%. The effects of vapour uptake on the osmotic concentration of the blood is to reduce this from 452 to 373 mosmol liter^{-1} which is less

203

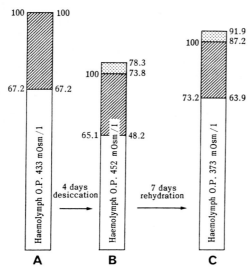

Fig. 97. Analysis of the water content (blank areas) and blood osmotic pressure in *Arenivaga* nymphs. *A*: before dehydration; *B*: after dehydration; *C*: after rehydration. *Numbers to right of each column*: arbitrary weight units, *to the left*: percentages of total wet weight at the times concerned. *Stippled areas*: faecal material. During rehydration there is a net movement of water from the air into the insects. (From Edney, 1966)

than the expected reduction in the absence of regulation (Fig. 97). Further work (Edney, 1968 b) showed that regulation during dehydration and rehydration occurs as the result of removal from or addition to the blood of osmotically active substances. (Chap. 6.C.V.2)

Finally, critical equilibrium humidities present a very wide range—from 94% R.H. in some *Ixodes* species, to 47.5% in *Ctenolepisma terebrans* (Edney, 1971) and even 45% R.H. in *Thermobia*. The thysanurans seem to be particularly good at absorbing from low humidities, but there is no obvious correlation between C.E.H. and phyletic position.

D. The Energetics of Absorption

The uptake of water vapour sometimes takes place against very steep gradients, and this raises the question of the amount of energy involved. The process is certainly an active one, since it ceases at death (Kalmus, 1936; Edney, 1947), is inhibited by anaesthesia (Lees 1946 a; Browning, 1954) or by poisoning (Lees, 1946 a; Belozerov and Seravin, 1960).

One of the earliest estimates was that of Lees (1948 b) who found that the actual amount of energy required is small compared with the energy resources of the ticks concerned (about 1 part in 3,800 of stored fat per day), and Ramsay (1964) reached a similar conclusion for the larval meal worm, *Tenebrio molitor*.

Kanungo (1965) compared oxygen uptake by mites which were or were not absorbing water and found no significant difference. Calculation also showed that such a difference would be very small, for the observed oxygen uptake was $0.17 \mu l$ $O_2 h^{-1}$ by a standard female mite *(Laelaps echidnina)*, while the oxygen necessary to supply the energy needed to concentrate water from a_v 0.90 to a_w 0.99 is only $0.0006 \mu l$ per mite h^{-1}.

For *Arenivaga*, calculations based on the well-known relationship $G = -RT \ln P_2/P_1$ show that a 100-mg insect, absorbing 6 mg water day^{-1} expends about $0.036 cal day^{-1}$ in doing so, while its basal metabolic rate is $2.6 cal day^{-1}$. The extra energy necessary for uptake is, therefore, relatively very small.

Maddrell (1971) considered the energetics of uptake of water vapour by *Thermobia*, and calculated that to move $5 \mu g min^{-1}$ of water from a_v 0.5 (in the air) to a_w 0.99 (in the blood) takes $0.006 cal h^{-1}$. Even at 10% efficiency this represents only $18 \mu g h^{-1}$ of glucose, less than 0.07% of the body weight. For comparison, a flying insect uses up to 20% of its body weight h^{-1} (Weis-Fogh, 1967).

Thus, although the gradient is often very steep, the energy required to concentrate the quantities involved is comparatively small. However, this is not to say that the problem is solved, for although the energy is available to the whole insect, releasing it in the very small mass of living tissue concerned in the energy consuming process of absorption raises difficult thermodynamic problems, as Brodsky *et al.* (1955) pointed out in connection with the standing gradient hypothesis for solute coupled water transport. The mechanism for water vapour uptake is still an open question.

E. The Site of Absorption

In order to discover the mechanism involved we must first be clear about the site or sites where uptake occurs. Mellanby (1932c) was inclined to implicate the tracheae, but most subsequent workers have favored the whole cuticle or elsewhere (see below), as the site of uptake. The reasons for this are compelling in the case of those mites and ticks where a tracheal system is absent, but such reasoning need not be extrapolated to arthropods which possess a tracheal system. Lees (1946a) observed absorption in *Ixodes* (a tick which does have tracheae) even when the spiracles were blocked, and he also observed that damage by abrasion to one part of the cuticle puts the whole cuticle out of action so far as uptake is concerned. In *Arenivaga*, successful blocking of the spiracles leads to death, so that experiment is inconclusive; but abrasion, which of course leads to a great increase in water loss and might therefore mask absorption, does not damage the cuticle as a whole because covering the abrasion with nail polish immediately restores net absorption (Edney, 1966).

In fact it is difficult to see how tracheoles, which usually contain some fluid probably nearly in equilibrium with the blood (a_w 0.99), could absorb water vapour from air down to a_v 0.5. More recently Noble-Nesbitt (1970b, 1973, 1975) has found that uptake by *Thermobia domestica* (originally named

Lepismodes inquilinus), which may occur from air at a_v 0.45, is prevented by blocking the anus. If one of these insects is supported in such a way that both the anterior and the posterior ends may be enclosed in separate cells at known humidities, uptake occurs only when the posterior end is in high humidity (Fig. 98), and the effect is reversible. The rectum in this insect consists

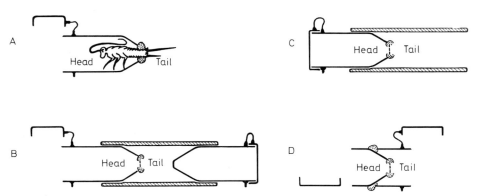

Fig. 98 A–D. A double chamber array for investigating water vapour uptake in *Thermobia*. (A) Basic component, formed from a capped polythene capsule. (B) Complete array, showing head-end open. (C) Complete array, showing tail-end open. (D) Modification for use on electrobalance. Cross-hatching indicates polythene or perspex tubing, and stippling indicates wax used to seal the insect in place and to cement the parts of the array together. The position of the insect is indicated in (A). (From Noble-Nesbitt, 1975)

of two parts, the posterior part forming an anal chamber or anal sac, that opens through three movable anal valves to the outside. When an insect is fully hydrated the anal valves are closed, but if a dehydrated insect is exposed to humidities above a_v 0.45, the valves open and close rhythmically every 1 to 2 s, permitting tidal ventilation of the anal chamber. Furthermore, given the volume of the chamber, and the amount of water vapour contained in air of known humidity, calculation suggests that the maximum rate of water absorption would demand ventilation at the observed frequency. Such ventilation prevents the air in the chamber from equilibrating in a_v with the absorbing mechanism (whatever this may be), and consequently aids in the diffusion of water vapour between the air and the rectal epithelium. The fine structure of the latter is strongly indicative of some form of activated transport (Noirot and Noirot-Timothée, 1971), and we shall return to this later (Chap. 9.G.).

Okasha (1971, 1973) raised an objection to the above proposal on the grounds that uptake continues until the insects' original volume is reached (at which time the water content percent may of course be much higher than it was originally). He suggested that blocking the anus of *Thermobia* may in some way interfere with a sensory pathway that switches on vapour uptake. However, there is no confirmatory evidence to support this alternative, and the evidence in favour of Noble-Nesbitt's mechanism is strong.

The case for rectal absorption is additionally strengthened by an observation of Slifer and Sekhon (1970) made in a quite different context. These authors noted that when *Ctenolopisma lineata pilifera* (another thysanuran) is kept in a culture where damp filter paper is available, the insects when resting always have their posterior ends on the paper, even though the remainder of the body may be away from it.

Tenebrio, as we saw above (Chap. 6.C.VII.7.), can reduce its faecal water content to a level in equilibrium with a_v 0.9 by absorption of water vapour from the posterior part of the rectum, and Noble-Nesbitt (1973) found in corroboration of this, that anal blockage prevents uptake in this insect, the effect being reversible if the block is removed within a few days (otherwise inhibition of uptake may be permanent). Dunbar and Winston (1975), after ligation experiments, also reached the conclusion that uptake occurs in the rectum. They found, as did Machin (1975), that uptake in high humidities may continue until the insects' water content is far higher than the original (pre-desiccation) level. Dunbar and Winston suggest that reabsorption of liquid water from the rectal contents is the biologically important process—uptake of water vapour (in *Tenebrio*) being a necessary, but adaptively unimportant, consequence. Machin views the process as being adaptive in itself, pointing out that fasting larvae that have absorbed grow faster than those that have not when both are subsequently fed, and that the rate of uptake (which increases at higher humidities) leads to a more rapid gain than net uptake from the food at 92% and above. It is strange that these larvae do not avoid abnormally high water contents as a result of absorption. In other arthropods the process ceases at an appropriate level. Perhaps regulation is achieved in nature by moving away from relative humidities of 88% or above, and this could be checked experimentally.

Further measurements by Machin (1976) strongly suggest that the rectum is the site of uptake in *Tenebrio* larvae, and that a fluid compartment, where the solute concentration may be controlled by the absorptive mechanism, exists between the rectal lumen and the blood. Water exchange rates between this compartment and the atmosphere during periods of uptake are much greater than those through the cuticle at other times. The rectum is probably closed off from the air by anal valves when ambient humidity is less than a_v 0.9 (the equilibrium humidity), and this may isolate the proposed fluid compartment from potentially desiccating conditions.

Thus the rectum seems to be the site for water vapour uptake in some insects at least. There are recent reports, however, that this is not so in ticks. Mc Enroe (1972) found that blocking the anus in *Dermacentor variabilis* has no effect on water vapour uptake, and he is inclined to believe that the cuticle is implicated. However, Rudolph and Knülle (1974) and McMullen *et. al.* (1976) found that blocking the mouthparts of the tick *Amblyomma variegatus* prevents uptake, while blocking the anus, or the whole dorsal surface, does not. Furthermore, ticks with mouthparts free but projecting through a sealed film into a small compartment containing a 1 μl drop of KNO_3 solution (which produces a_v 0.93), reduce the volume of the drop by absorption of vapour through the mouthparts. These authors also have evidence that secretion of a highly concentrated saliva in equilibrium with a_v 0.75 (as is a saturated solution of sodium chloride) may be involved in this process. It is pertinent to recall here that the salivary glands of many ixodid ticks are known to be involved in water balance (Chap. 6.C.III.). Another case of oral absorption has recently been postulated by O'Donnell (1977)— this time for *Arenivaga*. When these insects are absorbing, but not otherwise, two bladder-like vesicles, diverticula of the hypopharynx, are protracted. Blocking the anus or the dorsal surface of the insects does not entirely inhibit absorption, while blocking the mouthparts does. Further development of these interesting discoveries will be awaited with interest.

F. The Rate of Absorption

Another area in which information is needed before we can understand the mechanism is the rate at which uptake occurs, and the relationship of this to water loss by transpiration. It seems to be generally true that uptake is faster than loss. For example, Knülle and Spadafora (1969) found that in *Liposcelis* the ratio (rate in to rate out) is 38:1. In *Arenivaga* the ratio is up to 20:1. Noble-Nesbitt (1969) and Locke (1964) found similar ratios for *Thermobia* and *Tenebrio* larvae, respectively, but of course the integument is probably not involved in these species.

If we assume that in some cases uptake does occur across the integument, then the net effect that is usually measured is the algebraic sum of at least three contemporaneous processes: passive efflux, passive influx and activated influx; and it is important that the pathways and rates of each of these processes be identified and measured. The recent work of Wharton and his associates in this field has been innovative and useful (Wharton and Devine, 1968; Knülle and Spadafora, 1969; Knülle and Devine, 1972; Wharton and Arlian, 1972; Devine and Wharton, 1973; Arlian and Wharton, 1974).

By using tritiated water, Wharton and Devine (1968) showed that in the spiny rat mite *Laelaps*, under conditions above the C.E.H., where no change in weight occurs, (1) the time taken to turn over half the total water content is 18.6 h, (2) transpiration and sorption both occur at the same rate, according to first-order kinetics, with a coefficient $k_T = k_S = 0.0373\,h^{-1}$ (i.e., transpiration and sorption both proceed at a rate of 3.73% of the total water mass per hour). For mites weighing 100 μg, this works out at a turnover rate of 7.46 μg h^{-1}, and the conclusion is clear that measurements of net water movement lead to serious underestimation of cuticle permeability.

It has for long been a debatable question as to whether the mechanism involved in uptake at higher humidities ceases to function below the C.E.H. (which of course varies from one species to another), or whether it continues to function, and, although overshadowed by the much greater rate of transpiration at lower humidities, nevertheless restricts net water loss to a lower rate than would otherwise be the case. Lees (1947), Edney (1957), Winston and Nelson (1965), and Knülle (1967) have taken the view that restriction of water loss

208

by living insects (as compared with loss from dead ones) below the C.E.H. may indeed reflect the continued action of the uptake mechanism.

Using tritiated water, Knülle and Devine (1972) measured the net water movement and its components in the tick *Dermacentor variabilis* over a wide range of humidities, and reached the conclusion that transpiration occurs at about $1\% \, h^{-1}$ throughout the humidity range from 0% to 92.5%, while uptake occurs at rates which are nearly proportional to ambient a_v. Thus they argue that the reason why net loss increases with decreasing humidity is not that the rate of movement of water molecules outwards increases, but rather that the rate of diffusion of water molecules inwards decreases.

By comparing rates of uptake in carbon dioxide-anaesthetized and control ticks, they obtained data suggesting that inward movement occurs by passive diffusion over the whole humidity range, but is reinforced by an active uptake mechanism (inactivated by carbon dioxide) at or above 85% R.H. (a_v 0.85). The combined rate is then $0.200 \, \mu g \, h^{-1} \, larva^{-1}$, of which $0.156 \, \mu g$ is passive absorption and $0.044 \, \mu g$ is active.

The answer to the initial question is that in these animals the uptake mechanism is inactive below the C.E.H.

Meanwhile Noble-Nesbitt (1969) obtained confirmatory evidence, by different methods, with *Thermobia*. He found that net water loss from this insect at humidities below its C.E.H. is proportional to $1-a_v$, which is a function of vapour pressure deficit, and not to $0.5-a_v$ as it should be if the active uptake mechanism reduces a_w at the insects' surface to equilibrium with 50% R.H., i.e., to a_w 0.5. Noble-Nesbitt's data are compatible with the ideas of water movement proposed by Knülle and Devine, but of course they are also compatible with rectal uptake if this process is inactivated by $a_v < 0.5$. It would be very interesting to measure the components of water movement through the cuticle in both directions by means of a radioactive tracer.

Recently, Devine and Wharton (1973) obtained further data about water exchange in *Laelaps*, this time in conditions other than those ensuring zero net weight change. Their results suggest that, as in *Dermacentor* (Knülle and Devine, 1972), transpiration in *Laelaps* is independent of ambient a_v while uptake is affected by a_v and takes place through the cuticle. No distinction was looked for between passive inward diffusion and active uptake.

The physiological significance of a decrease in transpiration rate with time, which they also observed, is not clear. Devine and Wharton (1973) propose that it is related to a progressive decrease in water mass in the arthropod. Decreases in rates of water loss with time during anything but short periods of desiccation have often been observed (Edney, 1951 b; Bursell, 1955; Loveridge, 1968 a; among others). Such decreases have been ascribed to changes in the permeability of one or another layer of the cuticle, or of the apical membrane of the epidermal cells, or to exhaustion of hygroscopic water in the cuticle and failure to replace it owing to a constant high impermeability of the apical membrane (Berridge, 1970). Any of these mechanisms could play a part in determining the effects observed, and further work is necessary to identify and measure the factors involved.

G. Possible Mechanisms

Several mechanisms to explain uptake through the integument have been proposed. Locke (1964) notes that the secretion of lytic enzymes by the epidermal cells into the endocuticle occurs during preparation for moulting, and when insects are starved. These enzymes depolymerize glycoproteins of the endocuticle, converting them to smaller molecules such as amino acids and glucosamine units. Such smaller molecules, Locke proposes, generate local osmotic concentrations high enough to cause the absorption of water molecules from outside the insect. These concentrated solutions could then be taken into the cells by pinocytosis (a process which normally forms part of the preparation for a moult) and subsequent repolymerization of the solutes would release water. A highly impermeable apical membrane would prevent water from moving out of the cell in the wrong direction.

One difficulty is that release and solution of molecules the size of even small amino acids would probably generate no more than a 1 or 2 osmolar solution, and yet insects are known to take up water from air at a_v 0.5, against a 55.5 osmolar gradient.

Beament's hypothesis uses the fact that protein molecules, if moved away from their isoelectric point, expose hydratable polar groups which have a high attraction for water molecules. If cyclic changes in the pH of the cuticle could be effected, water would be absorbed onto the protein chains during one phase of the cycle and released at another. This, combined with a valve permitting the movement of water inward but not out, would ensure that water was released to cells and not to the outside.

Such valves may exist in the fine capillaries observed by Locke in electron micrographs of the epicuticle of the caterpillar *Calpodes*. These so called wax canals are much narrower than pore canals, down to about 30 Å in *Calpodes*, and about 80 Å in *Thermobia* (Noble-Nesbitt, 1967). They are the equivalent, perhaps, of the epicuticular canals observed by Wigglesworth (1975) in *Rhodnius* (Chap. 3.B.VI.). During one phase of the cycle, water would be withdrawn by suction from such a capillary, a meniscus would form, at the surface of which there would be a low a_w. Calculations by Beament (1964) and Noble-Nesbitt (1969) based on extrapolations from the properties of larger capillaries, and therefore not necessarily applicable, suggest that water at the meniscus of a 30 Å capillary may have a_w 0.5, and could thus cause condensation of water molecules from air with $a_v > 0.5$. During the other phase of the cycle, when suction is released, water in the capillary would rise, flatten at the surface, and become covered by a monolayer of lipid molecules, thus preventing rapid evaporation. As the suction increased again, water would be drawn down the capillary and the lipid monolayer would be broken, so that water molecules could move rapidly into the capillary and condense on the meniscus surface (Fig. 23).

This model has certain advantages. It accounts for a lower impedance to flow inward than outward, and could also explain why the limiting a_v for uptake is species specific, if indeed the size of the microcapillaries differs between species in an appropriate manner. Noble-Nesbitt suggests that the model might also explain why transpiration increases in dead insects, on the grounds that withdrawal of water from the capillaries as the cuticle dries out would lead to rupture of the monolayer over the capillary opening. However, this remains to be demonstrated.

The weaknesses of the model are that it is not clear either how a cyclic change in pH of the cuticle could be brought about, or whether the surface lipids behave in the manner required. It is also rather inconvenient for the hypothesis (though by no means fatal) that the fine capillaries necessary for uptake from a_v 0.5 (30 Å) exist in *Calpodes* (which has never been shown to absorb water vapour), while in *Thermobia* (which does absorb from a_v 0.5) the capillaries are 80 Å in diameter.

Turning now to rectal uptake: two different systems so far are known, one in *Tenebrio molitor* larvae, the other in *Thermobia* (Chap. 9.E.), and in both, clearly observed structural features go some way to explaining the mechanisms involved.

The rectal complex of *Tenebrio* has already been described (Chap. 6.C.VII.7.) The epidermal cells are quite unspectacular (Grimstone *et al.*, 1968), apart from a small amount of apical folding and a somewhat more pronounced basal infolding. Solute coupled transport, based on the secretion and recycling of common ions within the rectal epidermal cells does not seem to be a likely mechanism for moving water from a_v 0.85 from the rectum because the appropriate fine structure is absent, and in any case this mechanism would involve solutions of very high osmotic pressure (equivalent to saturated potassium chloride, for example). A more likely material is the protein of M.W. 10,000–12,000 that occurs in the posterior perirectal space (Grimstone *et al.*, 1968). But even if this is directly involved in water vapour uptake, there are still some obstinate problems to be solved, particularly those concerning the release of water from the material, perhaps continuously, into the appropriate compartment. Machin's (1976) work on *Tenebrio* is very relevant and is referred to above (Chap. 6.E.).

In *Thermobia*, on the other hand, the fine structure of the epidermal cells lining the rectal sacs is very elaborate (Noirot and Noirot-Timothée, 1971) (Figs. 99 and 100). The apical membrane of these cells is very deeply and regularly infolded, the spaces between the infolds being occupied by long, rod-like mitochondria in hexagonal array. A thin (15 nm) coating of particles about 150 Å long covers the cytoplasmic surface of the infolded plasma membrane and is also quite close to the mitochondria. (Such particles are also present in Malpighian tubes and salivary glands.) The apical infolds are in communication with a distinct sub-cuticular space (reminiscent of the situation in *Periplaneta*). The basal membrane is infolded to a much smaller extent, and separated from the blood by a thin basement membrane. Intercellular tight junctions are present apically between adjacent lateral membranes, but the intercellular spaces (which are feebly enlarged here and there to form small sinuses) are open basally.

It seems highly probable that this kind of organization is associated with transport in some form. The long narrow channels could be suitable for the establishment of standing gradients, energized perhaps by the closely associated mitochondria; but it is by no means clear how this would work, particularly

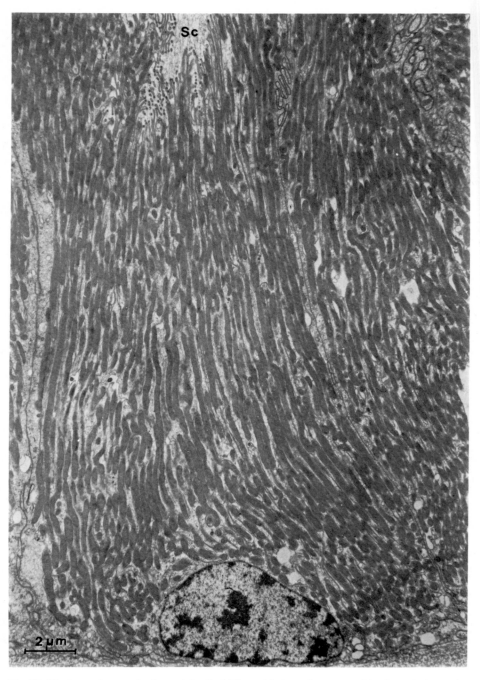

Fig. 99. Electron micrograph of a rectal cell of *Thermobia* in a plane normal to the apical margin. There is a large basal nucleus, but much of the volume of the cell is occupied by extensive apical infoldings (intercellular canals) continuous with the sub-cuticular space *(Sc)*. Numerous elongated mitochondria are associated with the canals. × 6,600. (From Noirot and Noirot-Timothée, 1971)

212

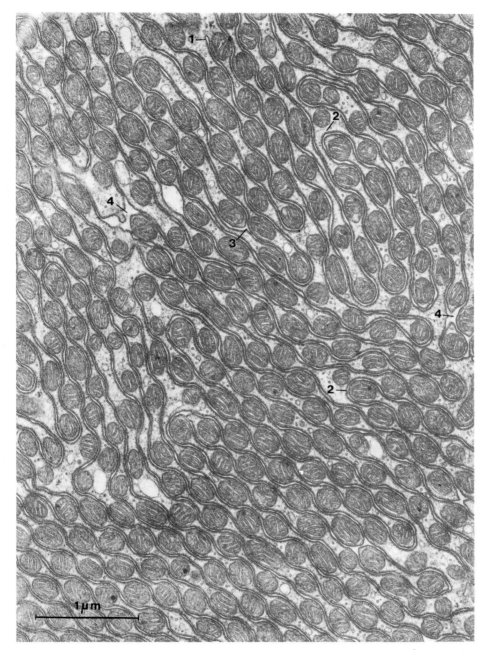

Fig. 100. Electron micrograph of a rectal cell of *Thermobia* parallel to the basal surface, showing the regular arrangement of cytoplasmic processes bounded by intercellular canals. 1: blind end of a canal; 2: a fold; 3: a bifurcation; 4: anastomosis of canals. The long rod-like mitochondria in hexagonal array are cut transversely. × 26,000. (From Noirot and Noirot-Timothée, 1971)

because the maximum a_v generated by a saturated solution of sodium chloride is about 0.75, and water vapour at a_v 0.45 has to be moved. A material such as glycerol, which is mutually soluble with water, might do better. Or perhaps the lining layer of particulate material may be involved. But further speculation is not appropriate at present.

H. Conclusions

It would be gratifying to be able to synthesize the available information into an acceptable explanation of water vapour uptake, but such cannot as yet be done, at least by the present writer. The process has been known for 40 years, but the mechanism is still virtually a mystery. The crux of the problem is how to generate a water activity as low as 0.5 somewhere in the cuticle or elsewhere to bring about the movement of water inwards from $a_v < 0.5$ outside, and then to release it for general use at a_w 0.99. The existence of fine capillaries by themselves does not solve the problem; there must be suction forces to remove water from them to generate appropriately low water activities within them.

There is no reason why all the known cases of water vapour uptake should involve the same mechanism. Indeed the evidence now suggests that several different mechanisms may be involved. Not only are the structural features believed to be associated with the processes quite various (there are two kinds of rectal systems, one or more oral and possibly several integumental ones), but large physiological differences, sometimes associated with structural differences, also exist, as exemplified by uptake from a_v 0.45 in *Thermobia* but only from a_v 0.9 in several ticks.

It is no more than trite to observe that we need more information. But particular attention should be paid to verifying sites of uptake and to describing the underlying structures. No structure, either gross or fine, has ever been associated with integumental uptake; and in this connection it would be appropriate to look at the integuments of atracheate mites and also to compare those of insects, particularly closely related ones, that do and do not absorb water vapour. The adult male (which does not absorb) and female (which does) of *Arenivaga*, might form useful material. Likewise, *Xenopsylla* larvae and prepupae (which have different parameters for uptake) might be interesting. But probably the most helpful work will be done by means of electron microprobe techniques. It would be especially interesting to use these to look at ion concentrations across the rectal complex of *Tenebrio* and the rectal sac wall of *Thermobia*.

This is a highly interesting field in which a defined, unsolved problem exists. It should provide excellent opportunities for cooperative work between biologists and biophysicists.

Chapter 10

Water Balance in Eggs

A. Structure

The eggs of most arthropods contain a relatively large amount of yolk covered, at the time of laying, by a thin vitelline membrane (produced by the oocyte itself) and a more or less tough chorion, or shell, which is produced largely by the follicular cells of the mother. Unlike the cuticle of later stages, the chorion contains no chitin; it is more akin to "cuticulin"—a stabilized lipo-protein complex (Chap. 3.B.VI.). The chorion takes different forms in different insects (refs in Wigglesworth, 1972), but in all it has to provide mechanical support and protection, permit the exchange of respiratory gases, and prevent dehydration, flooding and loss of valuable solutes.

In *Rhodnius*, where the structure has been closely studied, the chorion consists of several layers (Fig. 101). In many species the surface is intricately sculptured and provides many small, interconnected, strongly hydrophobic air spaces (the aeropyle) (Fig. 102). Films of air thus tightly held, known as plastrons, function as gaseous gills, permitting the uptake of oxygen even when the egg is covered with water (Thorpe, 1950; Hinton, 1963, 1968, and other papers). According to Hartley (1971) the fine mesh-work of the chorion in *Locusta* may fill with

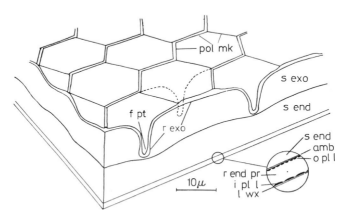

Fig. 101. Diagrammatic representation of a fragment of the completed egg shell of *Rhodnius*. Note the polygonal markings of the shell surface (*pol mk*), each with a follicular pit at its center (*f pt*). The resistant exochorion layer is thickened at the base of each pit and at the surface ridges (*r exo*). *s exo*: soft exochorion layer; *s end*: soft endochorion layer. *Inset*: The detailed structure of the resistant endochorion layers. *amb*: amber layer; *i pl l*: inner polyphenol layer; *o pl l*: outer polyphenol layer; *r end pr*: resistant endochorion protein layer; *l wx*: primary wax layer. (From Beament, 1946)

215

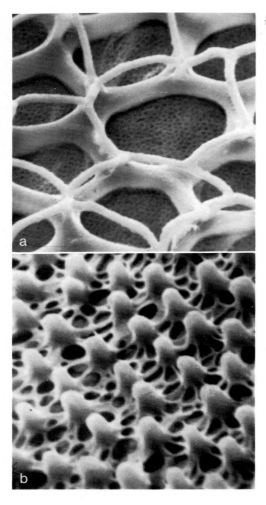

Fig. 102. The surface of the chorion of (a) a fly, *Fannia armata*; (b) a mosquito, *Culex modestus*, showing the intricately sculptured plastrons. Air is held in the meshwork whose surface is strongly hydrophobic. (a) × 6,800, (b) × 7,500. (From unpublished micrographs kindly furnished by H. E. Hinton)

air or with water; in other words its surface is hydrophilic, and chorionic sculpturing of the eggs of various sandfly species may prove to be related to differences in moistness of their various oviposition sites (Ward and Ready, 1975).

Equally important is the prevention of dehydration, and this is achieved in various ways. The eggs may be enclosed in oothecae, as in cockroaches (Roth and Willis, 1955a, b), or in frothy protective capsules of one form or another as in acridids; but more usually protection from dehydration is achieved by waterproofing one or more layers of the chorion, or of the serosal cuticle after this has been laid down. In *Rhodnius* the chorion itself is freely permeable to water, and impermeability is conferred just before the egg is laid, by a layer of lipid material, the "primary wax layer" about 0.5 μm thick probably secreted by the oocyte through the vitelline membrane and closely attached to the inside of the chorion (Beament, 1946). Primary wax layers have also been reported in *Melanoplus* (Slifer, 1948; Salt, 1952), *Lucilia* (Davies, 1948), and *Bombyx* (Takahashi, 1959), and a secondary wax layer (as part of the cuticle) by Beament (1946) for *Rhodnius* and by Slifer (1948) for *Melanoplus differentialis*.

After the zygote nucleus has divided several times, and the nuclei have migrated to the egg surface to form the serosa, or blastoderm, the latter begins to lay down a serosal cuticle below the existing vitelline membrane, and this may contain a secondary wax layer. The serosal cuticle in *Melanoplus* closely resembles an ordinary insect integument with a secondary wax layer near its surface. It consists of a thin "yellow cuticle" which is probably a tanned lipo-protein, overlying a white cuticle, composed largely of chitin (Slifer, 1938; Salt, 1952)

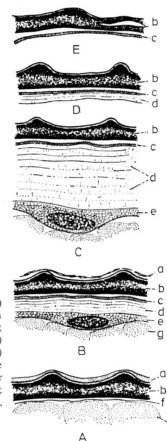

Fig. 103 A–E. Egg membranes in *Melanoplus differentialis*. (A) egg newly laid; (B) egg incubated 11 days at 25°C (the serosa has laid down the yellow cuticle and part of the white cuticle); (C) egg incubated 21 days (serosal cuticle fully formed); (D) egg 3 days before hatching (serosal cuticle much reduced); (E) egg ready to hatch (nothing remains but the chorion and the yellow part of the serosal cuticle); *a*: temporary secretory layer over the chorion; *b*: chorion; *c*: yellow cuticle; *d*: white cuticle; *e*: serosa; *f*: vitelline membrane; *g*: yolk. (From Wigglesworth, 1972; after Slifer)

(Fig. 103). According to Jahn (1936) the yellow cuticle is responsible for impermeability to inorganic ions. The wax layers behave rather like those of the epicuticle, showing in some cases rather sharp transition phenomena. In *Rhodnius* and in *Ornithodorus* the primary wax layer is much more permeable above 43°C and 45°C respectively (Beament, 1946). In ticks a superficial wax layer is secreted

217

over the egg shell by Gene's organ and there is no primary wax layer (Lees and Beament, 1948), while in *Tetranychus* eggs there is both a superficial and a primary layer. (Further references are given by Beament, 1949; Edney, 1957.)

Early in development, some of the serosal cells group together along an anterior-posterior axis to form the germ band, which sinks into the yolk below the serosa and ultimately forms the embryo proper.

Unfortunately, the terminology of the stages of development has been confused. Wheeler (1893) coined the terms "anatrepsis," "diapause" and "katatrepsis" to refer to three stages during the process of blastokinesis. Anatrepsis (literally, turning or twisting up) is now used not so much as a component of blastokinesis, but as a period of time leading up to revolution or blastokinesis. Katatrepsis, likewise, is used to refer to the period after blastokinesis and also to blastokinesis itself; while diapause, which was originally a resting period in the middle of blastokinesis, has been enormously widened in scope to mean almost any temporary cessation of development in any organism. Blastokinesis may itself be a prolonged and complicated process whereby the developing embryo undergoes devious turnings and migrations, usually amounting to a movement up around the yolk to a dorsal position and a subsequent return. Alternatively, very little in the way of morphogenetic gyrations may occur. In any case it is a term that has generally been accepted and used by insect embryologists and we shall use it here to mean the series of movements resulting in the embryo assuming the position in which it is finally hatched. It seems unfortunate to use "trepsis" for a period of time during which processes not forming a part of blastokinesis may occur, but to introduce completely different terms (e.g., antekinesis and postkinesis) at this stage would probably lead to even more confusion. Consequently we shall follow Lees (1976) and use *anatrepsis* to mean development up to blastokinesis and *katatrepsis* to mean development after blastokinesis.

In *Melanoplus differentialis* (Slifer, 1938), *Melanoplus bivittatus* (Salt, 1952), *Locustana* (Matthée, 1951) and in many other acridids, the serosa, when it lays down the serosal cuticle, also produces, at the posterior end of the egg, a small area known as the hydropyle, where the white cuticle is thin and the yellow cuticle thicker, but where both seem to be relatively permeable to water. Figure 104 shows the hydropylar area of a grasshopper *(Eyprepocnemis plorans)* together with the subterminal ring of micropyles for the passage of sperm. As its name implies, the function of this structure is believed to be water uptake in some eggs (Chap. 10.C.). Cricket eggs, however, take up water without having a hydropyle. For a fuller description of morphological changes during acridid egg development, see Shulov and Pener (1963).

B. Water Loss by Evaporation

As a result of the presence of the wax layers, and perhaps of the serosal cuticle, transpiration from arthropod eggs is often very slow (see Table 13), an effect that is of course highly desirable when the eggs are laid in dry places. The permeabilities of some egg membranes (e.g., *Rhodnius*, *Lucilia*) are in fact among the lowest recorded for any animal membrane. However, permeability may vary considerably during the course of development in eggs that absorb water

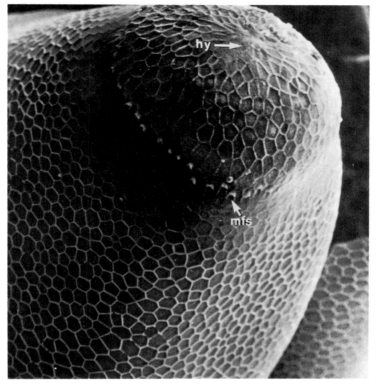

Fig. 104. Posterior pole of an egg of the grasshopper *Eypreprocnemis plorans* to show the hydropylar region, the ring of micropyles and the sculpturation of the chorionic surface. *hy*: hydropyle area; *mp*: micropyles. (From Luca and Viscuso, 1974)

(Chap. 10.E.). Eggs that are laid in damp soil, such as those of *Phyllopertha* and *Tipula* (Laughlin, 1958), *Teleogryllus commodus* (Browning, 1953), or *Ocypus olens* (Lincoln, 1961), lose water much more rapidly (although at a lower rate than that from a free water surface) and seem not to be strongly waterproofed, particularly when young (Browning, 1967).

There is general agreement that incorporation of lipids, either as a discrete layer, or by impregnation of other layers, combined with quinone tanning of the serosal cuticle, can reduce transpiratory water loss to satisfactory levels. The mechanisms and sites of water absorption seem to be more various and controversial, and to these we shall now turn.

C. Water Uptake

Many arthropod eggs can develop in unsaturated air. *Rhodnius* provides a well known example of this, but it is probably true also of many species whose eggs are laid in exposed places. In others, although development may proceed up to a point in dry air, contact with moisture is obligatory at some time. In *Melanoplus bivittatus*, for example, development proceeds until anatrepsis is complete, but further development is inhibited unless water is available.

In many species it seems that water absorption, if it occurs at all, must take place at a particular stage of embryogenesis, although the stage may differ from one insect to another (Browning, 1967). However, this is not always true, as the following examples show. *Chorthippus brunneus* has an obligate diapause just before blastokinesis. Water is absorbed during anatrepsis according to the usual S-shaped pattern, falls off during diapause, but may proceed again at a rapid rate after diapause (Moriarty, 1969a, b). But in this species as well as in several other species of *Chorthippus* and other acridids, Moriarty (1969) found that development will proceed as long as sufficient water is taken in—the developmental stage at which this occurs being unimportant. Thus development can continue without any extra absorption after diapause is broken by chilling, provided that enough water has been taken in before diapause. Furthermore, termination of diapause does not depend on water absorption (as it does in *Melanoplus differentialis*), because *C. brunneus* may absorb water after chilling, but without any further morphogenetic development.

Similarly, in the field *Camnulla pellucida* eggs absorb water as soon as the temperature rises sufficiently after the winter, and absorption is almost complete before development begins. In other conditions, however, these eggs may develop very slowly without water absorption. Such restraint may be broken by chilling, after which water absorption and development both proceed rapidly, and hatching occurs some ten days later (Pickford, 1975).

In the cricket *Teleogryllus*, diapause intervenes in the development of some eggs, but water absorption always occurs before diapause, at the time when the embryo becomes separated from the serosa. If the eggs of this insect are held at the low temperature of 13 °C for a month after being laid, diapause is avoided, and the eggs on return to higher temperatures, develop to completion. In these eggs, the amount of water absorbed is independent of temperature, but the rate of absorption, and hence the time taken to complete it, is affected, as shown in Figure 105 (Browning, 1953), and the same is true of *Chortoicetes* (Wardhaugh, 1973). The relationships between diapause, stage of development, and water uptake in some other orthopterans are shown in Table 35.

In some insect eggs, specialized organs are involved in water absorption. Aleurodid eggs, for example, are laid with a short process in the form of a thin walled bladder inserted into a leaf, and this absorbs water at a rate fast enough to balance evaporation from the rest of the egg (Weber, 1931). In several acridids, including *Melanoplus* and *Locustana*, the hydropyle at the posterior pole (see above) has been thought to absorb water. Slifer (1938) found that if the hydropyle in *Melanoplus differentialis* is sealed, no water is absorbed

220

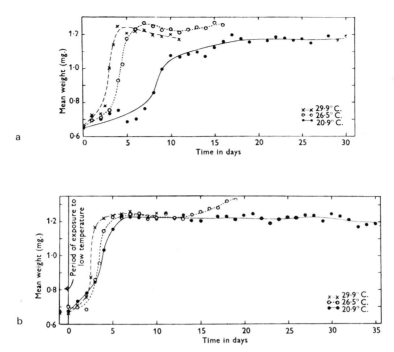

Fig. 105a and b. Changes in weight of eggs of the cricket *Teleogryllus* as a result of the uptake of water as early development proceeds. In (a) the effect of lower temperatures is to delay uptake (and a proportion of the eggs go into diapause). In (b) the eggs have been previously exposed to 13°C for various periods of time, after which uptake is rapid and further diapause is avoided. (From Browning, 1953)

and development is arrested, and Salt (1952) came to a similar conclusion for *M. bivittatus*. In *Melanoplus* Slifer (1946) found that the permeability of the hydropyle (and hence the termination of diapause) is controlled by lipid deposition and removal. The evidence is that treatment with xylol permits development to proceed and water to be absorbed (Slifer, 1946). Andrewartha (1952) suggested that such resumption of development may be a general "wound" response and Hinton (1957) urged the need to distinguish between water required to complete diapause and that required for further development. In later experiments Slifer (1958) showed that treatment of eggs with a relatively harmless mineral oil also permits water absorption and development.

Mathur (1944) observed uptake of water only at the posterior tip of the egg in *Schistocerca*, but does not claim that this is necessarily through a hydropyle. He excluded the micropylar channels, which are sub-terminal, as sites for water uptake (Fig. 104). Shulov and Pener (1963) believe that at least the later stages of water uptake in *Schistocerca* occur through such a hydropyle. Moloo (1971), however, has evidence to the contrary, for he found that when various fractions of the egg surface are covered with water-impermeable material, satisfactory absorption occurs through the uncovered areas, even when the posterior third

221

Table 35. The relationship between stage of development, diapause (if this occurs) and water absorption in various orthopteran arthropods

Species	Period of diapause (if this occurs)	Period of water uptake	Reference
Acheta domesticus		Days 3–4 at 27°C, before katatrepsis	McFarlane and Furneaux (1964), McFarlane (1970)
Camnulla pellucida (in lab)	Late anatrepsis— broken by chilling	Late anatrepsis— after diapause	Pickford (1975)
Camnulla pellucida (in field)		When temp. rises but before development begins	Pickford (1975)
Chorthippus brunneus (and other acridids)	After anatrepsis before blastokinesis	During anatrepsis and again after diapause	Moriarty (1969a, b, 1970)
Chortoicetes terminifera (in lab)	Diapause usually absent. Dev. arrested by lack of water at anatrepsis	During anatrepsis	Wardhaugh (1970), Lees (1976)
Chortoicetes terminifera (in field)	Near end of anatrepsis	Before diapause	Wardhaugh (1973)
Locusta migratoria	After anatrepsis	During anatrepsis (days 5–6 at 30°C)	Browning (1969b)
Locustana pardalina	Begins on 8th day	During anatrepsis (days 3–6)	Matthée (1951)
Melanoplus differentialis	Early katatrepsis, before water absorption	Early katatrepsis but after diapause	Slifer (1938)
Melanoplus bivittatus	Late katatrepsis	During anatrepsis before diapause	Salt (1949, 1952)
Schistocerca gregaria	Dev. stopped at anatrepsis if no water available	During anatrepsis and early katatrepsis	Shulov (1952), Shulov and Pener (1963)
Teleogryllus commodus	During katatrepsis after water absorption	During anatrepsis before diapause	Browning (1965)

(including the hydropyle) has been covered. Similarly Hartley (1961) found that the whole egg surface of *Chorthippus brunneus* takes up water readily (particularly if some water is already present), and this was confirmed by Moriarty (1969a). In *Chorthippus parellelus*, according to Hartley, the whole of the chorionic surface is hydrofuge except for a ring of small micropyles near the posterior pole, through which water readily enters the egg.

Ewer (pers. comm.) has found that in *Schistocerca* water uptake occurs according to the usual S-shaped curve during anatrepsis, but only if the posterior pole (which eventually contains the hydropyle) of the egg is in contact with liquid water. However, he reports further that provided water is taken up through the posterior pole for 24 h, the remainder of the necessary uptake may occur through the whole chorion, and he suggests that initial posterior pole uptake may stimulate changes in egg membrane permeability, and thus permit rapid uptake over the whole surface. The early uptake referred to cannot occur through a hydropyle, for this is not present during the first few days. Loss of water by transpiration is also enhanced at this time. In *Chorthippus brunneus* too, Moriarty (1966a) obtained strong, though indirect evidence that eggs which have started to absorb water become dehydrated (if exposed to dry air) much more rapidly than those that have not. The advantage of such a system is plain: chorionic impermeability is maintained, and thus the risk of dehydration is reduced, until free water is available and the absorption process, whatever this may be, is triggered. If the posterior pole alone is capable of absorption from the beginning while the chorion only becomes so a little later, this might help to explain discrepancies in earlier authors' conclusions.

In most eggs that absorb water, the process is very slow at first, becomes much more rapid, and finally returns to a slow rate, to give the characteristic S-shaped curve (Browning, 1967). Even in cockroaches, where embryogenesis occurs while the eggs are in oothecae, water uptake may follow such a curve. It does so in those species such as *Nauphoeta cinerea* which are viviparous and in which the ootheca is retained within the female's body, and may do so even when the ootheca is only partly retained, as in *Lophoblatta brevis* (Roth, 1968c), where water is absorbed from the female through the permeable, anterior (retained) end of the ootheca. In cockroaches such as *Blatta* and *Periplaneta*, where the oothecae are shed early in the process of embryogenesis, water uptake is restricted to that which exists as a liquid film inside the ootheca (Roth and Willis, 1955b). For further discussion of water balance in cockroach eggs, and of its phylogenetic implications see Roth and Willis (1955a, b, 1958), Roth (1968a, b) and other papers by these authors. Other insects that show the typical S-shaped curve are referred to by Browning (1953), and Moriarty (1970) found this pattern in several species of *Chorthippus* and other acridids.

Water is absorbed only from a liquid surface. Absorption of water vapour by eggs of arthropods has never been demonstrated, even when it has been looked for (Shulov, 1952), although in a few instances, uptake appears to take place against a small osmotic gradient. In *Locustana*, for example, Matthée observed uptake from a 625 mosmol liter^{-1} glucose solution, and Moloo (1971) similarly demonstrated uptake by *Schistocerca* eggs from a 700 mosmol liter^{-1} glucose solution, their own internal osmotic pressure being 450 mosmol liter^{-1}. The fact that the increase in rate of absorption with rising temperature in *Melanoplus* has a Q_{10} of 10 or above argues for an energy consuming process, for if osmosis alone were involved the rate would be proportional to the absolute temperature. In *Teleogryllus*, normally absorbing eggs immediately (but reversibly) cease development and absorption if they are transferred to a nitrogen atmosphere (Browning, 1965). A similar but less marked effect was found in *Schistocerca* by Moloo (1971), who suggests that the absorption that he observed in eggs under nitrogen may have been the result of anaerobiosis. In other cases, absorption occurs only from very dilute solutions, possibly by osmosis. For example, the eggs of *Chorthippus brunneus* absorb water from 0.1 M NaCl solution, but not from 0.4 M or above, and the development of *Teleogryllus* eggs is inhibited when water uptake is due to commence if the eggs are kept in 0.5 M sucrose solution (Browning, 1965). The eggs of *Chortoicetes* continue to absorb water and to develop when kept in nitrogen, and this, as Lees (1976) points out, raises the question of the extent of anaerobiosis in other species, a question that deserves to be studied for its solution might throw light on several perplexing problems.

D. The Control of Water Uptake

When water absorption occurs only at particular times, interest has centered on the mechanisms that initiate and terminate the process, and this has led to a consideration of the properties, particularly as regards permeability and tensile strength, of egg shells. For a useful discussion of the earlier work in this field see Browning (1967) and McFarlane (1970). A good deal of the work has been done with acridids or gryllids, and the assumption is usually made that water uptake occurs through the whole shell rather than through the hydropyle.

In *Teleogryllus commodus*, water uptake commences as soon as the serosa is complete, follows an S-shaped curve during anatrepsis, and ceases at blastokinesis (Browning, 1953) (Fig. 105). Experiments using deuterium (Browning and Forrest, 1960) suggested that permeability of the shell of *Teleogryllus* eggs does not vary during development in a way that would account for the water absorption curve. McFarlane objected on the grounds that the radioisotope used might have lodged in the shell without penetrating the egg; that the results could have been ascribed to hydrogen ion exchange, and on other grounds. Further work (Browning, 1969a) with tritiated water showed that radioactivity was in fact largely in the egg and not in the shell. Experimenting further with *Locusta migratoria* (for the sake of its larger eggs) Browning (1969b) investigated shell permeability by cutting off one or other end of the egg and using the shell as an osmometer. These experiments showed again that at all stages of development the shells are permeable in both directions and that the presence or absence of the hydropyle makes little difference (in this case its function remains unexplained). Parallel experiments with *Teleogryllus* gave similar (though less complete) results. Diapausing eggs of *Locusta* did, however, become much less permeable than developing eggs, and since water absorption ceases at the end of anatrepsis when diapause begins, there is to this extent a correlation between shell permeability and water absorption.

Osmotic changes in *Teleogryllus* eggs cannot be held to explain changes in absorption rate, for most estimates show that the O.P. of orthopteran eggs lies between 360 and 540 mosmol liter^{-1}, and does not vary greatly during development (Matthée, 1951; Laughlin, 1957; McFarlane and Kennard, 1960; Browning, 1965). However, these values do mean that if the shell were a semipermeable membrane the egg would develop an internal hydrostatic pressure of about 8 to 12 ats., or 87.6 to 124 g mm^{-2}.

As a result, Browning (1967) proposes that in *Teleogryllus* at least the shell functions as a semipermeable membrane, influx of water during the first few days of anatrepsis being prevented by an internally developed hydrostatic pressure. As soon as the serosa is complete, however, (about 2 days at 25° to 30°C) the endochorion undergoes a change in structure (presumably as a result of serosal action) and becomes fragmented (Fig. 106), thus permitting the shell to enlarge and water to enter osmotically. An alternative proposal would be that the cracks and shears observed in the endochorion are the result of internal hydrostatic pressure generated by water uptake.

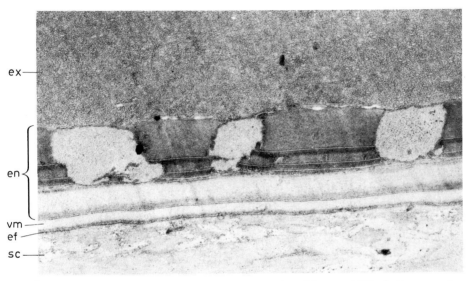

Fig. 106. Electron micrograph (× 80,000) of a transverse section of the egg of *Teleogryllus commodus*, after 9 days development at 27°C. The endochorion has fragmented horizontally and radially, however the vertical gaps do not extend completely through the endochorion. *ex*: exochorion; *en*: endochorion; *ef*: serosal epicuticle; *sc*: serosal cuticle; *vm*: vitelline membrane. (From an unpublished micrograph kindly furnished by T. O. Browning)

If Browning's proposal is correct, the egg shell should at first be able to withstand the necessary pressure without splitting, and indeed he showed that *Teleogryllus* eggs can withstand externally applied pressures equivalent to 137 g mm^{-2}. However, as Browning points out, other orthopteran eggs split at much lower pressures (see below), and so this explanation cannot be universally true.

McFarlane (1970) has further criticized Browning's interpretation of the situation in *Locusta* on the grounds that the experimental manipulations involved may have affected the properties of the egg shell, and that the rate of water movement in and out of the balloons in fact shows them to be rather impermeable. McFarlane attaches more importance to changing properties of the serosal cuticle, and does not accept internal wall pressure as a controlling factor initially preventing uptake. Until the internal hydrostatic pressure of *Teleogryllus* eggs can be measured, this question will probably remain unsettled.

E. Uptake by Eggs of the Locust *Chortoicetes*

More recently Lees (1976) has developed a technique for inserting small capillaries into the eggs of the grasshopper *Chortoicetes*, and thereby measuring the normal internal hydrostatic pressure as well as the breaking pressure of these eggs. In *Chortoicetes terminifera* water uptake follows the usual S-shaped curve.

Newly laid, untanned egg shells resist about $1.5 \, \text{g mm}^{-2}$, while those with tanned cuticle resist $5.0 \, \text{g mm}^{-2}$ ($10 \, \text{g mm}^{-2}$ if they are supported in their normal egg pod foam) and up to $20 \, \text{g mm}^{-2}$ as they become older. But the osmotic concentration of their contents represents hydrostatic pressures of more than $77 \, \text{g mm}^{-2}$, so that wall pressure would be quite incapable of preventing water influx if the shell acted as a semipermeable membrane. However, measured internal hydrostatic pressures were $<0.6 \, \text{g mm}^{-2}$ in preabsorptive eggs, and reached only 5 g and $3 \, \text{g mm}^{-2}$ in absorptive and postabsorptive eggs, respectively.

Accordingly, Lees proposes the following mechanism governing water uptake in *Chortoicetes terminifera*. Water is excluded in the very young egg by a permeability barrier in the chorion. This could be the primary wax layer. After about 5 days, the serosal layer is completed and begins to secrete a cuticle. At the same time, yolk material cleaves (becomes divided into cellular compartments) and is mobilized for cuticle formation, resulting in the appearance of smaller molecules and to a rise in O.P.$_i$ from about 400 to 500 mosmol liter^{-1}. At this time influx occurs and the water content is more than doubled. After a further 5 days or so, anatrepsis is complete, blastokinesis occurs, the serosa is detached from the cuticle, O.P.$_i$ falls to about 285 mosmol liter^{-1}, and water uptake ceases. During the uptake phase, influx is effected by osmosis, for subjection of the eggs to an iso-osmotic environment completely (but reversibly) inhibits uptake. Subjection of the eggs to a nitrogen atmosphere does not prevent uptake (as it does in *Teleogryllus*), so that, as in *Locustana* (Matthée, 1951) the process is unlikely to be an active (energy consuming) one, as Browning believes it to be in *Locusta*.

An increase in O.P.$_i$ from 400 to 500 mosmol liter^{-1} is certainly incapable of explaining the almost all or nothing type of difference between the preabsorptive and absorptive egg, since the gradient tending to move water in increases only by $5:4$, or 25%. It therefore seems that there is a reduction in resistance, and this could be the result of serosal cell activity affecting the nature of the primary wax layer (as suggested by Lees). In this connection we note that in *Locusta* too, Moloo emphasizes the point that water uptake commences immediately the serosa is complete, but not before.

At blastokinesis a fall in O.P.$_i$ to 285 mosmol liter^{-1} is again not a sufficient explanation of the cessation of water influx. A further barrier to movement must be created, and this could possibly be the completion of the serosal cuticle

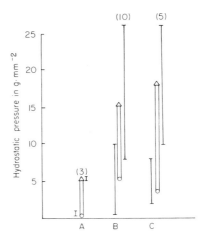

Fig. 107. The mean internal hydrostatic pressure *(circles)* and the mean breaking pressure *(triangles)* for eggs and their chorions of the locust *Chortoicetes*. *A*: two-day-old pre-absorptive phase; *B*: 5–8 days old, during the absorption of water; *C*: 11 days old, post-absorptive phase. Bars on the left of each column represent the range of measured values in normal eggs—those on the right, the breaking pressures of the same eggs. Numbers in brackets refer to the numbers of eggs used. (Redrawn from Lees, 1976)

together with a secondary wax layer (or lipid permeation of one or another layer). McFarlane (1970) suggests a mechanism rather similar to this for the cricket *Acheta domesticus*.

Lees (1976) has found a smaller, but certainly measurable, amount of water absorption by *Chortoicetes* eggs after blastokinesis (Fig. 107), which might be thought to lead to excess tension in the shell. However, this does not occur because the shell becomes "leaky," and the development of temporary pinholes or cracks permits the exit of a sufficient amount of liquid to preserve a reasonable internal hydrostatic pressure. The phenomenon has also been recorded by Salt (1952) in *Melanoplus bivittatus*.

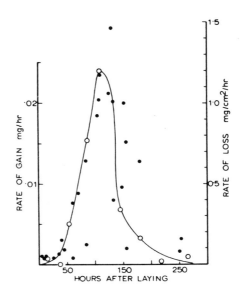

Fig. 108. The uptake of water *(open circles)* and the loss of water in dry air *(closed circles)* by eggs of *Phyllopertha* at different times after laying. (From Bursell, 1974; after Laughlin, 1957)

227

There is a good deal of evidence that the rate of loss of water by transpiration through the shell is greater during periods when water absorption would normally be going on. Matthée (1951) found this for *Locustana*, and in *Chortoicetes* the rate of transpiration varies greatly at different developmental stages, being very high at the time of maximum water absorption (Lees, 1976) (see above). In *Phyllopertha*, outward permeability increases dramatically (from 3 to 60 μg cm^{-2} h^{-1} mm Hg^{-1}) during the period of water uptake (Laughlin, 1957) (Fig. 108), as it does in *Melanoplus bivittatus* (Salt, 1952). It is difficult to resist the conclusion that an increase in shell permeability is generally associated with the period of water uptake, although this is not true of *Teleogryllus*. There is demonstrable variability between species, but there is no compelling evidence that at any one developmental stage, the shell of an egg (excluding living tissues) is more permeable to water inwards than outwards.

F. Conclusions

The proposals advanced by Lees (1976) for *Chortoicetes* can probably be accepted as a working hypothesis for the mechanism of water uptake in many orthopteran eggs at least. Certain anomalies are left outstanding, including the function of the hydropyle if indeed this is not involved in water absorption. *Teleogryllus* is a somewhat aberrant insect, for water uptake is inhibited by a nitrogen atmosphere (suggesting that it is an active process), and uptake is normally prevented during the early stages of anatrepsis by internal hydrostatic pressure and a very strong chorion. A test of this hypothesis (for *Teleogryllus*) would be to measure the actual internal hydrostatic pressure of the early egg. If, unlike *Chortoicetes*, this is indeed as high as 100 to 120 g mm^{-2} ($\equiv$ 10–12 bars), Browning's proposal would be strongly supported. The *Teleogryllus* type of water mechanism may be restricted to small eggs, because the strength of a cylindrical shell necessary to resist a given internal hydrostatic pressure is proportional to its diameter. As it is, the chorion of *Teleogryllus* is much thicker than that of *Chortoicetes*, although the egg is only about one-fourth the size.

As regards the hydropyle, the careful work of Slifer, Pener, Shulov and others certainly indicates that this structure (or at least the posterior portion of the egg) has to do with water absorption. Yet other acridids possess hydropyles, and although their function is unknown, they do not seem to play an important part in water balance. Lees (pers. comm.) suggests that it may be involved in oxygen absorption. It hardly seems likely that a well-developed organ such as this has a specific function in one insect yet a completely different one (or none) in other closely related forms. Perhaps the fact that many egg chorions possess systems of well-developed air cavities (aeropyles) at and below the surface, through which water might diffuse from one part of the cuticle to another, may be relevant to this question, but solution of the puzzle will (as always) await further experimentation.

Apart from some early useful work on *Rhodnius*, mites and ticks, there has been very little work on water balance mechanisms in arthropod eggs other than those of orthopterans. Yet the egg stage is clearly a vulnerable one, and owing to their very small size, part of this vulnerability surely stems from the need to regulate water loss or water uptake. Judging by the work already done, this is a challenging field, and we can hope for some important advances in the future. Particularly valuable would be work on the fine structure of the serosa and other membranes, particularly during different phases of water absorption. The fact that in many cases water absorption begins promptly on serosal completion, and ceases at blastokinesis as the serosa is destroyed, is very suggestive, and points to the need for further investigation of the function of this important epithelium. However, we should not forget that many, perhaps most, eggs do not absorb water. They develop in relatively dry air, and the means by which they absorb oxygen while maintaining a high degree of impermeability to water deserves further consideration.

Chapter 11

Conclusions

A. General

This chapter will serve in part as a summary by recalling the main features of the earlier sections, but its chief purpose is to identify areas that are currently active or that may offer particular promise for future development. A case will be made for the need to integrate the results of physiological studies with reference to the whole animal in a natural setting.

Terrestrial arthropods comprise such a wide variety of forms and modes of life that most biological "problems" including those associated with water affairs, affect them in vastly different, often directly opposite ways. There is not one water balance problem to be solved in a general sense, there are many; and the so-called solutions that involve responses by one species in one direction imply action in the opposite direction in another. Potassium may have to be eliminated by one species, conserved by another; water may be in excess or in extremely short supply. Even in one individual, regulation may have to switch "signs" from time to time. In a way, the foregoing chapters have simply exemplified the above generalization—diversity of problems met by diversity of solutions. However, there are of course certain regularities or consistencies that aid in the appreciation of the whole picture, and we shall attempt to identify these.

B. The Components of Water Loss

I. Transpiration and Cuticle Structure

Unfortunately there has been relatively little exchange of information and ideas between animal and plant eco-physiologists; even some of the terms used by these two groups are not mutually understood. Water potential (Ψ) in bars or atmospheres is a very useful measure, so far almost entirely confined to the plant literature. (See Chap. 1 for the relationship between Ψ and other measures of water concentration.) Psi is applicable to both liquid and vapour phases and includes the effects of solute concentration, hydrostatic pressure and temperature. Water activity in the liquid phase (a_w) and the vapour phase (a_v) is a dimensionless ratio—a useful measure that includes osmotic concentration in the liquid phase. It is proportional to relative humidity in the vapour phase, and its use, by permitting the expression of body fluid concentration and ambient humidity in comparable terms, draws attention to the size of water activity gradients within organisms and between them and their environments.

Plant physiologists also use terms for expressing permeability to water, and its reciprocal, resistance, in a simple form not so far frequently used by animal physiologists namely, "cm s^{-1}" and "s cm^{-1}," respectively. Use of these terms draws attention to the fact that the common expression "mg cm^{-2}h^{-1}" is not a measure of permeability, but of transpiration. The now more usual, but still cumbersome form, mg cm^{-2}h^{-1} mm Hg^{-1}, is a measure of permeability, and a quick rule of thumb for its conversion to cm s^{-1} (both at 25°C) is given by the relationship:

$$2.94 \times (\mu g \, cm^{-2} h^{-1} \, mm \, Hg^{-1}) = (cm \, s^{-1}) \times 10^4$$

Resistances of arthropod cuticles range from 10^4 to 12.6 s cm^{-1} for tsetse-fly pupae and centipedes respectively; and these compare with a range of about 30 to 200 s cm^{-1} or more for plant cuticles (Nobel, 1974).

Probably the best way to present information is to cite mass, surface area, water content and permeability in cm s^{-1} from which most physiologically significant data can be obtained. If ecological effects are in question, information about the lowest (or highest) tolerable water content is also necessary.

The functions of arthropod cuticles are numerous, but it seems probable that water affairs have played a large part in determining the various forms and structures observed. In the past, interest has centered on the nature and disposition of epicuticular lipids in determining cuticle permeability, and there is little doubt that the epicuticle is involved in water conservation, although it is possible that superficial abrasion of the epicuticle leads to increased permeability not as a direct result of the destruction of a thin impermeable layer, but because such injury involves the epidermal cells in repair activities which might lead to a temporary interference with the function of their apical membranes as water barriers (Chap. 3.B.IX.).

The classical idea that an oriented monolayer of lipid molecules on or near the surface of the epicuticle is largely responsible for impermeability to water, has been questioned on several grounds: that the predominant lipids are hydrocarbons which are non-polar and do not form tightly packed monolayers; that alcohols and phospholipids that do form good monolayers are virtually absent (however, fatty acids are present and could perform this function); and that an oriented monolayer has never been satisfactorily demonstrated. As an alternative, a bulk distribution of mixtures of various classes of lipids throughout the epicuticle has been proposed. These perhaps change from crystalline to liquid crystalline states at higher temperatures, and because each lipid has its own transition point, both the position and shape of a temperature: permeability curve may depend on the nature and degree of heterogeneity of the lipids present (Chap. 3.B.VII.). However, the evidence is inconclusive, and the monolayer theory has not been disproved.

Whatever the mechanisms involved, cuticle permeability is certainly correlated in a general way with the environmental conditions in which different arthropods live. Small size, immobility (as in pupae or eggs), high environmental temperatures, and low humidities, tend to be correlated with low cuticle permeabilities in the arthropods concerned. These are mainly permanent adaptations resulting from the evolutionary history of species, not physiological adaptations that vary from time to time within each individual.

Recent investigations on cuticle structure and function have been encouraging (Neville, 1975). So far as water affairs are concerned, further research will probably concentrate on the relation between cuticle lipid species and permeability properties—particularly as these vary between species and from time to time; on the significance of epidermal cell membranes for permeability control, and on the details of epicuticular development and structure, particularly as regards the puzzling movements of materials through pore canals and epicuticular channels.

Until recently, the integument has been thought of as possessing a fixed permeability; but recent evidence is beginning to show that this is wrong, and that integumental permeability may be variable within limits and even controlled by the insect in an adaptive manner. Undoubtedly we shall hear more of this in the future, and again work will probably concentrate on the physical effects of dehydration on cuticle structure as well as the possibility of hormonal control.

In fact the possibility of a relationship between hormonal action, apical membrane function, plasticization, cuticular water content and permeability, offers an attractive field for exploration, and this is discussed further in Chapter 3.B.IX.

II. Respiratory Water Loss

Membranes that are permeable to oxygen but not to water are unknown. Thus, for a particular metabolic rate in terms of oxygen consumption, there is probably a minimal obligatory water loss from the respiratory surfaces of any arthropod, although this may vary from one species to another. Most terrestrial forms have some means of closing the entrance to their respiratory systems, so that while the capacity for rapid oxygen uptake (and consequent rapid water loss) exists, this is not utilized to the full except in times of need.

Several points of interest emerge. The typical tracheal system of so many land arthropods puts a long diffusion path between the outside air and the moist respiratory surfaces. Whether or not this has a beneficial effect on water loss per unit oxygen uptake is not clear (Chap. 4.B.), but the advantages of spiracular control are manifest, and there is a lot of evidence showing, not only that spiracular closure reduces water loss, but that there is a regulatory process whereby water loss can be related to the immediate water status of the individual (Chap. 4.D.).

Increased oxygen uptake, although it certainly involves greater spiracular opening (whether wider or more frequent opening is unknown), does not, it seems, involve the animal in proportionately greater water loss. This is a point of considerable interest, but the mechanism is unknown. Perhaps a lower P_{O_2} in the metabolizing cell increases the oxygen gradient between cell and outside air without increasing the water gradient, but a firm answer must await further research.

Similarly the intermittent release of carbon dioxide is said to conserve water, but the evidence so far is inconclusive despite the elegant work that has been done on intermittent releases in some insects (Chap. 4.E.). Physiological mechanisms involved in spiracular control have been investigated, and in general these seem to be highly adaptive for water balance; for example, higher concentration of carbon dioxide is necessary to cause spiracle opening when an animal is short of water.

In arthropods other than insects very little indeed is known about spiracular control, and not very much about any aspect of respiratory water loss, so that information in this area would be useful.

In view of the above, it is impossible to compare respiratory with cuticular water loss except in specific circumstances. Two important variables are cuticle permeability and the level of metabolic activity. Clearly, cuticular loss would form a larger fraction of total loss in a resting centipede than in a resting tsetse-fly; but in even partially active tsetse-flies, spiracular loss increases from 25% to 57% (Chap. 4.F.). In general, spiracular loss is minimal while the animals are at rest, but may increase dramatically as soon as they become active. Spiders, which have lung books, are unable to maintain high activity for long, probably because their blood does not transport oxygen rapidly enough. The tracheal system of insects is independent of blood circulation and highly efficient, permitting metabolic energy release in wing muscles, for example, at one of the highest rates for any known living tissue.

III. Evaporative Cooling

Evaporative cooling for long periods is out of the question for most arthropods because they are too small. However, for short periods, particularly in species with adequate water supplies, beneficial evaporative cooling does occur. Such species include the larger isopods (with permeable integuments), tsetse-flies (with adequate water supplies in their blood meal, and spiracular control of water loss), and large leaf-eating caterpillars such as those of the saw-fly, *Perga*. Further examples are reported from time to time, and it may turn out that evaporative cooling is a good deal more common among arthropods than has been thought. Perhaps water loss by evaporative cooling is obligatory if an animal is necessarily exposed to high temperatures; but if such temperatures can be avoided by behavioural means, this is undoubtedly done, and to that extent water loss by evaporative cooling is controllable.

IV. Nitrogen Excretion

Water may be lost with nitrogen excretion and with the elimination of excess inorganic and other ions. This latter will be referred to further below.

Of the three commonly observed forms in which nitrogen is eliminated, ammonia (NH_3) contains most hydrogen and least energy; urea is intermediate, and uric acid (or guanine, the alternative used by arachnids), contains least hydrogen and most energy. Uric acid and guanine, because they are insoluble in water, are preferred by most land arthropods, and they pose no problem so far as water conservation is concerned, although this form of excretion obliges an insect to eat more than it otherwise would to obtain the requisite energy (Chap. 6.B.I.). Curiously enough, land isopods eliminate nitrogen in the form of gaseous ammonia, and this involves water loss, not only because of the high $H:N$ ratio, but because a moist integument is necessary for ammonia release (Chap. 6.B.II.).

C. The Components of Water Gain

I. Eating and Drinking

Eating and drinking are both potential sources of water gain, and there is no doubt that the latter is a regulatory mechanism, because some arthropods, both larval and adult, are known to drink free water to an extent necessary to replenish their body water. Again, less information is available about non-insectan groups.

Because "dry" food contains some water it would seem that extra consumption, beyond the energetic or nutritional requirements, would be advantageous when water is in short supply. But this has never been conclusively demonstrated. Insects are known to eat more moist food than dry food (Chap. 7.A.), but this of course does not prove that they eat excess moist food for what water it contains. In fact, when locusts are fed on dry food (5% water content) the rate of water loss with the faeces is greater than the rate of gain with the food, and the insects then reduce food consumption and in this way save water (Chap. 6.C.V.7.).

Questions about the amount and effects of liquid water absorbed with food at first sight look simple enough, but turn out to be complex and rather interesting. The process is in principle regulatory, and unlike the situation with regard to metabolic water discussed below, no question of excess energy arises if the food ingested is not digested or absorbed while its water content is. However, far too little is known about the processes involved, and further work using labelled food to permit the accurate measurement of ingestion rates would be well worth while.

If the food is blood (as it is in mosquitoes, tsetse-flies, ticks and others) the problem is different and involves the elimination of excess water. Very efficient regulatory mechanisms are available to ensure that the appropriate amount of water is retained, having regard to the animals' previous water status. Such elimination usually takes place through Malpighian tubule activity, but in ixodid ticks, elimination is achieved by salivary glands, and in argassid ticks by coxal glands.

Current problems with regard to drinking concern the nature of the stimulation and control mechanisms. Blood volume seems to be involved in *Phormia*, blood chloride concentration in *Lucilia*, and the situation in locusts is at present being investigated. The whole problem of initiation, rate and termination of drinking and feeding is proving to be highly complex (some aspects of it are discussed in Chap. 7.B.III.). Behavioural mechanisms are of central importance for this problem but these aspects are peripheral to our somewhat arbitrarily defined interests. For a review of work in this currently active field see Browne (1975).

Absorption of liquid water through the cuticle has been postulated but not rigorously confirmed. However, uptake certainly does take place via the rectum and other special organs (Chap. 7.D.). These processes are almost certainly regulatory, although there is hardly any information about the amounts involved and their relationship to needs. This is a neglected field, and one in which it should not be too difficult to garner useful information.

II. Metabolic Water

Oxidative metabolism is a source of water in all organisms. Here we are interested essentially in two questions: do arthropods regulate the amount of metabolic water they produce, and how important is this source compared with other components of the water budget?

The first question is a double one. There is some evidence that *Ephestia* and some other stored products insects metabolize more food in dry situations (not necessarily at a faster rate, but in terms of the whole period of their development) and produce more metabolic water as a result (Chap. 8.B.). On the other hand, the parallel suggestion that animals may augment their water supplies by switching from carbohydrate to fat metabolism is denied by the physiological fact that for equal energy production, fats actually yield less water than carbohydrates (Chap. 8.C.). When it is in plentiful supply, water can perhaps be stored by synthesis of fat from carbohydrate, to be subsequently liberated during water shortage. Evidence tending to support this suggestion has been found in cockroaches (Chap. 8.C.). But in this case the fate of the excess energy produced in liberating the water poses a problem, and this is discussed in Chapter 8.B.

The significance of metabolic water in relation to other sources varies with circumstances. For arthropods living in dry stored products metabolic water is an essential and a large component (Chap. 8.B.); for locusts eating fresh grass this is insignificant, but it again becomes relatively, though not absolutely, of great importance if the insects have only dry food. During flight, when neither food nor water is available, metabolism is the only source of water, and whether or not this is sufficient to maintain an insect in water balance poses an interesting question. The answer is at least indicated for migratory locusts, where Weis-Fogh's (1967) tentative model shows the combined effects of ambient humidity, temperature and radiation on water loss, and compares these with water gain by metabolism. There is also some useful information for aphids. But clearly this is an area in which information, although it is technically difficult to obtain, would contribute greatly to our understanding of water affairs.

III. Absorption of Water Vapour

The last source of water gain to be considered is the absorption of water vapour. This is an intriguing process that is apparently restricted to certain groups of arthropods. It never appears in winged forms or stages and only appears in animals without access to free water (but by no means in all such animals). Originally believed to take place through the cuticle, it now seems that in some species of insects at least, the rectum is the site of uptake, while in ixodid ticks and *Arenivaga* the mouth may be involved (Chap. 9.E.).

The process is regulatory in effect—seldom leading to overhydration—and seems to be limited by ambient water vapour activity rather than vapour pressure. The process consumes little energy in relation to the total metabolic output of the animal, although the necessary concentration of energy in the few cells involved poses a problem. The ecological significance of the process is probably very great and is discussed further above (Chap. 9.B.). The mechanisms underlying this important process are not known.

Needless to say, research in this field is very active, and particular needs have already been referred to (Chap. 9.H.). Firm identification of the sites of absorption in a range of different arthropods is clearly desirable, and quantitative measurements of flux rates are necessary, bearing in mind that net movement across the cuticle is but the algebraic sum of possibly much larger component fluxes, some of which may be energy consuming.

D. Water Balance

To summarize so far, we may identify the channels of gain and loss that are obligatory, and those that may be regulated. Cuticle permeability has been determined, within limits, by evolutionary processes. A minimum respiratory loss is obligatory; above this, loss may be regulated by spiracular apparatus and by ventilation in large forms. Excretory water loss is negligible so far as nitrogen is concerned (except for isopods), but may be used as an important regulatory mechanism by the production of moist faeces. When necessary, faecal water content may be reduced to about 10% of total weight by rectal reabsorption. Loss by salivation is probably not great and it may be obligatory; it is, however, regulatory in ixodid ticks where it forms the main avenue for the elimination of excess water.

Gain of water with the food is obligatory insofar as feeding is obligatory. It has been shown to be regulatory in the sense that moist food may be chosen over dry and that the intake of dry food is reduced below the optimal nutrient level if its consumption would lead to net water loss. Gain by drinking is certainly regulatory, and gain of liquid water by absorption through special regions or special organs (probably not including the general cuticle) is also regulatory. Gain by metabolism is largely obligatory; it may be regulatory in terms of the total amount eaten over long periods, but changing from carbohydrate to fat metabolism is not advantageous for water conservation. Finally, water vapour uptake is clearly regulatory, although by no means common to all arthropods.

E. Osmotic and Ionic Regulation

Osmotic and ionic regulation may involve several different processes, some of which may lead to overall loss of water, others to overall gain, but all are, of course, regulatory in a wide sense, and in fact they are central to the whole concept of water balance. This is a very active field, and since it has been admirably reviewed quite recently by Berridge (1970), Maddrell (1971), Stobbart and Shaw (1974) and others, comment here may be reasonably brief.

Central to the whole complex of processes is the Malpighian tubule–rectal wall partnership. In simple terms, it is supposed that Malpighian tubules produce an iso-osmotic fluid, rich in potassium or sodium, and containing proportionate samples of all save the largest molecules in the blood, at a rate that is governed by blood O.P. This fluid is then modified by selective absorption of water and ions in the rectum to extents that are necessary to maintain a constant composition, osmolarity and volume of the blood.

Recent work has shown that this view is essentially sound even though certain modifications are in order. Such modifications include the facts (1) that other parts of the alimentary canal are involved in regulation, e.g. the salivary glands of ixodid ticks, the labial glands of saturniids, and the mid- and hind-guts of various arthropods (Chap. 6.C.V.9.); (2) that the tubule fluid is not always iso-osmotic; (3) that several kinds of ions in addition to sodium and potassium are actively secreted into the Malpighian tubule (Chap. 6.C.V.3.); (4) that rectal fluid may be rendered hyper-osmotic by the secretion of solutes into it rather than by reabsorption of water from it (Chap. 6.C.V.6.), and (5) that large molecules may cross the tubule wall by diffusion along intercellular pathways.

But interest has mainly centered on further exploration of the extent of osmotic and ionic regulation, and of the mechanisms that lie behind these processes, particularly the involvement and mode of action of hormones (Chap. 6.D.) and of the mechanisms concerned with iso-osmotic tubule fluid production, transepithelial movements of water against osmotic gradients, and the selective secretion and reabsorption of ions both for solute coupled water transport and for the maintenance of ionic composition.

Comparison of the Malpighian tubules of *Rhodnius* (where the essentials were first worked out) with those of other insects shows that there are differences in the extent to which different regions of the tubules are specialized structurally and, presumably, functionally. Further exploration of this structure-function relationship will be rewarding.

As regards the extent of regulation, work so far suggests that in some insects, in which blood acts as a water store, water movements in and out of the blood are not accompanied by proportional osmotic changes (Chap. 6.C.V.2.). In other words there is evidence of osmotic regulation of the blood. This suggests that ions may be sequestered or rendered inactive in some way, and there is a suggestion that the fat body of cockroaches may be involved as an ion sink (Chap. 6.B.I.). The ability to move water out of the rectum against impressive gradients has been verified in several insects, and this facility is, of course, a corner stone for the whole fabric of water balance. In this way locusts and cockroaches can raise the osmotic concentration of their rectal fluids to 1 M or more (depending on the water status of the insect concerned), and with the aid of cryptonephridial equipment, *Tenebrio* can bring its rectal contents into equilibrium with a_w 0.9 or less, which is equivalent to 6.1 osmol liter^{-1} (Chap. 6.C.VII.7.).

The cellular mechanisms involved are of the greatest interest, and the fine structure of several epithelia has been investigated, including Malpighian tubules (and their cryptonephric derivations), mid-gut, salivary glands, rectal walls and filter chambers (Chap. 6.C.VII.). In most of these there is an overall basic similarity in structure involving fine channels and spaces, with large surface areas, derived from infoldings of the basal, apical or lateral membranes of the cells concerned. Hypotheses based on either double membrane or standing gradient theory have been proposed to account for the function of all these structures, and some evidence in support has come from micro-sampling of intercellular fluids. However, there are still many unresolved questions in this fast growing area (Chap. 6.C.VII.3.); active investigation continues, and the picture will certainly be revised in the light of further information. Microprobe techniques now being developed will contribute much to the solution of these problems by permitting the measurement of ionic concentrations over very small intracellular regions.

In spite of some excellent pioneering work on the function of cryptonephridia, the mechanism of these systems is incompletely understood. In *Tenebrio* their activity is probably associated with water vapour uptake as well as withdrawal of water from the faecal material, for they reduce the water content of the latter to equilibrium with air at a_v 0.9. On the other hand water vapour uptake occurs in *Thermobia* from much lower a_v's without the aid of a cryptonephric system. Investigation of the nature and significance of the high molecular weight material concentrated in the posterior perinephric space of *Tenebrio*, and of the "particulate" material in the cells of *Thermobia*, may throw considerable light on this mysterious process (Chap. 6.C.VII.7.). In addition, there is certainly a need for studies on the cryptonephric systems of other insects, and Ramsay's recent work on lepidopteran larvae (referred to in Chap. 6.C.VII.7.) is particularly useful in this regard.

Perhaps related to cryptonephric structures are the filter chamber systems found in sap-feeding insects, where rapid concentration of food and rapid elimination of excess water is necessary (Chap. 6.C.VI.2.). The problem for these insects is indeed the reverse of that for dry-living species, but the general structural

238

relationships of both systems have elements in common. The study of filter chamber physiology has only recently been commenced: it should prove to be a fertile field, and the application to it of micro-analytical techniques now available will be particularly valuable.

Hormonal participation in excretion and osmoregulation is a field that has already yielded highly interesting results. The existence and function, if not the precise chemical nature, of both diuretic and anti-diuretic hormones are known in several insects and arachnids (Chap. 6.D.), and work is concentrating on the mode of action of these materials. It has recently become possible to localize neurohaemal release sites precisely, and exploitation of this development can be expected to yield new insights. Already there is some indication of a relationship between specific activity, release sites and target areas on the one hand and feeding habits and general ecology on the other hand. It is important from the point of view of general biology that these kinds of relationships should be identified.

The recent indication that brain hormones may affect cuticular permeability is intriguing, and the relationship of this to other areas of interest is referred to in Chapter 3.B.IX.

Finally, most of the work on osmoregulation has involved insects. But there are some notable exceptions (Chap. 6.C.III.), and judged by the interesting results of work on ticks, for example, further study of non-hexapods should be rewarding.

F. The Water Affairs of Eggs

Being small, sometimes very small, eggs epitomize some of the water balance problems facing all arthropods that live in dry regions, as well as those that are subject to flooding.

Unlike pupae (where intake of water by feeding is also impossible), some eggs absorb a lot of water, as much as their initial weight or more, directly from the environment. The biological advantage of this is clear: it releases the female parent from the obligation of using a lot of water to provide for her eggs. To put it another way, she can lay more eggs for an equal cost in terms of water. Furthermore, the necessity to absorb water for development to proceed to completion effectively synchronizes the latter with abundant (if temporary) water in the environment and thereby increases the probability that larvae when they hatch will find abundant food. In the light of this, the fact that water absorption has been recorded only in a very few arthropods is somewhat surprising: perhaps further research will discover further examples. It would certainly be most desirable for work to be done on the water balance of eggs other than those of hemipterans, gryllids and acridids. Locusts and crickets have provided almost all the information so far; yet they are not representative, even of insects, by reason of their size and of their water absorbing capacities.

There are certainly some fascinating problems outstanding in connection with acridid egg development. One concerns the function of the hydropyle—

whether or not this structure is necessary for water absorption. Another concerns the means for switching the absorption process, whatever this may be, on and off, and a third concerns the relationship between water absorption, development and diapause.

G. Water Balance and the Whole Animal

From one point of view, the purpose of physiological analysis is to illuminate ecological facts, or even to make ecological predictions possible. Be this as it may, some attempt should be made to apply the information now available about water balance mechanisms to real situations. It is very difficult to make such integrations with due scientific rigor (this no doubt is the reason why it has rarely been done), because a real situation is always exceedingly complex and usually unstable, so that even if the very large number of internal and external variables involved in water balance could be measured and their combined effect on water exchange between an organism and its environment at a particular instant predicted, such prediction would be invalid a short time later as the variables alter.

However, this is not to say that such "modeling" is valueless. If the model is approximately correct it will at least serve to indicate the kinds of variables that are involved, and more importantly, the relative importance of each, and the limits within which regulation by the living animal is possible. In this way the significance of water in the lives of arthropods may become a little clearer, and the intellectual gap between field ecologists and laboratory physiologists may be narrowed, with benefit to both.

Some references to ecological situations have already been made in the course of the previous chapters. These have included the effects of evaporative cooling on water loss in field conditions (Chap. 5.C.), the relationship between cuticle permeability and habitat (Chap. 3.C.), the relationship between spiracular control and water status (Chap. 4.D.), and the significance of metabolic water during flight (Chap. 8.D.). Weis-Fogh's model for water balance in flying locusts is a good example of the kind of integrative approach that we are discussing, but apart from this, most ecological applications have dealt with relationships between one environmental component and one avenue of water loss or gain.

There have been several reasonably successful attempts to integrate avenues of heat gain and loss in the determination of body temperature in field conditons [e. g. Hadley (1970a) on desert beetles and scorpions, Edney (1971b) on tenebrionid beetles, Henwood (1975, 1976) on desert beetles and Heinrich (1975 and earlier papers) on several other insects], but there is very little work that attempts to do the same for all avenues of water gain and loss. Yet this approach is absolutely necessary if we are to appreciate the relative importance of the avenues concerned, the means available for control of each, and the limits within which regulation is possible.

The following two examples of such attempts at integration—incompletely documented though the second is—will serve to illustrate the need for research rather than to demonstrate principles.

The first example concerns an insect about which a good deal is known—*Locusta migratoria*. The main components of water loss and gain in these insects have already been referred to; cuticular and spiracular loss in Chapter 4.F., faecal loss in Chapter 6.C.V.7., gain by feeding in Chapter 7.A., drinking in Chapter 7.B.III., and by the production of metabolic water in Chapter 8.C. Each of these components except the last can be varied by the insect, and is thus, in principle at least, regulatory.

Using his own data, Loveridge (1975) has drawn up two water balance sheets applicable to adult, non-flying locusts in air at 50% R.H. and 30°C, while feeding either on fresh or on dry grass (see also Chap. 6.C.V.7.). His summary is shown in Figure 109, and is very illuminating.

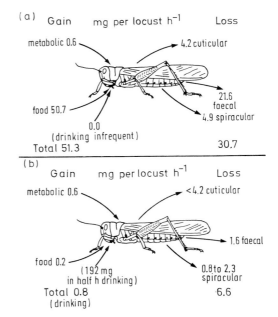

Fig. 109a and b. The water balance of a "standard" locust (1.6 g size) at 50% R.H., 30°C. (a) Locust feeding on fresh, green grass. (b) Locust feeding on dry food. Components of water gain and total gain are listed on the left, and components of loss and total loss are listed on the right in each case. (From Loveridge, 1975)

When fed on fresh grass, water taken in with the food amounts to 50.7 mg per locust h^{-1} and is overwhelmingly the largest component of gain. Faecal loss at 21.6 mg per locust h^{-1} is the largest component on the loss side. In these circumstances there is an overall gain of water which could, if necessary, be balanced by increasing ventilation and faecal loss. When the insects feed on dry grass, however, the intake of food is severely curtailed and water gained in this way is minimal while metabolic water production remains at 0.6 mg per locust h^{-1}. Cuticular loss is slightly reduced, faecal loss is greatly reduced (both absolutely and in proportion to total loss), and spiracular loss is considerably reduced proportionately, but not by very much absolutely; giving a total loss of 6.6 compared with a total gain of 0.8 mg per locust h^{-1}. In these circumstances, if water becomes available, locusts drink avidly (up to 192 mg per locust in half an hour according to Loveridge's figures). If water is not available (and in a field situation such availability is very unlikely) the insects' water balance cannot be maintained.

The second example concerns an insect about which comparatively little is known, and the conclusions must be correspondingly tentative. *Arenivaga investigata* lives in desert sand dunes, where the surface temperature rises to 70°C on occasion. However, the insects are found several centimeters below the surface in the daytime, and they become active just below the surface only at night (Edney, 1974) (Fig. 91). Measurements of temperatures and humidity in their environment during a hot day in summer are shown in Figure 92 (Edney *et al.*, 1974). These data show that even when the surface microclimate is least tolerable, equable temperatures, and humidities of 82% or so (the threshold level for vapour absorption) are available not far below.

The nymphs and adult females can absorb water vapour from air down to a$_v$ 0.825 (82.5% R.H.); and knowing the rate of this (Chap. 9.D.) together with the rate of transpiration, of faecal water loss and metabolic water production (Edney, 1966), it is possible to construct a tentative water balance sheet representing the situation at 88% R.H. and in "dry" air (Table 36).

Table 36. Water exchange in *Arenivaga* nymphs. Data in mg/(100 mg)/day at 25°C for a 320 mg insect. [Data (which are approximations) from Edney, 1966]

	"Dry" air	88% R.H.
Loss		
Faeces	0.19	0.19
Cuticular and spiracular	5.43	0.65
Total	5.62	0.84
Gain		
Food	0.22	0.44
Metabolism	0.87	0.87
Vapour absorption	0	2.14
Total	1.09	3.45

These data suggest that even though no free water is available and the food contains only 20% water, *Arenivaga* can remain in water balance provided that air at 82% R.H. or above is available, so that they can absorb water vapour. In dry air, loss is very much greater than gain. It seems that the ability to absorb water vapour contributes significantly to water balance and may be critical for survival.

The fact that the conclusions in the above two examples have to be tentative underlines the need for more work in which physiological data on various aspects of water balance can be seen in relation to each other and, in particular, to actual field conditions. Even the scanty data now available make it clear that each of the avenues of water loss and gain may be the most important one in particular circumstances, and it is not possible (for example) to say that transpiratory water loss is more (or less) important than faecal water loss in any general sense. However, it is probably going to be possible to say, for any one ecologically consistent group of arthropods (e.g. graminivores) that certain avenues are always more important than others as regulatory mechanisms. The justification for this is that in graminivores, and probably in other trophic categories, the absolute amount of water taken in with the food, and the absolute amount lost with the faeces, can both vary over a much wider range than the absolute amounts gained or lost through other channels (although the relative changes in rates through other channels may be equally large or larger). Thus for graminivores, and probably for several others, food intake and faecal loss are by far the most important channels for the regulation of body water content.

H. Concluding Remarks

Land arthropods are so various in their structure, physiology and behaviour, that any attempt to generalize is hazardous. There are three aspects of the problem (already referred to several times in passing) which should perhaps be re-emphasized in these concluding remarks.

The first is the obvious paucity of information in all areas about arthropods other than insects. It is much to be hoped that future work will be concerned with such attractive experimental animals as the larger spiders (eurypelmids, for example), myriapods, scorpions; the smaller arachnids including ticks and mites, and last but not least with those ubiquitous, highly successful (though ammonotelic) land crustaceans, the isopods.

Secondly, there is a tendency among biologists interested in water affairs to think in stereotypes—the arthropod stereotype is an insect-like animal forever inventing "strategies" to conserve energy and to conserve water. This is not the place to discuss the energy fallacy, but we should remember that in so many situations water is over abundant and arthropods are faced with the problem of getting rid of it, or living in spite of it. We have not had much to do with the aeropyles of eggs, for example, but these are probably as important as hydropyles because they prevent death by anoxia.

Thirdly, and perhaps most importantly, the time has surely come when physiologists should take serious thought about the opportunities to be had by becoming acquainted with real ecological situations and the associated problems. One way of moving in this direction, within the field of water affairs, is to attempt to integrate the effects of the various physiological processes whose rates are known, and to apply the results to whole animals in real life. The advantages of this sort of exercise have been discussed above and the paucity of material available for discussion there underlines the present plea for further work along these lines. There are good examples of integrative work on water balance in vertebrates (Schmidt-Nielsen, 1964, 1975) and it would seem that arthropods should now be treated in like manner.

Judged by any standards, the arthropod level of organization is highly successful for land life, and perhaps a large part of this success may be put down to their effective water balance mechanisms which permit rigorous conservation while allowing for rapid elimination of excess water when this is necessary. The study of arthropod water affairs has already yielded information and ideas of general biological interest, and current indications point to a healthy future.

References

Abushama, F. T.: 1970. Loss of water from the grasshopper *Poecilocerus hieroglyphicus* (Klug), compared with the tree locust *Anacridium melanorhodon melanorhodon* (Walker). Z. Angew. Entomol. **66**, 160–167

Adams, P. A., Heath, J. E.: 1964. An evaporative cooling mechanism in *Pholus achemon* (Sphingidae). J. Res. Lepidoptera **3**, 69–72

Agren, L.: 1975. Fine structure of the thoracic salivary gland of the male *Bombus lapidarius* L. Hymenoptera Apidae. Z. **3**, 19–31

Ahearn, G. A.: 1970a. The control of water loss in desert tenebrionid beetles. J. Exp. Biol. **53**, 573–595

Ahearn, G. A.: 1970b. Water balance in the whipscorpion, *Mastigoproctus giganteus* (Lucas) (Arachnida, Uropygi). Comp. Biochem. Physiol. **35**, 339–353

Ahearn, G. A., Hadley, N. F.: 1969. The effects of temperature and humidity on water loss in two desert tenebrionid beetles, *Eleodes armata* and *Cryptoglossa verrucosa*. Comp. Biochem. Physiol. **30**, 739–749

Ahearn, G. A., Hadley, N. F.: 1976. Functional roles of luminal sodium and potassium in water transport across desert scorpion ileum. Nature (London) **261**, 66–68

Alexander, A., Ewer, D. W.: 1955. A note on the function of the eversible sacs of the onychophoran, *Opisthopatus cinctipes* Purcell. Ann. Natal Mus. **13**, 217–222

Alexander, P., Kitchener, J. A., Briscoe, H. V. A.: 1944. Inert dust insecticides. Ann. Appl. Biol. **31**, 143–159

Almquist, S.: 1971. Resistance to desiccation in some dune living spiders. Oikos **22**, 225–229

Altmann, G.: 1956. Die Regulation des Wasserhaushaltes der Honigbiene. Insectes Soc. **3**, 33–40

Amin, E. S.: 1960. The wax of the silkworm ecdysis skin. I. The fatty acids. J. Chem. Soc. London **1960**, 1410–1411

Anderson, S. O.: 1974. Cuticular sclerotization in larval and adult locusts, *Schistocerca gregaria*. J. Insect Physiol. **20**, 1537–1552

Anderson, S. O., Weis-Fogh, T.: 1964. Resilin. A rubber-like protein in arthropod cuticle. Adv. Insect Physiol. **2**, 1–65

Andrewartha, H. G.: 1952. Diapause in relation to the ecology of insects. Biol. Rev. Cambridge Phil. Soc. Rev. **27**, 50–107

Andrewartha, H. G., Birch, L. C.: 1954. The distribution and abundance of animals. Chicago: Univ. Chicago, 782 pp

Andrewartha, H. G., Birch, L. C.: 1960. Some recent contributions to the study of the distribution and abundance of insects. Ann. Rev. Entomol. **5**, 219–242

Araman, S. F., Said, A.: 1972. Biochemical and physiological studies of certain ticks (Ixodoidea): the ionic regulatory role of the coxal organs of *Argas (Persicargas) persicus* (Oken) and *A. (P.) arboreus* Kaiser, Hoogstraal and Kuhls (Argassidae). J. Parasitol. **58**, 348–353

Arlian, G. A.: 1975. Dehydration and survival of the European house dust mite, *Dermatophagoides pteronyssinus*. J. Medic. Entomol. **12**, 437–442

Arlian, L. G., Eckstrand, I. A.: 1975. Water balance in *Drosophila pseudoobscura*, and its ecological implications. Ann. Entomol. Soc. Amer. **68**, 827–832

Arlian, L. G., Wharton, G. W.: 1974. Kinetics of active and passive components of water exchange between the air and a mite, *Dermatophagoides farinae*. J. Insect Physiol. **20**, 1063–1077

Armold, M. T., Blomquist, G. J., Jackson, L. L.: 1969. Cuticular lipids of insects. III. The surface lipids of the aquatic and terrestrial life forms of the Big Stonefly, *Pteronarcys californica* Newport. Comp. Biochem. Physiol. **31**, 685–692

Arms, K., Feeny, P., Lederhouse, R. C.: 1974. Sodium: stimulus for puddling behavior by tiger swallowtail butterflies, *Papilio glaucus*. Science **185**, 372–374

Asperen, K. van, Esch, I. van.: 1954. A simple microtitration method for the determination of calcium and magnesium in the haemolymph of insects. Nature (London) **174**, 927

Asperen, K. van, Esch, I. van.: 1956. The chemical composition of the haemolymph in *Periplaneta americana* with special reference to the mineral constituents. Arch. Néerl. Zool. **11**, 342–360

Aston, R. J.: 1975. The role of adenosine 3′-5′ cyclic monophosphate in relation to the diuretic hormone of *Rhodnius prolixus*. J. Insect Physiol. **21**, 1873–1877

Atkinson, P., Gilby, A. R.: 1970. Autoxidation of insect lipids: inhibition on the cuticle of the American cockroach. Science **168**, 992

Babcock, S. M.: 1912. Metabolic water: its production and role in vital phenomena. Res. Bull. Wisconsin Agr. Exp. Sta. **22**, 87–181

Babers, F. H.: 1938. An analysis of the blood of the sixth-instar Southern Armyworm (*Prodenia eridania*). J. Agr. Res. **57**, 697–706

Baker, G., Peffer, J. H., Johnson, L. H., Hastings, E.: 1960. Estimation of the composition of the cuticular wax of the mormon cricket, *Anabrus simplex* Hold. J. Insect Physiol. **5**, 47–60

Balashov, Y. S.: 1958. The excretion processes and activity of Malpighian tubes in ticks. Parasitol. Sb., Akad. Nauk SSSR, Zool. Inst. **18**, 120

Balashov, Y. S.: 1972. Blood sucking ticks (Ixodoidea)—Vectors of disease of Man and animals. Entomol. Soc. Am., Misc. Publ. # 5, **8**, 161–376

Balshin, M., Phillips, J. E.: 1971. Active absorption of amino-acids in the rectum of the desert locust (*Schistocerca gregaria*). Nature New Biology **233**, 53–55

Barker, R. W., Burris, E., Sauer, J. R., Hair, J. A.: 1973. Composition of tick oral secretions obtained by three different methods. J. Med. Entomol. **10**, 198–201

Beadle, L. C., Shaw, J.: 1950. The retention of salt and the regulation of the non-protein nitrogen fractions in the blood of the aquatic larva, *Sialis lutaria*. J. Exp. Biol. **27**, 96–109

Beadly, D. J.: 1974. Fine structure of the integument of the ticks, *Boophilus decoloratus* Koch and *B. microplus* (Canastrini) (Acarina : Ixodidae). Intern. J. Insect Morphol. Embryol. **3**, 1–12

Beament, J. W. L.: 1945. The cuticular lipoids of insects. J. Exp. Biol. **21**, 115–131

Beament, J. W. L.: 1946. The formation and structure of the chorion of the egg in an hemipteran, *Rhodnius prolixus*. Quart. J. Microsc. Sci. **87**, 393–439

Beament, J. W. L.: 1948. The role of wax layers in the waterproofing of insect cuticle and egg-shell. Discus. Faraday Soc. **3**, 117–182

Beament, J. W. L.: 1949. The penetration of insect egg shells. II. The properties and permeability of the sub-corial membranes during development of *Rhodnius prolixus* Stahl. Bull. Entomol. Res. **39**, 467–488

Beament, J. W. L.: 1955. Wax secretion in the cockroach. J. Exp. Biol. **32**, 514–538

Beament, J. W. L.: 1958. The effect of temperature on the waterproofing mechanism of an insect. J. Exp. Biol. **35**, 494–519

Beament, J. W. L.: 1959. The waterproofing mechanism of arthropods. I. The effect of temperature on cuticle permeability in terrestrial insects and ticks. J. Exp. Biol. **36**, 391–422

Beament, J. W. L.: 1961. The water relations of the insect cuticle. Biol. Rev. Cambridge Phil. Soc. **36**, 281–320

Beament, J. W. L.: 1964. Active transport and passive movement of water in insects. Adv. Insect Physiol. **2**, 67–129

Beament, J. W. L.: 1965. The active transport of water: evidence, models, and mechanisms. Symp. Soc. Exper. Biol. Fogg, G. E. (ed.). Cambridge, pp. 273–298

Beament, J. W. L., Noble-Nesbitt, J., Watson, J. A. L.: 1964. The waterproofing mechanisms of arthropods. III. Cuticular permeability in the firebrat, *Thermobia domestica* (Packard). J. Exp. Biol. **41**, 323–330

Beatty, I. M., Gilby, A. R.: 1969. The major hydrocarbons of a cockroach cuticular wax. Naturwissenschaften **56**, 373–374

Beckel, W. E.: 1958. The morphology, histology, and physiology of the spiracular regulatory apparatus of *Hyalophora cecropia* (L.). Proc. 10th Intern. Congr. Entomol. **2**, 87–115

Bedford, J. J., Leader, J. P.: 1975. The composition of the haemolymph of the New Zealand centipede *Cormocephalus rubriceps* (Newport). Comp. Biochem. Physiol. **50 A**, 561–564

246

Belozerov, V. N., Seravin, L. N.: 1960. Water balance regulation in *Alectrobius tholozani* at different atmospheric humidities. Med. Parazitol. **29**, 309–313

Bennet-Clark, H. C.: 1961. The Mechanics of Feeding in the Bug *Rhodnius prolixus* Stal. Thesis, University of Cambridge

Bennet-Clark, H. C.: 1962. Active control of the mechanical properties of insect endocuticle. J. Insect Physiol. **8**, 627–633

Bergman, W.: 1938. The composition of the ether-extractives from exuviae of the silkworm *Bombyx mori*. Ann. Entomol. Soc. Am. **31**, 315–321

Bergold, G.: 1935. Die Ausbildung der Stigmen bei Coleopteren verschiedener Biotope. Z. Morphol. Oekol. Tiere **29**, 511–526

Bernays, E. A., Chapman, R. F.: 1972. Meal size in nymphs of *Locusta migratoria*. Entomol. Exp. Appl. **15**, 399–410

Bernays, E. A., Chapman, R. F.: 1974a. Changes in haemolymph osmotic pressure in *Locusta migratoria* larvae in relation to feeding. J. Entomol. **48**, 149–155

Bernays, E. A., Chapman, R. F.: 1974b. The effect of haemolymph osmotic pressure on the meal size of nymphs of *Locusta migratoria* L. J. Exp. Biol. **61**, 473–480

Bernays, E. A., Chapman, R. F.: 1974c. The regulation of food intake by acridids. In: Experimental Analysis of Insect Behaviour. Browne, L. B. (ed.). Berlin–Heidelberg–New York: Springer, pp. 48–59

Berridge, M. J.: 1965a. The physiology of excretion in the cotton stainer, *Dysdercus fasciatus* Signoret. I. Anatomy, water excretion and osmoregulation. J. Exp. Biol. **43**, 511–521

Berridge, M. J.: 1965b. The physiology of excretion in the cotton stainer, *Dysdercus fasciatus* Signoret. II. Inorganic excretion and ionic regulation. J. Exp. Biol. **43**, 523–533

Berridge, M. J.: 1965c. The physiology of excretion in the cotton stainer, *Dysdercus fasciatus* Signoret. III. Nitrogen excretion and excretory metabolism. J. Exp. Biol. **43**, 535–552

Berridge, M. J.: 1966a. Metabolic pathways of isolated Malpighian tubules of the blow fly functioning in an artificial medium. J. Insect Physiol. **12**, 1523–1538

Berridge, M. J.: 1966b. The physiology of excretion in the cotton stainer, *Dysdercus fasciatus* Signoret. IV. Hormonal control of excretion. J. Exp. Biol. **44**, 553–566

Berridge, M. J.: 1968. Unpublished experiments

Berridge, M. J.: 1969. Urine formation by the Malpighian tubules of *Calliphora*. II. Anions. J. Exp. Biol. **50**, 15–28

Berridge, M. J.: 1970. Osmoregulation in terrestrial arthropods. In: Chemical Zoology. Florkin, M., Scheer, B. T. (eds.). New York: Academic Press, Vol. V, Part A, pp. 287–320

Berridge, M. J., Gupta, B. L.: 1967. Fine structural changes in relation to ion and water transport in the rectal papillae of the blow fly, *Calliphora*. J. Cell Sci. **2**, 89–112

Berridge, M. J., Oschman, J. L.: 1969. A structural basis for fluid secretion by Malpighian tubules. Tissue Cell **1**, 247–272

Berridge, M. J., Patel, N. G.: 1968. Insect salivary glands: stimulation of fluid secretion by 5-hydroxy-tryptamine and adenosine-3,5-monophosphate. Science **162**, 462–463

Betts, A. D.: 1930. The ingestion of syrup by the honey bee. Bee World **9**, 85–90

Bialaszewicz, K., Landau, C.: 1938. Sur la composition minérale de l'hémolymphe des vers à soie et sur les changements qu'elle subit au course de la croissance et pendant le métamorphose. Acta Biol. Exp. **12**, 307

Binyon, J., Lewis, J. G. E.: 1963. Physiological adaptations of two species of centipede (Chilopoda: Geophilomorpha) to life on the shore. J. Mar. Biol. Ass. UK **43**, 49–55

Bishop, G. H., Briggs, A. P., Ronzoni, E.: 1925. Body fluids of the honey bee larva. II. Chemical constituents of the blood, and their osmotic effects. J. Biol. Chem. **66**, 77

Bliss, D. E.: 1968. Transition from water to land in decapod crustaceans. Am. Zool. **8**, 355–392

Bliss, D. E., Mantel, L. H.: 1968. Adaptations of crustaceans to land: A summary and analysis of new findings. Am. Zool. **8**, 673–685

Blower, G.: 1951. A comparative study of the chilopod and diplopod cuticle. Quart. J. Microsc. Sci. **92**, 141–161

Blower, J. G.: 1955. Millipedes and centipedes as soil animals. In: Soil Zoology. Kevan, D. K. McE. (ed.). London: Butterworths Scientific, pp. 138–151

247

Bodine, J. H.: 1921. Factors influencing the water content and the rate of metabolism of certain Orthoptera. J. Exp. Zool. **32**, 137–164

Bodine, J. H.: 1933. The effect of hypertonic solution on oxygen uptake by the eggs of *Melanoplus* (Orthoptera). Physiol. Zool. **6**, 150–158

Bohm, B. C., Hadley, N. F.: 1976. Tritium determined water flux in the free-roaming desert tenebrionid beetle, *Eleodes armata*. In prep.

Bolwig, N.: 1953. On the variation of the osmotic pressure of the haemolymph in flies. S. African Industr. Chem. **7**, 113–115

Boné, G. J.: 1943. Recherches sur les glandes coxales et la régulation du mileu interne chez l'*Ornithodorus moubata* Murray (Acarina, Ixodidae). Ann. Soc. Roy. Zool. Belge **74**, 16–312

Boné, G. J.: 1947. Regulation of the sodium potassium ratio in insects. Nature (London) **160**, 679

Boné, G. J.: 1964. La rapport sodium/potassium dans la liquide coelemique des insectes. 1. Ses relations avec le régime alimentaire. Ann. Soc. Roy. Zool. Belge **75**, 123–132

Bouligand, Y.: 1965. Sur une architecture torsadée répandue dans de nombreuses cuticles d'arthropodes. C. R. Acad. Sci., Paris **261**, 3665–3668

Bradley, T. J., Phillips, J. E.: 1975. The secretion of hyperosmotic fluid by the rectum of saline water mosquito larvae, *Aedes taeniorhynchus*. J. Exp. Biol. **63**, 331–342

Brady, J.: 1975. "Hunger" in the tsetsefly: the nutritional correlates of behaviour. J. Insect Physiol. **21**, 807–829

Brockway, A. P., Schneiderman, H. A.: 1967. Strain-gauge transducer studies on intratracheal pressure and pupal length during discontinuous respiration in diapausing silkworm pupae. J. Insect Physiol. **13**, 1413–1451

Brodsky, W. A., Rehm, W. S., Dennis, W. J., Miller, D. G.: 1955. Thermodynamic analysis of the intracellular osmotic gradient hypothesis of active water transport. Science **121**, 302–303

Browne, L. B.: 1964. Water regulation in insects. Ann. Rev. Entomol. **9**, 63–82

Browne, L. B.: 1968. Effects of altering the composition and volume of the haemolymph on water ingestion of the blow fly *Lucilia cuprina*. J. Insect Physiol. **14**, 1603–1620

Browne, L. B. (ed.): 1974. Experimental Analysis of Insect Behaviour. Berlin–Heidelberg–New York: Springer, 366 pp

Browne, L. B.: 1975. Regulatory mechanisms in insect feeding. Adv. Insect Physiol. **11**, 1–116

Browne, L. B., Dudzinski, A.: 1968. Some changes resulting from water deprivation in the blow fly, *Lucilia cuprina*. J. Insect Physiol. **14**, 1423–1434

Browne, L. B., Moorhouse, J. E., Gerwen, A. C. M. van: 1975a. Sensory adaptation and the regulation of meal size in the Australian plague locust, *Chortoicetes terminifera*. J. Insect Physiol. **21**, 1633–1639

Browne, L. B., Moorhouse, J. E., Gerwen, A. C. M. van: 1975b. An excitatory state generated during feeding in the locust, *Chortoicetes terminifera*. J. Insect Physiol. **21**, 1731–1735

Browne, L. B., Moorhouse, J. E., Gerwen, A. C. M. van: 1976. A relationship between weight loss during food deprivation and subsequent meal size in the locust *Chortoicetes terminifera*. J. Insect Physiol. **22**, 89–94

Browning, T. O.: 1953. The influence of temperature and moisture on the uptake and loss of water in the eggs of *Gryllulus commodus* Walker (Orthoptera–Gryllidae). J. Exp. Biol. **30**, 104–115

Browning, T. O.: 1954. Water balance in the tick *Ornithodoros moubata* Murray, with particular reference to the influence of carbon dioxide on the uptake and loss of water. J. Exp. Biol. **31**, 331–340

Browning, T. O.: 1965. Observations on the absorption of water, diapause and embryogenesis in the eggs of the cricket *Teleogryllus commodus* (Walker). J. Exp. Biol. **43**, 433–439

Browning, T. O.: 1967. Water, and the eggs of insects. In: Insects and Physiology. Beament, J. W. L., Treherne, J. E. (eds.). London: Oliver and Boyd, pp. 315–328

Browning, T. O.: 1969a. The permeability of the shell of the egg of *Teleogryllus commodus* measured with the aid of tritiated water. J. Exp. Biol. **51**, 397–405

Browning, T. O.: 1969b. Permeability to water of the shell of the egg of *Locusta migratoria migratorioides*, with observations on the egg of *Teleogryllus commodus*. J. Exp. Biol. **51**, 99–105

Browning, T. O., Forrest, W. W.: 1960. The permeability of the shell of the egg of *Acheta commodus* Walker (Orthoptera, Gryllidae). J. Exp. Biol. **37**, 213–217

248

Broza, M., Borut, A., Pener, M.: 1976. Osmoregulation in the desert tenebrionid beetle *Trachyderma philistina* Reiche during dehydration and subsequent rehydration. Israel J. Med. Sci. **12**, 868–871

Buck, J. B.: 1953. Physical properties and chemical composition of insect blood. In: Insect Physiology. Roeder, K. D. (ed.). New York: Wiley, pp. 147–190

Buck, J. B.: 1958. Cyclic CO_2 release in insects. IV. A theory of mechanism. Biol. Bull. **114**, 118–140

Buck, J.: 1962. Some physical aspects of insect respiration. Ann. Rev. Entomol. **7**, 27–66

Buck, J.: 1965. Hydration and respiration in chironomid larvae. J. Insect Physiol. **11**, 1503–1516

Buck, J., Friedman, S.: 1958. Cyclic CO_2 release in diapausing pupae—III. CO_2 capacity of the blood: carbonic anhydrase. J. Insect Physiol. **2**, 52–60

Buck, J., Keister, M.: 1955. Cyclic CO_2 release in diapausing *Agapema* pupae. Biol. Bull. **109**, 144–163

Bullock, J. A., Smith, P. H.: 1971. The relation between dry and fresh weight in some caterpillars. Entomol. Exp. Appl. **14**, 125–131

Burkett, B. N., Schneiderman, H. A.: 1968. Coordinated neuromuscular activity in insect spiracles at sub-zero temperatures. Nature (London) **217**, 75–96

Burkett, B. N., Schneiderman, H. A.: 1974a. Roles of oxygen and carbon dioxide in the control of spiracular function in cecropia pupae. Biol. Bull. Woods Hole **147**, 274–293

Burkett, B. N., Schneiderman, H. A.: 1974b. Discontinuous respiration in insects at low temperatures: intra-tracheal pressure changes and spiracular valve behavior. Biol. Bull. Woods Hole **147**, 294–310

Bursell, E.: 1955. The transpiration of terrestrial isopods. J. Exp. Biol. **32**, 238–255

Bursell, E.: 1957a. Spiracular control of water loss in the tsetse fly. Proc. Roy. Entomol. Soc. London, Ser. A **32**, 21–29

Bursell, E.: 1957b. The effect of humidity on the activity of tsetse flies. J. Exp. Biol. **34**, 42–51

Bursell, E.: 1958. The water balance of tsetse pupae. Phil. Trans. Roy. Soc. London, Ser. B **241**, 179–210

Bursell, E.: 1959a. The water balance of tsetse flies. Trans. Roy. Entomol. Soc. London **111**, 205–235

Bursell, E.: 1959b. Metabolic rate and water loss during flight. Ann. Rep. E. Afr. Tsetse Trypanosom. Res. Org. **1958**, 32–35

Bursell, E.: 1960. Loss of water by excretion and defaecation in the tsetse fly. J. Exp. Biol. **37**, 689–697

Bursell, E.: 1964. Nitrogenous waste products of the tsetse fly. In: Proc. 12th Intern. Congr. Entomol. **797**. Freeman, P. (ed.). Roy. Entomol. Soc. (London)

Bursell, E.: 1967. Excretion of nitrogen in insects. Adv. Insect Physiol. **4**, 33–67

Bursell, E.: 1970. An Introduction to Insect Physiology. New York: Academic Press, 276 pp

Bursell, E.: 1974a. Environmental aspects—temperature. In: The Physiology of Insecta. Rockstein, M. (ed.). New York: Academic Press, Vol. II, pp. 1–41

Bursell, E.: 1974b. Environmental aspects—humidity. In: The Physiology of Insecta. Rockstein, M. (ed.). New York: Academic Press, Vol. II, pp. 44–84

Bursell, E., Clements, A. N.: 1967. The cuticular lipids of the larva of *Tenebrio molitor* L. (Coleoptera). J. Insect Physiol. **13**, 1671–1678

Butcher, R. W., Sutherland, E. W.: 1962. Adenosine 3′,5′-phosphate in biological materials. I. Purification and properties of cyclic 3′,5′-nucleotide phosphodiesterase, and use of this enzyme to characterise adenosine 3′,5′-phosphate in human urine. J. Biol. Chem. **237**, 1244–1250

Butler, C. G.: 1940. The choice of drinking water by the honey bee. J. Exp. Biol. **17**, 253–261

Buxton, P. A.: 1924. Heat, moisture and animal life in deserts. Proc. Roy. Soc. London, Ser. B **96**, 123–131

Buxton, P. A.: 1930. Evaporation from the meal-worm (*Tenebrio*: Coleoptera) and atmospheric humidity. Proc. Roy. Soc. London, Ser. B **106**, 560–577

Buxton, P. A.: 1932a. Terrestrial insects and the humidity of their environment. Biol. Rev. Cambridge Phil. Soc. **7**, 275–320

Buxton, P. A.: 1932b. The relation of adult *Rhodnius prolixus* (Reduviidae, Rhynchota) to atmospheric humidity. Parasitology **24**, 429–439

Buxton, P. A.: 1955. The Natural History of Tsetse-Flies. London: Lewis, 816 pp

Buxton, P. A., Lewis, D. J.: 1934. Climate and tsetse-flies: laboratory studies upon *Glossina submorsitans* and *tachinoides*. Phil. Trans. Roy. Soc. London, Ser. B **224**, 175–240

Byzoua, Y. B.: 1974. Carbonic anhydrase in the hemolymph of the beetle *Blaps scutellata*. J. Evol. Biochem. Physiol. **9**, 530–531

Carrington, C. B., Tenney, S. M.: 1959. Chemical constituents of haemolymph and tissue in *Telea polyphemus* Cram. with particular reference to the question of ion binding. J. Insect Physiol. **3**, 402–413

Casey, T. M.: 1976. Flight energetics in sphinx moths: heat production and heat loss in *Hyles lineata* during free flight. J. Exp. Biol. **64**, 545–560

Chapman, D., Leslie, R. B.: 1970. Structure and function of phospholipids in membranes. In: Membranes of Mitochondria and Chloroplasts. Racker, E. (ed.). New York: Van Nostrand Reinhold, pp. 91–126

Chefurka, W., Pepper, J. H.: 1955. Studies on the cuticle of the grasshopper *Melanoplus bivittatus* (Say) (Orthoptera : Acrididae). Can. Entomol. **87**, 145–171

Cheng, L.: 1974. Notes on the Ecology of the Oceanic Insect *Halobates*. Mar. Fisheries Rev. **36**, 1–7

Cheung, W. W. K., Marshall, A. T.: 1973a. Water and ion regulation in cicadas in relation to xylem feeding. J. Insect Physiol. **19**, 1801–1816

Cheung, W. W. K., Marshall, A. T.: 1973b. Studies on water and ion transport in homopteran insects: ultrastructure and cytochemistry of the cicadoid and cercopoid midgut. Tissue Cell **5**, 651–669

Church, N. S.: 1960. Heat loss and the body temperatures of flying insects. I. Heat loss by evaporation of water from the body. J. Exp. Biol. **37**, 171–185

Cloudsley-Thompson, J. L.: 1950. The water relations and cuticle of *Paradesmus gracilis* (Diplopoda, Strongylosomidae). Quart. J. Microsc. Sci. **91**, 453–464

Cloudsley-Thompson, J. L.: 1956. Studies in diurnal rhythms. VI. Bioclimatic observations in Tunisia and their significance in relation to the physiology of the fauna, especially woodlice, centipedes, scorpions and beetles. Ann. Mag. Nat. Hist. **9**, 305–329

Cloudsley-Thompson, J. L.: 1957. Studies in diurnal rhythms. V. Nocturnal ecology and water-relations of the British cribellate spiders of the genus *Ciniflo* Bl. Linnean Soc. J. Zool. **43**, 134–152

Cloudsley-Thompson, J. L.: 1958. Notes on Arachnida 29: drinking by scorpions. Entomol. Mon. Mag. **94**, 229

Cloudsley-Thompson, J. L.: 1959. Studies in diurnal rhythms. IX. The water-relations of some nocturnal tropical arthropods. Entomol. Exp. Appl. **2**, 249–256

Cloudsley-Thompson, J. L.: 1961. Observations on the natural history of the "Camel-Spider" *Galeodes arabs* C. L. Koch (Solifugae : Galeodidae) in the Sudan. Entomol. Mon. Mag. **97**, 145–152

Cloudsley-Thompson, J. L.: 1962. Some aspects of the physiology of *Buthotus minax* (Scorpiones : Buthidae) with remarks on other African scorpions. Entomol. Mon. Mag. **98**, 243–246

Cloudsley-Thompson, J. L.: 1963. Some aspects of the physiology of *Buthotus minax* (Scorpiones, Buthidae) with remarks on other African scorpions. Entomol. Mon. Mag. **98**, 243–246

Cloudsley-Thompson, J. L.: 1967. The water-relations of scorpions and tarantulas from the sonoran desert. Entomol. Mon. Mag. **103**, 217–220

Cloudsley-Thompson, J. L.: 1970. Terrestrial invertebrates. In: Comparative Physiology of Thermoregulation. Whittow, G. C. (ed.). New York: Academic Press, Vol. I, pp. 15–77

Cloudsley-Thompson, J. L.: 1973. Climatic factors affecting the nocturnal emergence of woodlice and other arthropods. Entomol. Mon. Mag. **109**, 123–124

Cloudsley-Thompson, J. L.: 1975. Adaptations of Arthropoda to arid environments. Ann. Rev. Entomol. **20**, 261–283

Cloudsley-Thompson, J. L., Crawford, C. S.: 1970. Water and temperature relations, and diurnal rhythms of scolopendromorph centipedes. Entomol. Exp. Appl. **13**, 187–193

Cochran, D. G.: 1975. Excretion in insects. In: Insect Biochemistry and Function. Candy, D. J., Kiley, B. A. (eds.). London: Chapman and Hall, pp. 177–314

Cochran, D. G.: 1976. Excreta analysis on additional cockroach species and the house cricket. Comp. Biochem. Physiol. **53A**, 79–81

Cockbain, A. J.: 1961a. Fuel utilization and duration of tethered flight in *Aphis fabae* Scop. J. Exp. Biol. **38**, 163–174

Cockbain, A. J.: 1961b. Water relationships of *Aphis fabae* Scop. during tethered flight. J. Exp. Biol. **38**, 175–180

Colosi, I. de S.: 1933. L'assunzione dell'acqua per via cutanea. Publ. Staz. Zool. Napoli **13**, 12–38

250

Cooper, A. F., Van Gundy, S. D.: 1971. Senescence, quiescense and cryptobiosis. In: Plant Parasitic Nematodes. Zuckerman, B. M., Mai, W. F., Rohde, R. A. (eds.). New York: Academic Press, pp. 297–318

Copeland, E.: 1968. Fine structure of salt and water uptake in the land crab *Gecaricenus lateralis*. Am. Zool. **8**, 417–432

Cottrell, C. B.: 1962. The imaginal ecdysis of blowflies. Observations on the hydrostatic mechanisms involved in digging and expansion. J. Exp. Biol. **39**, 431–448

Crawford, C. S.: 1972. Water relations in a desert millipede, *Orthoporus ornatus* (Girard) (Spirostreptidae). Comp. Biochem. Physiol. **42A**, 521–535

Crawford, C. S., Cloudsley-Thompson, J. L.: 1971. Water relations and desiccation-avoiding behavior in the vinegaroon *Mastigoproctus giganteus* (Arachnida: Uropygi). Entomol. Exp. Appl. **14**, 99–106

Crawford, C. S., Wooten, R. C.: 1973. Water relations in *Diplocentrus spitzeri*, a semimontane scorpion from the Southwestern United States. Physiol. Zool. **6**, 218–229

Croghan, P. C.: 1959. The interstitial soil-water habitat and the evolution of terrestrial arthropods. Proc. Roy. Phys. Soc. Edinburgh **27**, 103–104

Crowe, J. H.: 1971. Anhydrobiosis: an unsolved problem. Am. Natur. **105**, 563–574

Crowe, J. H.: 1972. Evaporative water loss by tardigrades under controlled relative humidities. Biol. Bull. Wood's Hole **142**, 407–616

Crowe, J. H., Cooper, A. F.: 1971. Cryptobiosis. Sci. Am. **225**, 30–36

Curran, P. F.: 1960. Na, Cl and water transport by rat ileum in vitro. J. Gen. Physiol. **43**, 1137–1148

Curry, A.: 1974. The spiracle structure and resistance to dessication of centipedes. Symp. Zool. Soc. London **32**, 365–382

Danielli, J. F., Davson, H.: 1935. A contribution to the theory of permeability of thin films. J. Cell. Comp. Physiol. **5**, 495–508

Davies, I., King, P. E.: 1975. The structure of the rectal papilla in a parasitoid hymenopteran *Nasonia vitripennis* Hymenoptera Pteromalidae. Cell Tissue Res. **161**, 413–419

Davies, L.: 1948. Laboratory studies on the egg of the blowfly, *Lucilia sericata* (Mg.). J. Exp. Biol. **25**, 71–85

Davies, M. E., Edney, E. B.: 1952. The evaporation of water from spiders. J. Exp. Biol. **29**, 571–582

Davis, M. T. B.: 1974a. Critical temperature and changes in cuticular lipids in the rabbit tick, *Haemaphysalis leporispalustris*. J. Insect Physiol. **20**, 1087–1100

Davis, M. T. B.: 1974b. Changes in critical temperature during nymphal and adult development in the Rabbit Tick, *Haemaphysalis leporispalustris* (Acari: Ixodides: Ixodidae). J. Exp. Biol. **60**, 85–94

Dejours, P.: 1975. Principles of Comparative Respiratory Physiology. Amsterdam: Elsevier, 253 pp

Delachambre, J.: 1975. Variations in structure of the abdominal cuticle in *Tenebrio molitor* Insecta Coleoptera. Tissue Cell **7**, 669–676

Delphin, R.: 1963. Histology and possible functions of neurosecretory cells in the ventral ganglia of *Schistocerca gregaria* Forsk. Nature (London) **200**, 913–915

Delye, G.: 1969. Permeabilité du tegument et resistance aux temperatures elevées de quelques Arthropodes sahariens. Bull. Soc. Entomol. France **74**, 51–55

Den Boer, P. J.: 1961. The ecological significance of activity patterns in the woodlouse *Porcellio scaber* Latr. (Isopoda). Arch. Neer. Zool. **14**, 283–409

Dennell, R.: 1946. A study of an insect cuticle: the larval cuticle of *Sarcophaga falculata* Pand. (Diptera). Proc. Roy. Soc. London, Ser. B **133**, 348–373

Dennell, R., Malek, S. R. A.: 1955. The cuticle of the cockroach *Periplaneta americana*. II. The epicuticle. Proc. Roy. Soc. London, Ser. B **143**, 239–257

Dennell, R., Malek, S. R. A.: 1956. The cuticle of the cockroach *Periplaneta americana*. VI. The composition of the cuticle as determined by quantitive analysis. Proc. Roy. Soc. London, Ser. B **145**, 249–258

Dethier, V. G.: 1974. Sensory input and the inconstant fly. In: Experimental Analysis of Insect Behaviour. Browne, B. L. (ed.). Berlin–Heidelberg–New York: Springer, pp. 21–31

Dethier, V. G., Bodenstein, D.: 1958. Hunger in the blowfly. Z. Tierpsychol. **15**, 129–140

Dethier, V. G., Evans, D. R.: 1961. The physiological control of water ingestion in the blowfly. Biol. Bull. Woods Hole **121**, 108–116

Devine, T. L., Wharton, G. W.: 1973. Kinetics of water exchange between a mite, *Laelaps echidnina*, and the surrounding air. J. Insect Physiol. **19**, 243–254

Diamond, J. M., Bossert, W. H.: 1967. Standing gradient osmotic flow. A mechanism for coupling of water and solute transport in epithelia. J. Gen. Physiol. **50**, 2061–2083

Diamond, J. M., Bossert, W. H.: 1968. Functional consequences of ultrastructural geometry in "backwards" fluid-transporting epithelia. J. Cell Biol. **37**, 694–702

Diehl, P. A.: 1973. Paraffin synthesis in the oenocytes of the desert locust. Nature (London) **243**, 468–470

Dinsdale, D.: Excretion in an oribatid mite *Phthiracarus* sp. (Arachnida : Acari). J. Zool. London **177**, 225–231

Djajakusumah, T., Miles, P. W.: 1966. Changes in the relative amounts of soluble protein and amino acid in the haemolymph of the locust, *Chortoicetes terminifera* Walker (Orthoptera : Acrididae), in relation to dehydration and subsequent hydration. Aust. J. Biol. Sci. **19**, 1081–1094

Dodds, S. E., Ewer, D. W.: 1952. On the rate of water loss of *Peripatopsis moseleyi* (Wood-Mason). Ann. Natal Mus. **12**, 275–278.

Downs, J. A.: 1973. Lepidoptera feeding at puddle-margins, dung and carrion. J. Lepidop. Soc. **27**, 89–99

Drach, P.: 1939. Mue et cycle d'intermue chez les crustacés décapodes. Ann. Inst. Oceanog. Paris (N. S.) **19**, 103–392

Drach, P.: 1944. Étude preliminaire sur le cycle d'intermue et son conditionnement hormonal chez *Leander serratus* (Pennant). Bull. Biol. France Belge **78**, 40–62

Dresel, I. B., Moyle, V.: 1950. Nitrogenous excretion of amphipods and isopods. J. Exp. Biol. **27**, 210–225

Dreyer, W. A.: 1935. Water content of insects in relation to temperature and humidity. Biol. Bull. **69**, 338–339

Dreyer, W. A.: 1938. Seasonal weight and total water content of the mound building ant, *Formica exsecoides* Forel. Ecology **19**, 38–49

Drummond, F. H.: 1953. The eversible vesicles of *Campodea* (Thysanura). Proc. Roy. Entomol. Soc. London, Ser. A **28**, 145–148

Duchâteau, G., Florkin, M., Leclercq, J.: 1953. Concentrations des bases fixes et types de composition de la base totale de l'hémolymphe des insects. Arch. Intern. Physiol. Biochim. **61**, 518

Duffey, E.: 1962. A population study of spiders in limestone grassland—the field layer fauna. Oikos **13**, 15–34

Dunbar, B. S., Winston, P. W.: 1975. The site of active uptake of atmospheric water in larvae of *Tenebrio molitor*. J. Insect Physiol. **21**, 495–500

Ebeling, W.: 1961. Physico-chemical mechanisms for the removal of insect wax by means of finely divided powder. Hilgardia **30**, 531–564

Ebeling, W.: 1974. Permeability of insect cuticle. In: The Physiology of Insecta. Rockstein, M. (ed.). New York: Academic Press, Vol. VI, pp. 271–343

Edney, E. B.: 1947. Laboratory studies on the bionomics of the rat fleas, *Xenopsylla brasiliensis*, Baker, and *X. cheopis*, Roths. II. Water relations during the cocoon period. Bull. Entomol. Res. **38**, 263–280

Edney, E. B.: 1951a. The body temperature of woodlice. J. Exp. Biol. **28**, 271–280

Edney, E. B.: 1951b. The evaporation of water from woodlice and the millipede *Glomeris*. J. Exp. Biol. **28**, 91–115

Edney, E. B.: 1953. The temperature of woodlice in the sun. J. Exp. Biol. **30**, 331–349

Edney, E. B.: 1954. Woodlice and the land habitat. Biol. Rev. **29**, 185–219

Edney, E. B.: 1957. The water relations of terrestrial arthropods. Monographs in Experimental Biology, 5. 109 pp. Cambridge

Edney, E. B.: 1960a. Terrestrial adaptations. In: The Physiology of Crustacea. Waterman, T. H. (ed.). New York: Academic Press, Vol. 1, pp. 367–393

Edney, E. B.: 1960b. The survival of animals in hot deserts. Smithsonian Reports for 1959 407–425

Edney, E. B.: 1961. The water and heat relationships of Fiddler crabs (*Uca* spp.). Trans. Roy. Soc. S. Afr. **36**, 71–91

Edney, E. B.: 1966. Absorption of water vapour from unsaturated air by *Arenivaga* sp. (Polyphagidae, Dictyoptera). Comp. Biochem. Physiol. **19**, 387–408

Edney, E. B.: 1968a. Transition from water to land in isopod crustaceans. Am. Zool. **8**, 309–326

Edney, E. B.: 1968b. The effect of water loss on the haemolymph of *Arenivaga* sp. and *Periplaneta americana*. Comp. Biochem. Physiol. **25**, 149–158

Edney, E. B.: 1971. Some aspects of water balance in tenebrionid beetles and a thysanuran from the Namib Desert of South Africa. Physiol. Zool. **44**, 61–76

Edney, E. B.: 1971. The body temperature of tenebrionid beetles in the Namib desert of Southern Africa. J. Exp. Biol. **55**, 253–272

Edney, E. B.: 1974. Desert arthropods. In: Desert Biology. Brown, G. W. (ed.). New York: Academic Press, Vol. II, pp. 311–384

Edney, E. B.: 1975. Absorption of water vapour from unsaturated air. In: Physiological Adaptation to the Environment. Vernberg, J. (ed.). New York: Intext Educational, pp. 77–97

Edney, E. B., Barrass, R.: 1962. The body temperature of the tsetse-fly *Glossina morsitans* Westwood (Diptera Muscidae). J. Insect Physiol. **8**, 469–481

Edney, E. B., Haynes, S., Gibo, D.: 1974. Distribution and activity of the desert cockroach *Arenivaga investigata* (Polyphagidae) in relation to microclimate. Ecology **55**, 420–427

Edney, E. B., McFarlane, J.: 1974. The effect of temperature on transpiration in the desert cockroach, *Arenivaga investigata*, and in *Periplaneta americana*. Physiol. Zool. **47**, 1–12

Edney, E. B., Spencer, J. O.: 1955. Cutaneous respiration in woodlice. J. Exp. Biol. **32**, 256–269

Edwards, J. S.: 1962. A note on water uptake and gustatory discrimination in a predatory reduviid (Hemiptera). J. Insect Physiol. **8**, 113–115

Edwards, L. B., Patton, R. L.: 1967. Carbonic anhydrase in the house cricket, *Ancheta domesticus*. J. Insect Physiol. **13**, 1333–1341

Evans, A. C.: 1934. Studies on the influence of the environment on the sheep blow fly, *Lucilia sericata* Meig. I. The influence of temperature and humidity on the egg. Parasitology **26**, 366–377

Evans, A. C.: 1944. Observations on the biology and physiology of wireworms of the genus *Agriotes* Esch. Ann. Appl. Biol. **31**, 235–250

Evans, D. R.: 1961. Control of the responsiveness of the blowfly to water. Nature (London) **190**, 1132–1133

Evans, D. R., Mellon, D.: 1961. Electrophysiological studies of a water receptor associated with the taste sensilla of the blowfly. J. Gen. Physiol. **45**, 487–500

Ewer, D. W., Ewer, R. F.: 1942. The biology and behaviour of *Ptinus tectus* (Coleoptera, Ptinidae) a pest of stored products. III. The effect of temperature and humidity on oviposition, feeding, and duration of life cycle. J. Exp. Biol. **18**, 290–305

Falke, H.: 1931. Beiträge zur Lebensgeschichte und zur Partembryonal-Entwicklung von *Ixodes ricinus*. Z. Morph. Ökol. Tiere **21**, 567–607

Farquharson, P. A.: 1974a. A study of the Malpighian tubules of the pill millipede, *Glomeris marginata* (Villers). I. The isolation of the tubules in a ringer solution. J. Exp. Biol. **60**, 13–28

Farquharson, P. A.: 1974b. A study of the Malpighian tubules of the pill millipede, *Glomeris marginata* (Villers). II. The effect of variations in osmotic pressure and sodium and potassium concentrations on fluid production. J. Exp. Biol. **60**, 29–39

Farquharson, P. A.: 1974c. A study of the Malpighian tubules of the pill millipede, *Glomeris marginata* (Villers). III. The permeability characteristics of the tubule. J. Exp. Biol. **60**, 41–51

Fayadh, L.: 1969. Salt and water balance in the desert locust *Schistocerca gregaria* Forskal. M. Sc. Thesis, University of Newcastle-upon-Tyne

Filshie, B. K.: 1970a. The resistance of epicuticular components of an insect to extraction with lipid solvents. Tissue Cell **2**, 181–190

Filshie, B. K.: 1970b. The fine structure and deposition of the larval cuticle of the sheep blowfly (*Lucilia cuprina*). Tissue Cell **2**, 479–489

Florkin, M., Jeuniaux, C.: 1974. Hemolymph: composition. In: The Physiology of Insecta. Rockstein, N. (ed.). New York: Academic Press, Vol. V, pp. 256–307

Fraenkel, G., Blewett, M.: 1943. The Vitamin B-complex requirements of several insects. Biochem. J. **37**, 686–692

Fraenkel, G., Blewett, M.: 1944. The utilisation of metabolic water in insects. Bull. Entomol. Res. **35**, 127–139

Frayha, G. J., Dajani, R. M., Almaz, O., Sweatman, G. K.: 1974. Chemical composition of the coxal fluid of the argasid tick *Ornithodoros savignyi*. J. Med. Entomol. **11**, 168–172

Free, J. B.: Spencer-Booth, Y.: 1958. Observations on the temperature regulation and food consumption of honey bees (*Apis mellifera*). J. Exp. Biol. **35**, 930–937

Free, J. B., Spencer-Booth, Y.: 1962. The upper lethal temperatures of honey bees. Entomol. Exp. Appl. **5**, 249–254

Friauf, J. J., Edney, E. B.: 1969. A new species of *Arenivaga* from desert sand dunes in southern California. Proc. Entomol. Soc. Washington **71**, 1–7

Frick, J. H., Brown, D. J., Sauer, J. R.: 1974. Determining the extracellular fluid volume (inulin space) of salivary glands of the lone star tick, *Amblyomma americanum* (L.). Ann. Entomol. Soc. Am. **67**, 994–996

Füller, H.: 1966. Elektronenmikroskopische Untersuchungen der Malpighischen Gefäße von *Lithobius forficatus* (L.). Z. Wiss. Zool. **173**, 191–217

Galbreath, R. A.: 1975. Water balance across the cuticle of an insect. J. Exp. Biol. **62**, 115–120

Gee, J. D.: 1975a. Diuresis in the tsetse-fly *Glossina austeni*. J. Exp. Biol. **63**, 381–390

Gee, J. D.: 1975b. The control of diuresis in the tsetse-fly *Glossina austeni*: a preliminary investigation of the diuretic hormone. J. Exp. Biol. **63**, 391

Gee, J. D.: 1976. Active transport of sodium by the Malpighian tubules of the tsetse-fly *Glossina morsitans*. J. Exp. Biol. **64**, 357–368

Gese, P. K.: 1950. The concentration of certain inorganic constituents in the blood of Cynthia pupa *Samia walkeri* Felder and Felder. Physiol. Zool. **23**, 109–113

Gibbs, K. E., Morrison, F. O.: 1959. The cuticle of the two-spotted spider mite, *Tetranychus telarius* (Linnaeus) (Acarina : Tetranychidae). Can. J. Zool. **37**, 633–637

Gifford, C. A.: 1968. Accumulation of uric acid in the land crab *Cardisoma guahumi*. Am. Zool. **8**, 521–539

Gilbert, L. I.: 1967a. Lipid metabolism and function in insects. In: Advances in Insect Physiology. Beament, J. W. L., Treherne, J. E., Wigglesworth, V. B. (eds.). Amsterdam: Elsevier, Vol. IV, pp. 69–211

Gilbert, L. I.: 1967b. Biochemical correlations in insect metamorphosis. In: Comprehensive Biochemistry. Florkin, M. F., Stotz, E. H. (eds.). Amsterdam: Elsevier, Vol. XXVIII, pp. 199–252

Gilbert, L. I., O'Connor, J. D.: 1970. Lipid metabolism and transport in arthropods. Chem. Zool. **5A**, 229–253

Gilby, A. R.: 1965. Lipids and their metabolism in insects. Ann. Rev. Entomol. **10**, 141–160

Gilby, A. R., Cox, M. E.: 1963. The cuticular lipids of the cockroach, *Periplaneta americana* (L.). J. Insect Physiol. **9**, 671–681

Gilby, A. R., McKellar, J. W.: 1970. The composition of the empty puparia of a blowfly. J. Insect Physiol. **16**, 1517–1529

Gluud, A.: 1968. Zur Feinstruktur der Insectencuticula. Ein Beitrag zur Frage des Eigengiftschutzes der Wanzencuticula. Zool. Jb. Anat. **85**, 191–227

Goodchild, A. J. P.: 1963. Some new observations on the intestinal structures concerned with water disposal in sap-sucking Hemiptera. Trans. Roy. Entomol. Soc. London **115**, 217–237

Goodchild, A. J. P.: 1966. Evolution of the alimentary canal in the Hemiptera. Biol. Rev. Cambridge Phil. Soc. **47**, 97–140

Goodrich, B. S.: 1970. Cuticular lipids of adults and puparia of the Australian blowfly *Lucilia cuprina* (Wied.). J. Lipid Res. **11**, 1–6

Gouranton, J.: 1968a. Composition, structure, et mode de formation des concrétions minérales dans l'intestine moyen des homoptères cercopides. J. Cell Biol. **37**, 316–328

Gouranton, J.: 1968b. Étude ultrastructurale du système cryptonéphridien de *Cicadella viridis* L. (Homoptera, Jassidae). C. R. Acad. Sci., Paris **266**, 1403–1406

Gouranton, J.: 1968c. Ultrastructures en rapport avec un transit d'eau. Etude de la "chambre filtrante" de *Cicadella viridis* L. (Homoptera, Jassidae). J. Microscopie **7**, 559–574

Graham-Smith, G. S.: 1934. The alimentary canal of *Calliphora erythrocephala* L., with special reference to its musculature and to the proventriculus, rectal valve and papillae. Parasitology **26**, 176–248

Gregson, J. D.: 1967. Observations on the movement of fluids in the vicinity of the mouth parts of naturally feeding *Dermacentor andersoni* Stiles. Parasitology **57**, 1–8

Grimstone, A. V., Mullinger, A. M., Ramsay, J. A.: 1968. Further studies on the rectal complex of the mealworm, *Tenebrio molitor* L. (Coleoptera, Tenebrionidae). Phil. Trans. Roy. Soc. London **253**, 343–382

Gross, W. J.: 1955. Aspects of osmotic regulation in crabs showing the terrestrial habit. Am. Natur. **89**, 205–222

254

Gross, W. J.: 1964. Trends in water and salt regulation among aquatic amphibious crabs. Biol. Bull. **127**, 447–466

Gross, W. J., Lasiewski, R. C., Dennis, M., Rudy, P.: 1966. Salt and water balance in selected crabs of Madagascar. Comp. Biochem. Physiol. **17**, 641–660

Gubb, D.: 1975. A direct visualisation of helicoidal architecture in *Carcinus maenas* and *Halocynthia papillosa* by scanning electron microscopy. Tissue Cell **7**, 19–32

Gunn, D. L.: 1933. The temperature, and humidity relations of the cockroach *Blatta orientalis*. I. Desiccation. J. Exp. Biol. **10**, 274–285

Gunn, D. L.: 1942. Body temperature in poikilothermal animals. Biol. Rev. **17**, 293–314

Gunn, D. L., Notley, F. B.: 1936. The temperature and humidity relations of the cockroach. IV. Thermal death-point. J. Exp. Biol. **13**, 28–34

Gupta, B. L.: Water movement in cells and tissues. (In press)

Gupta, B. L., Berridge, M. J.: 1966a. A coat of repeating subunits on the cytoplasmic surface of the plasma membrane in the rectal papillae of the blowfly, *Calliphora erythrocephala* (Meig.), studies in situ by electron microscopy. J. Cell Biol. **29**, 376–382

Gupta, B. L., Berridge, M. J.: 1966b. Fine structural organization of the rectum in the blowfly *Calliphora erythrocephala* (Meig.) with special reference to connective tissue, tracheae and neurosecretory innervation in the rectal papillae. J. Morph. **120**, 23–82

Gupta, M.: 1961. Rectal glands in woodlice. Nature (London) **191**, 406–407

Hackman, R. H.: 1971. The integument of arthropods. In: Chemical Zoology. Florkin, M., Scheer, B. T. (eds.). New York: Academic Press, Vol. VI, B, pp. 1–62

Hackman, R. H.: 1974. Chemistry of the insect cuticle. In: The physiology of Insecta. Rockstein, M. (ed.). New York: Academic Press, Vol. VI, pp. 216–270

Hackman, R. H., Goldbert, M.: 1975. *Peripatus*: its affinities and its cuticle. Science **190**, 582

Hadley, N. F.: 1970a. Micrometeorology and energy exchange in two desert arthropods. Ecology **51**, 434–444

Hadley, N. F.: 1970b. Water relations of the desert scorpion, *Hadrurus arizonensis*. J. Exp. Biol. **53**, 547–558

Hadley, N. F.: 1971. Water uptake by drinking in the scorpion, *Centruroides sculpturatus* (Buthidae). The Southwestern Naturalist **15**, 504–505

Hadley, N. F.: 1972. Desert species and adaptation. Am. Scient. **60**, 338–347

Hadley, N. F.: 1974. Adaptational biology of desert scorpions. J. Arachnol. **2**, 11–23

Hafez, M., El-Ziady, S., Hefnawy, T.: 1970. Biochemical and physiological studies of certain ticks (Ixodoidea). Cuticular permeability of *Hyalomma* (H.) *dromedarii* Koch (Ixodidae) and *Ornithodoros* (O.) *savigny* (Audouin) (Argasidae). J. Parasitol. **56**, 154–168

Hagvar, S., Ostbye, E.: 1974. Oxygen consumption, caloric values, water and ash content of some dominant terrestrial arthropods from alpine habitats at Finse, South Norway. Nor. Entomol. Tidsskr. **21**, 117–126

Hair, J. A., Sauer, J. R., Durham, K. A.: 1975. Water balance and humidity preference in three species of ticks. J. Med. Entomol. **12**, 37–47

Hamdy, B. H.: 1972. Biochemical and physiological studies of certain ticks (Ixodoidea). Nitrogenous excretory products of *Argas (Persicargas) arboreus* Kaiser, Hoogstral and Kohls, and of other argassid and ixodid species. J. Med. Entomol. **9**, 346–350

Hamilton, A. G.: 1964. The occurrence of periodic or continuous discharge of carbon dioxide by male desert locusts (*Schistocerca gregaria* Forskål) measured by an infra-red gas analyser. Proc. Roy. Soc. London, Ser. B **160**, 373–395

Hamilton, W. J., Seely, M. K.: 1975. Fog basking by the Namib Desert beetle *Onymacris unguicularis*. Nature (London) **262**, 284–285

Harmsen, R.: 1966. The excretory role of pteridines in insects. J. Exp. Biol. **45**, 1–13

Hartenstein, R.: 1968. Nitrogen metabolism in the terrestrial isopod, *Oniscus asellus*. Am. Zool. **8**, 507–519

Hartley, G. S.: 1948. Contribution to a discussion of asymmetry. Disc. Faraday Soc. **3**, 223

Hartley, J. C.: 1961. The shell of acridid eggs. Quart. J. Microsc. Sci. **102**, 249–255

Harvey, W. R., Nedergaard, S.: 1964. Sodium-independent active transport of potassium in the isolated midgut of the cecropia silkworm. Proc. Nat. Acad. Sci. U.S. **51**, 757–765

Hassan, A. A. G.: 1944. The structure and mechanism of the spiracular regulatory apparatus in adult Diptera and certain other groups of insects. Trans. Roy. Entomol. Soc. London **94**, 103–153

Haverty, M. I., Nutting, W. L.: 1976. Environmental factors affecting the geographical distribution of two ecologically equivalent termite species in Arizona, USA. Amer. Midland Naturalist **95**, 20–27

Hawke, D. D., Farley, R. D.: 1973. Ecology and behavior of the desert burrowing cockroach *Arenivaga* sp. (Dictyoptera, Polyphagidae). Oecologia **11**, 263–279

Hecker, J., Diehl, P. A., Aeschlimann, A.: 1969. Recherches sur l'ultrastructure et l'histochimie de l'organe coxal d'*Ornithodorus moubata* (Murray) (Ixodoidea Argasidae). Acta Trop **26**, 346–360

Hefnawy, T.: 1970. Biochemical and physiological studies of certain ticks (Ixodidea). Water loss from the spiracles of *Hyalomma* (H.) *dromedarii* Koch and *Ornithodorus* (O.) *savignyi* (Audouin). J. Parasitol. **56**, 362–366

Heinrich, B.: 1974. Thermoregulation in endothermic insects. Science **185**, 747–756

Heinrich, B.: 1975. Thermoregulation and flight energetics of desert insects. In: Environmental Physiology of Desert Organisms. Hadley, N. (ed.). Stroudsburg, Penn: Dowden, Hutchinson and Ross, pp. 90–105

Henneberry, T. J., Adams, J. R., Cantwell, G.: 1965. Fine structure of the integument of the two spotted spider mite *Tetranychus telarius* (Acarina : Tetranychidae). Ann. Entomol. Soc. Am. **58**, 532–535

Henwood, K.: 1975. Infrared transmittance as an alternative thermal strategy in the desert beetle *Onymacris plana*. Science **189**, 993–994

Henwood, K.: 1976. A field tested thermoregulation model for two diurnal Namib Desert tenebrionid beetles. Ecology **56**, 1329–1342

Hepburn, H. R., Joffe, I.: 1974. Locust solid cuticle: a time sequence of mechanical properties. S. Afr. J. Sci. **70**, 83

Herreid, C. F.: 1969a. Water loss of crabs from different habitats. Comp. Biochem. Physiol. **28**, 829–839

Herreid, C. F.: 1969b. Integumental permeability of crabs and adaptation to land. Comp. Biochem. Physiol. **29**, 423–429

Hewitt, P. H., Nel, J. J. C., Schoeman, I.: 1971. Influence of group size on water imbibition by *Hodotermes mossambicus* alate termites. J. Insect Physiol. **17**, 587–600

Hill, A. E.: 1975a. Solute-solvent coupling in epithelia: a critical examination of the standing-gradient osmotic flow theory. Proc. Roy. Soc. London, Ser. B **190**, 99–114

Hill, A. E.: 1975b. Solute-solvent coupling in epithelia: an electro-osmotic theory of fluid transfer. Proc. Roy. Soc. London, Ser. B **190**, 115–134

Himmer, A.: 1932. Die Temperaturverhältnisse bei den sozialen Hymenopteren. Biol. Rev. Cambridge Phil. Soc. **7**, 224–253

Hinton, H. E.: 1957. Some aspects of diapause. Sci. Progress **178**, 307–320

Hinton, H. E.: 1960. A fly larva that tolerates dehydration and temperatures of $-270°$ to $+102°C$. Nature (London) **188**, 336–337

Hinton, H. E.: 1963. The respiratory system of the egg-shell of the blowfly, *Calliphora erythrocephala* Meig., as seen with the electron microscope. J. Insect Physiol. **9**, 121–129

Hinton, H. E.: 1968. Structures and protective devices of the egg of the mosquito *Culex pipiens*. J. Insect Physiol. **14**, 145–161

Hinton, H. E.: 1973. Some recent work on the colours of insects and their likely significance. Proc. British Entomol. Soc. **6**, 43–54

Hitchcock, L. F.: 1955. Studies on the non-parasitic stages of the cattle tick *Boophilus microplus* (Canestrini) (Acarina : Ixodidae). Austral. J. Zool. **3**, 295–311

Hochrainer, H.: 1942. Der Wasserhaushalt bei Insekten und die Faktoren, die denselben bestimmen. Zool. Jahrb. Allg. Zool. Physiol. **60**, 387–436

Hodson, A. C.: 1937. Some aspects of the role of water in insect hibernation. Ecol. Monog. **7**, 271–315

Holdgate, M. W., Seal, M.: 1956. The epicuticular wax layers of the pupa of *Tenebrio molitor* L. J. Exp. Biol. **33**, 82–106

Hook Van, R. I.: 1971. Energy and nutrient dynamics of spider and orthopteran populations in a grassland ecosystem. Ecol. Monog. **41**, 1–26

Hook Van, R. I., Deal, S. L.: 1972. Tritium uptake and elimination by tissue-bound and body-water components in crickets (*Acheta domesticus*). J. Insect Physiol. **19**, 681–687

Hopkins, C. R.: 1967. The fine structural changes observed in the rectal papillae of the mosquito *Aedes aegypti* L. and their relation to the epithelial transport of water and inorganic ions. J. Roy. Microsc. Soc. **83**, 235–252

Horne, F. R.: 1969. Purine excretion in five scorpions, a uropygid and a centipede. Biol. Bull. Woods Hole **137**, 155–160

Horowitz, M.: 1970. The water balance of the terrestrial isopod *Porcellio scaber*. Entomol. Exp. Appl. **13**, 173–178

House, C. R.: 1974. Water Transport in Cells and Tissues. Baltimore: Williams and Wilkins. 562 pp

Hoyle, G.: 1954. Changes in the blood potassium concentration of the African migratory locust (*Locusta migratoria migratorioides* R. & F.) during food deprivation, and the effect on neuromuscular activity. J. Exp. Biol. **31**, 260–270

Hoyle, G.: 1960. The action of carbon dioxide gas on an insect spiracular muscle. J. Insect Physiol. **4**, 63–79

Hoyle, G.: 1961. Functional contracture in a spiracular muscle. J. Insect Physiol. **7**, 305–314

Hsu, M. H., Sauer, J. R.: 1974. Sodium, potassium, chloride and water balance in the feeding lone star tick *Amblyomma americanum* (Linnaeus) (Acarina, Ixodidae). J. Kansas Entomol. Soc. **47**, 536–537

Hsu, M. H., Sauer, J. R.: 1975. Ion and water balance in the feeding lone star tick. Comp. Biochem. Physiol. **52A**, 269–276

Humphreys, W. F.: 1975. The influence of burrowing and thermoregulatory behaviour on the water relations of *Geolycosa godeffroyi* (Araneae: Lycosidae), an Australian wolf spider. Oecologia **21**, 291–311

Humphries, D. A.: 1967. Uptake of atmospheric water by the hen flea *Ceratophyllus gallinae* (Schrank). Nature (London) **214**, 426

Hurley, D. E.: 1959. Notes on the ecology and environmental adaptations of the terrestrial amphipoda. Pacific Sci. **13**, 107–129

Hurley, D. E.: 1968. Transition from water to land in amphipod crustaceans. Am. Zool. **8**, 327–353

Hurst, H.: 1941. Insect cuticle as an asymmetric membrane. Nature (London) **147**, 388–389

Hurst, H.: 1948. Asymmetrical behavior of insect cuticle in relation to water permeability. Discuss. Faraday Soc. **3**, 193–210

Irvine, H. B.: 1966. In vitro rectal transport and rectal ultrastructure in the desert locust, *Schistocerca gregaria*. M. Sc. Thesis, University of British Columbia, Vancouver, B. C.

Irvine, H. B., Phillips, J. E.: 1971. Effects of respiratory inhibitors and ouabain on water transport by isolated locust rectum. J. Insect Physiol. **17**, 381–383

Ito, S.: 1961. The endoplasmic reticulum of gastric parietal cells. J. Biophys. Biochem. Cytol. **11**, 333–348

Jack, R. W.: 1939. Studies on the physiology and behaviour of *Glossina morsitans* Westwood. Mem. Dept. Agric. Salisbury, Rhodesia. **1**, 1–203

Jackson, C. H. N.: 1937. Water and fat content of tsetse flies. Nature (London) **139**, 674–675

Jackson, L. L.: 1970. Cuticular lipids of insects: II. Hydrocarbons of the cockroaches *Periplaneta australasiae*, *Periplaneta brunnea*, and *Periplaneta fuliginosa*. Lipids **5**, 38–41

Jackson, L. L., Armold, M. T., Regnier, F. E.: 1974. Cuticular lipids of adult fleshflies, *Sarcophaga bullata*. Insect Biochem. **4**, 369–379

Jackson, L. L., Baker, G. L.: 1970. Cuticular lipids of insects. Lipids **5**, 239–246

Jahn, T. L.: 1936. Studies on the nature and permeability of the grasshopper egg membranes: III. Changes in electrical properties of the membranes during development. J. Cell. Comp. Physiol. **8**, 289–300

Jakovlev, V., Kruger, F.: 1953. Vergleichende Untersuchungen zur Physiologie der Transpiration der Orthopteren. Zool. Physiol. Tiere **64**, 391–428

Jeuniaux, C.: 1971. Hemolymph—Arthropoda. In: Chemical Zoology. Florkin, H., Scheer, B. T. (eds.). New York: Academic Press, Vol. VI, pp. 63–118

Jochum, F.: 1956. Changes in the reaction chains in the insect organism caused by diethyl-nitrophenyl thiosulphate. Hofchen-Briefe **9**, 289–384

Johnson, C. G.: 1942. Insect survival in relation to the rate of water loss. Biol. Rev. Cambridge Phil. Soc. **17**, 151–177

Jones, B. M.: 1954. On the role of the integument in acarine development and its bearing on pupa formation. Quart. J. Microsc. Sci. **95**, 169–181

Jones, E. W.: 1951. Laboratory studies on the moisture relations of *Limonius* (Coleoptera : Elateridae). Ecology **32**, 284–293

Jones, R. E.: 1975. Dehydration in an Australian rockpool chironomid larva *(Paraborniella tonnoiri)*. J. Entomol. **49 A**, 111–119

Kafatos, F. C.: 1968. The labial gland : a salt secreting organ of saturniid moths. J. Exp. Biol. **48**, 435–453

Kalmus, H.: 1936. Die Verwendung der Tracheenblasen der Corethralarve als Mikrohygrometer. Z. Wiss. Mikrosk. **53**, 215–219

Kanungo, K.: 1965. Oxygen uptake in relation to water balance of a mite *(Echinolaelaps echidninus)* in unsaturated air. J. Insect Physiol. **11**, 557–568

Kanwisher, J. W.: 1966. Tracheal gas dynamics in pupae of the cecropia silkworm. Biol. Bull. Woods Hole **130**, 96–105

Kaufman, W. R., Phillips, J. E.: 1973 a. Ion and water balance in the ixodid tick *Dermacentor andersoni*. I. Routes of ion and water excretion. J. Exp. Biol. **58**, 523–536

Kaufman, W. R., Phillips, J. R.: 1973 b. Ion and water balance in the ixodid tick *Dermacentor andersoni*. II. Mechanism and control of salivary secretion. J. Exp. Biol. **58**, 537–547

Kaufman, W. R., Phillips, J. E.: 1973 c. Ion and water balance in the ixodid tick *Dermacentor andersoni*. III. Influence of monovalent ions and osmotic pressure on salivary secretion. J. Exp. Biol. **58**, 549–564

Kaufman, Z. S.: 1962. Structure and development of stigmata in *Lithobius forficatus* L. (Chilopoda, Lithobiidae). Ent. Obozr. **41**, 223–225

Kennedy, J. S., Fosbrooke, I. H. M.: 1972. The plant in the life of an aphid. Symp. Roy. Entomol. Soc. London **6**, 129–140

Kessel, R. G.: 1970. The permeability of dragonfly Malpighian tubule cells to protein using horseradish peroxidase as a tracer. J. Cell Biol. **47**, 299–303

Kevan, D. K. McE.: 1962. Soil animals. London: Witherby, 237 pp

Kilby, B. A.: 1963. The biochemistry of the insect fat body. In: Advances in Insect Physiology. Beament, J. W. L., Treherne, J. E., Wigglesworth, V. B. (eds.). London–New York: Academic Press, 1963, Vol. 1, pp. 111–174

King, G.: 1944. Permeability of keratin membranes. Nature (London) **154**, 575–576

Kirkland, W. L.: 1971. Ultrastructural change in the nymphal salivary glands of the rabbit tick, *Haemaphysalis leporispalustris*, during feeding. J. Insect Physiol. **17**, 1933–1946

Kitaoka, S., Morii, T.: 1970. Ionic and water balance in the feeding process of ixodid ticks. Nat. Inst. Anim. Hlth. Quart. **10**, 34–41

Knülle, W.: 1962. Die Abhängigkeit der Luftfeuchtereaktionen der Mehlmilbe *(Acarus siro)* vom Wassergehalt des Körpers. Z. Vergl. Physiol. **45**, 233–246

Knülle, W.: 1965. Equilibrium humidities and survival of some tick larvae. J. Med. Entomol. **2**, 335–338

Knülle, W.: 1967. Significance of fluctuating humidities and frequence of blood meals on the survival of the spiny rat mite, *Echinolaelaps echidninus* (Berlese). J. Med. Entomol. **4**, 322–325

Knülle, W.: 1967. Physiological properties and biological implications of the water vapour sorption mechanism in larvae of the Oriental rat flea, *Xenopsylla cheopis* (Roths.). J. Insect Physiol. **13**, 333–357

Knülle, W., Devine, T. L.: 1972. Evidence for active and passive components of sorption of atmospheric water vapour by larvae of the tick *Dermacentor variabilis*. J. Insect Physiol. **18**, 1653–1664

Knülle, W., Spadafora, R. R.: 1969. Water vapour sorption and humidity relationships in *Liposcelis* (Insecta : Psocoptera). J. Stored Prod. Res. **5**, 49–55

Knülle, W., Wharton, G. W.: 1964. Equilibrium humidities in arthropods and their ecological significance. Acarologia **6**, 299–306

Koidsumi, K.: 1934. Experimentelle Studien über die Transpiration und den Wärmehaushalt bei Insekten. Mem. Fac. Sci. Agric. Taihoku **12**, 1–380

Krafsur, E. S.: 1971 a. Behavior of thoracic spiracles of *Aedes* mosquitoes in controlled relative humidities. Ann. Entomol. Soc. Am. **64**, 93–97

Krafsur, E. S.: 1971 b. Influence of age and water balance on spiracular behavior in *Aedes* mosquitoes. Ann. Entomol. Soc. Am. **64**, 97–102

Krijgsman, B. J.: 1936. Beschouwingen over de voedingsreactie van carnivoren arthropoda. Vakblad Voor Biologen **18**, 58–70

Krishnan, G.: 1951. Phenolic tanning and pigmentation of the cuticle in *Carcinus maenas*. Quart. J. Microsc. Sci. **92**, 333–342

Krogh, A.: 1916. The Respiratory Exchange of Animals and Man. Monographs of Biochemistry. London: Longmans, Green

Krogh, A.: 1919. The diffusion of gases through animal tissues. J. Physiol. **52**, 391–408

Krogh, A.: 1920. Studien über Tracheenrespiration. II. Die Gasdiffusion in den Tracheen. Pflug. Arch. Ges. Physiol. **179**, 95–112

Kuemmel, G., Zerbst-Boroffka, I.: 1974. Elektronenmikroskopische und physiologische Untersuchungen an den Rectalpolstern von *Apis mellifica*. Cytobiologie **9**, 432–459

Kuenen, D. J.: 1959. Excretion and water balance in some land isopods. Entomol. Exp. Appl. **2**, 287–294

Lafon, M.: 1943. Recherches biochimiques et physiologiques sur le squelette tegumentaire des arthropodes. Ann. Sci. Natur. **11**, 113–146

Laird, T. B., Winston, P. W.: 1975. Water and osmotic pressure regulation in the cockroach, *Leucophaea maderae*. J. Insect Physiol. **21**, 1055–1060.

Laird, T. B., Winston, P. W., Braukman, M.: 1972. Water storage in the cockroach *Leucophaea maderae* F. Naturwissenschaften **59**, 515–516

Lamb, M. J.: 1968. Temperature and lifespan in *Drosophila*. Nature (London) **220**, 808–809

Lane, N. J., Treherne, J. E.: 1972. Studies on perineural junctional complexes and the site of uptake of microperoxidase and lanthanum in the cockroach central nervous system. Tissue Cell **4**, 427–436

Laughlin, R.: 1957. Absorption of water by the eggs of the garden chafer, *Phyllopertha horticola* L. J. Exp. Biol. **34**, 226–236

Laughlin, R.: 1958. Desiccation of eggs of the crane-fly *Tipula oleracea* L. Nature (London) **182**, 613

Laviolette, P., Mestres, G.: 1967. Les variatons du pH et de la pression osmotique de l'hémolymphe de *Galleria mellonella* L. Compt. Rend. Ser. D **265**, 979–982

Leclercq, J.: 1946. Des insectes qui boivent de l'eau. Bull Ann. Soc. Entomol. Belgique **82**, 71–75

Lee, R. M.: 1961. The variation of blood volume with age in the desert locust *(Schistocerca gregaria* Forsk.). J. Insect Physiol. **6**, 36–51

Lees, A. D.: 1946a. The water balance in *Ixodes ricinus* L. and certain other species of ticks. Parasitology **37**, 1–20

Lees, A. D.: 1946b. Chloride regulation and the function of the coxal glands in ticks. Parasitology **37**, 172–184

Lees, A. D.: 1947. Transpiration and the structure of the epicuticle in ticks. J. Exp. Biol. **23**, 379–410

Lees, A. D.: 1948a. The sensory physiology of the sheep tick, *Ixodes ricinus* L. J. Exp. Biol. **25**, 145–207

Lees, A. D.: 1948b. Passive and active water exchange through the cuticle of ticks. Disc. Faraday Soc. **3**, 187–192

Lees, A. D.: 1976. The role of pressure in controlling the entry of water into the developing eggs of the Australian plague locust *Chortoicetes terminifera* (Walker). Physiol. Entomol. **1**, 39–50

Lees, A. D., Beament, J. W. L.: 1948. An egg-waxing organ in ticks. Quart. J. Microsc. Sci. **89**, 291–333

Leslie, R. A., Robertson, H. A.: 1973. The structure of the salivary gland of the moth *Manduca sexta*. Z. Zellforsch. Mikrosk. Anat. **146**, 553–564

Lester, H. M. O., Lloyd, L.: 1928. Notes on the process of digestion in tsetse flies. Bull. Entomol. Res. **19**, 39–60

Levenbook, L.: 1950. The composition of horse bot fly *(Gastrophilus intestinalis)* larva blood. Biochem. J. **47**, 336–346

Levy, R. I., Schneiderman, H. A.: 1966a. Discontinuous respiration in insects. II. The direct measurement and significance of changes in tracheal gas composition during the respiratory cycle of silkworm pupae. J. Insect Physiol. **12**, 83–104

Levy, R. I., Schneiderman, H. A.: 1966b. Discontinuous respiration in insects. III. The effect of temperature and ambient oxygen tension in the gaseous composition of the tracheal system of silkworm pupae. J. Insect Physiol. **12**, 105–121

Levy, R. I., Schneiderman, H. A.: 1966c. Discontinuous respiration in insects. IV. Changes in intratracheal pressure during the respiratory cycle of silkworm pupae. J. Insect Physiol. **12**, 465–492

Lewis, J. G. E.: 1963. On the spiracle structure and resistance to desiccation of four species of geophilomorph centipedes. Entomol. Exp. Appl. **6**, 89–94

Licent, E.: 1912. Recherches d'anatomie et de physiologie comparées sur le tubule digestif des Homoptères supérieurs. Cellule **28**, 1–161

Lincoln, D. C. R.: 1961. The oxygen and water requirements of the eggs of *Ocypus olens* Müller (Staphylinidae, Coleoptera). J. Insect Physiol. **7**, 265–272

Lindauer, M.: 1954. Temperaturregulierung und Wasserhaushalt im Brenenstaat. Z. Vergl. Physiol. **36**, 391–432

Lindqvist, O. V.: 1968. Water regulation in terrestrial isopods, with comments on their behavior in a stimulus gradient. Ann. Zool. Fenn. **5**, 279–311

Lindqvist, O. V.: 1971. Evaporation in terrestrial isopods is determined by oral and anal discharge. Experientia **27**, 1496–1498

Lindqvist, O. V.: 1972. Components of water loss in terrestrial isopods. Physiol. Zool. **45**, 316–324

Lindqvist, O. V., Fitzgerald, G.: 1973. The gut as water source in the terrestrial isopod *Porcellio scaber* Latr. Am. Zool. **13**, 236

Lindqvist, O. V., Salminen, I., Winston, P. W.: 1972. Water content and water activity in the cuticle of terrestrial isopods. J. Exp. Biol. **56**, 49–55

Lindsay, E.: 1940. The biology of the silverfish, *Ctenolepisma longicaudata* Esch. with particular reference to its feeding habits. Proc. Roy. Soc. Victoria **52**, 35–83

Locke, M.: 1960. Cuticle and wax secretion in *Calpodes ethlius* (Lepidoptera, Hesperidae). Quart. J. Microsc. Sci. **101**, 333–338

Locke, M.: 1961. Pore canals and related structures in insect cuticle. J. Biophys. Biochem. Cytol. **10**, 589–618

Locke, M.: 1964. The structure and formation of the integument in insects. In: The Physiology of Insecta. Rockstein, M. (ed.). New York: Academic Press, Vol. III, pp. 379–470

Locke, M.: 1965. Permeability of insect cuticle to water and lipids. Science **147**, 295–298

Locke, M.: 1966. The structure and formation of the cuticulin layer in the epicuticle of an insect, *Calpodes ethlius* (Lepidoptera, Hesperiidae). J. Morph. **118**, 461–494

Locke, M.: 1974. The structure and formation of the integument of insects. In: The Physiology of Insecta. Rockstein, M. (ed.). New York: Academic Press, Vol. VI, pp. 124–213

Locke, M., Collins, J. V.: 1967. Protein uptake in multivesicular bodies in the molt-intermolt cycle of an insect. Science **155**, 467–469

Locke, M., Collins, J. V.: 1968. Protein uptake into multivesicular bodies and storage granules in the fat body of an insect. J. Cell Biol. **36**, 453–483

Locke, M., Krishnan, G.: 1971. The distribution of phenoloxidases and polyphenols during cuticle formation. Tissue Cell **3**, 103–126

Lockwood, A. P. M., Croghan, P. C.: 1959. Composition of the haemolymph of *Petrobius maritimus* Leach. Nature (London) **184**, 370–371

Lok, J. B., Cupp, E. W., Blomquist, G. J.: 1975. Cuticular lipids of the imported fire ants, *Solenopsis invicta* and *richteri*. Insect Biochem. **5**, 821–829

Londt, J. G. H., Whitehead, G. B.: 1972. Ecological studies of larval ticks in South Africa (Acarina : Ixodidae). Parasitology **65**, 469–490

Louw, G. N.: 1972. The role of advective fog in the water economy of certain Namib Desert animals. Symp. Zool. Soc. London **31**, 297–314

Louw, G. N., Hamilton, W. J., III.: 1972. Physiological and behavioural ecology of the ultra-psammophilous Namib Desert tenebrionid, *Lepidochora argentogrisea*. Madoqua **1**, 87–95

Loveridge, J. P.: 1967. Desiccation and ventilatory control of water loss in *Locusta migratoria migratorioides* R. and F. Nature (London) **214**, 1143–1144

Loveridge, J. P.: 1968a. The control of water loss in *Locusta migratoria migratorioides* R. and F. II. Water loss through the spiracles. J. Exp. Biol. **49**, 15–29

Loveridge, J. P.: 1968b. The control of water loss in *Locusta migratoria migratorioides* R. and F. I. Cuticular water loss. J. Exp. Biol. **49**, 1–13

Loveridge, J. P.: 1973. Age and the changes in water and fat content of adult laboratory-reared *Locusta migratoria migratorioides* R. and F. Rhodesian J. Agric. Res. **11**, 131–143

Loveridge, J. P.: 1974. Studies on the water relations of adult locusts. II. Water gain in the food and loss in the faeces. Trans. Rhodesia Sci. Assoc. **56**, 1–30

Loveridge, J. P.: 1975. Studies on the water balance of adult locusts. III. The water balance of non-flying locusts. Zoologica Africana **10**, 1–28

Loveridge, J. P., Bursell, E.: 1975. Studies on the water relations of adult locusts (Orthoptera, Acrididae). 1. Respiration and the production of metabolic water. Bull. Entomol. Res. **65**, 13–20

Lowry, W. P.: 1969. Weather and Life: An Introduction to Biometeorology. New York: Academic Press, 305 pp

Luca, V. de, Viscuso, R.: 1974. Morphologie de l'ootheque et du chorion de l'oeuf d'*Eyprepocnemis plorans* Charp. (Orthoptera : Acrididae). Acrida **3**, 267–275

Ludwig, D.: 1936. The effect of desiccation on survival and metamorphosis of the Japanese beetle (*Popillia japonica* Newman). Physiol. Zool. **9**, 27–42

Ludwig, D.: 1937. The effect of different relative humidities on respiratory metabolism and survival of the grasshopper *Chortophaga viridifasciata* De Geer. Physiol. Zool. **10**, 342–351

Ludwig, D.: 1950. The metabolism of starved nymphs of the grasshopper, *Chortophaga viridifasciata* De Geer. Physiol. Zool. **23**, 41–47

Ludwig, D., Anderson, J. M.: 1942. Effects of different humidities, at various temperatures, on the early development of four saturniid moths (*Platysamia cecropia* Linnaeus, *Telea polyphemus* Cramer, *Samia walkeri* Felder and Felder, and *Callosamia promethea* Drury) and on the weights and water contents of their larvae. Ecology **23**, 259–274

Ludwig, D., Landsman, H. M.: 1937. The effect of different relative humidities on survival and metamorphosis of the Japanese beetle (*Popillia japonica* Newman). Physiol. Zool. **10**, 171–179

Lüscher, M., Wyss-Huber, M.: 1966. Die Adenosine-Nukleotide im Fettkörper des adulten Weibchens von *Leucophaea maderae* im Laufe des Sexualzyklus. Rev. Suisse Zool. **71**, 183–194

Lundegärdh, H. G.: 1966. Plant Physiology. London: Oliver and Boyd, 549 pp

Luzzati, V., Husson, F.: 1962. The structure of the liquid crystal phases of lipid water systems. J. Cell Biol. **12**, 207–219

Machin, J.: 1975. Water balance in *Tenebrio molitor*, L. larvae: the effect of atmospheric water absorption. J. Comp. Physiol. **101**, 121–132

Machin, J.: 1976. Passive exchange during water vapour absorption in mealworms (*Tenebrio molitor*): a new approach to studying the phenomenon. J. Exp. Biol. **65**, 603–615.

Maddrell, S. H. P.: 1962. A diuretic hormone in *Rhodnius prolixus* Stål. Nature (London) **194**, 605–606

Maddrell, S. H. P.: 1963. Excretion in the blood-sucking bug, *Rhodnius prolixus* Stål. I. The control of diuresis. J. Exp. Biol. **40**, 247–256

Maddrell, S. H. P.: 1964a. Excretion in the blood sucking bug *Rhodnius prolixus* Stål. II. The normal course of diuresis and the effect of temperature. J. Exp. Biol. **41**, 163–176

Maddrell, S. H. P.: 1964b. Excretion in the blood sucking bug *Rhodnius prolixus* Stål. III. The control of release of diuretic hormone. J. Exp. Biol. **41**, 459–472

Maddrell, S. H. P.: 1966. The site of release of the diuretic hormone in *Rhodnius*. A new neurohaemal system in insects. J. Exp. Biol. **45**, 499–508

Maddrell, S. H. P.: 1967. Neurosecretion in insects. In: Insects and Physiology. Beament, J. W. L., Treherne, J. E. (eds.). Edinburgh: Oliver and Boyd, pp. 103–118

Maddrell, S. H. P.: 1969. Secretion by the Malpighian tubules of *Rhodnius*. The movements of ions and water. J. Exp. Biol. **51**, 71–97

Maddrell, S. H. P.: 1970. Neurosecretory control systems in insects. Symp. Roy. Entomol. Soc. London **5**, 101–116

Maddrell, S. H. P.: 1971. The mechanisms of insect excretory systems. Adv. Insect Physiol. **8**, 199–331

Maddrell, S. H. P., Casida, J. E.: 1971. Mechanism of insecticide-induced diuresis in *Rhodnius*. Nature (London) **231**, 55–56

Maddrell, S. H. P., Gardiner, B. O. C.: 1974. The passive permeability of insect Malpighian tubules to organic solutes. J. Exp. Biol. **60**, 641–652

Maddrell, S. H. P., Gardiner, B. O. C., Pilcher, D. E. M., Reynolds, S. E.: 1974. Active transport by insect Malpighian tubules of acidic dyes and of acylamides. J. Exp. Biol. **61**, 357–377

Maddrell, S. H. P., Gee, J. D.: 1974. Potassium-induced release of the diuretic hormones of *Rhodnius prolixus* and *Glossina austeni*: Ca dependence, time course and localization of neurohaemal areas. J. Exp. Biol. **61**, 155–171

Maddrell, S. H. P., Phillips, J. E.: 1975a. Secretion of hypo-osmotic fluid by the lower Malpighian tubules of *Rhodnius prolixus*. J. Exp. Biol. **62**, 671–683

Maddrell, S. H. P., Phillips, J. E.: 1975b. Active transport of sulphate ions by the Malpighian tubules of larvae of the mosquito *Aedes campestris*. J. Exp. Biol. **62**, 367–378

Maddrell, S. H. P., Pilcher, D. E. M., Gardiner, B. O. C.: 1969. Stimulatory effect of 5-hydroxytryptamine (serotonin) on secretion by Malpighian tubules of insects. Nature (London) **272**, 784–785

Maddrell, S. H. P., Pilcher, D. E. M., Gardiner, B. O. C.: 1971. Pharmacology of the Malpighian tubules of *Rhodnius* and *Carausius*: The structure-activity relationship of tryptamine analogues and the role of cyclic AMP. J. Exp. Biol. **54**, 779–804

Maddrell, S. H. P., Reynolds, S. E.: 1972. Release of hormones in insects after poisoning with insecticides. Nature (London) **236**, 404–406

Maelzer, D. A.: 1961. The effect of temperature and moisture on the immature stages of *Aphodius tasmaniae* Hope (Scarabaeidae) in the lower south-east of South Australia. Aust. J. Zool. **9**, 173–202

Major, P. W., Kilpatrick, R.: 1972. Cyclic AMP and hormone action. J. Endocrinol. **52**, 593–630

Mantel, L. H.: 1968. The foregut of *Gecarcinus lateralis* as an organ of salt and water balance. Am. Zool. **8**, 433–442

Manton, S. M.: 1964. Mandibular mechanisms and the evolution of arthropods. Phil. Trans. Roy. Soc. London, Ser. B **247**, 1–183

Manton, S. M.: 1970. Arthropods: introduction. In: Chemical Zoology. Florkin, M. F., Scheer, B. T. (eds.). New York: Academic Press, Vol. IV, Part A, pp. 1–34

Manton, S. M.: 1973. Arthropod phylogeny—a modern synthesis. J. Zool. **171**, 111–130

Manton, S. M., Heatley, N. G.: 1937. Studies on the Onycophora. II. The feeding, digestion and food storage of *Peripatopsis*. Phil. Trans. Roy. Soc. London, Ser. B **227**, 411–464

Marcuzzi, G.: 1955. Osservazioni fisico-chemiche sul sangue dei colleoteri tenebrionidi. I. La pressione osmotica nel *Tenebrio molitor* L. Atti Accad. naz. Lincei Rend. Ser. 8, **18**, 654–662

Marcuzzi, G.: 1956. L'osmoregulazione nel *Tenebrio molitor* L. (Col. Tenebrionidae). Atti. Accad. naz. Lincei Rend. Classe Sci. Fis. Mat. Nat. Ser. 18, **20**, 492–500

Marcuzzi, G.: 1960. Il bilancio idrico nei coleotteri tenebrionidi. Arch. Zool. Ital. **45**, 281–324

Marshall, A. T., Cheung, W. W. K.: 1973. Studies on water and ion transport in homopteran insects: Ultrastructure and cytochemistry of the cicadoid and cercopoid hindgut. Tissue Cell **5**, 671–678

Marshall, A. T., Cheung, W. W. K.: 1974. Studies on water and ion transport in homopteran insects: Ultrastructure and cytochemistry of the cicadoid and cercopoid Malpighian tubules and filter chamber. Tissue Cell **6**, 153–171

Marshall, A. T., Cheung, W. W. K.: 1975. Ionic balance of homoptera in relation to feeding site and plant sap composition. Entomol. Exp. Appl. **18**, 117–120

Mathur, C. B.: 1944. The site of the absorption of water by the egg of *Schistocerca gregaria*. Indian J. Entomol. **5**, 35–40

Matthée, J. J.: 1951. The structure and physiology of the egg of *Locustana pardalina* (Walk). Union S. Afr. Dept. Agric. Bull. **316**, 3–83

May, R. M.: 1935. La substance réductrice et le chlore du sang des Orthoptères. Bull. Soc. Chim. Biol. **17**, 1045

Mayes, K. R., Holdich, D. M.: 1976. The water content of muscle and cuticle of the woodlouse *Oniscus asellus* in conditions of hydration and desiccation. Comp. Biochem. Physiol. **53A**, 253–258

McEnroe, W. D.: 1961. The control of water loss by the two-spotted spider mite, *Tetranychus telarius*. Ann. Entomol. Soc. Am. **54**, 883–887

McEnroe, W. D.: 1963. The role of the digestive system in the water balance of the two-spotted spider mite. Advan. Acarol. **1**, 225–231

McEnroe, W. D.: 1971. Water balance and mortality in the adult female American dog tick, *Dermacentor variabilis* Say. (Acarina: Ixodidae). Amherst: Univ. Massachusetts, Res. Bull. **594**, pp. 3–15

McEnroe, W. D.: 1972. Equilibrium weights of *Dermacentor variabilis* Say. at near saturation (Acarina: Ixodidae). Acaralogia **14**, 365–367

262

McFarlane, J. E.: 1970. The permeability of the cricket egg shell. Comp. Biochem. Physiol. **37**, 133–141

McFarlane, J. E., Furneaux, P. J. S.: 1964. Revised curves for water absorption by the eggs of the house cricket *Acheta domesticus* (L.). Can. J. Zool. **42**, 239–243

McFarlane, J. E., Kennard, C. P.: 1960. Further observations on water absorption by the eggs of *Acheta domesticus* (L.). Can. J. Zool. **38**, 77–85

McMullen, H. L., Sauer, J. R., Burton, R. L.: 1976. Possible role in uptake of water vapour by ixodid tick salivary glands. J. Insect Physiol. **22**, 1281–1285

Mead-Briggs, A. R.: 1956. The effect of temperature upon the permeability to water of arthropod cuticles. J. Exp. Biol. **33**, 737–749

Megaw, M. W. J.: 1974. Studies on the water balance mechanism of the tick, *Boophilus microplus canestrini*. Comp. Biochem. Physiol. **48 A**, 115–125

Mellanby, K.: 1932a. The influence of atmospheric humidity on the thermal death point of a number of insects. J. Exp. Biol. **9**, 222–231

Mellanby, K.: 1932b. Effect of temperature and humidity on the metabolism of the fasting bed-bug *(Cimex lectularius)*, Hemiptera. Parasitology **24**, 419–428

Mellanby, K.: 1932c. The effect of atmospheric humidity on the metabolism of the fasting mealworm *(Tenebrio molitor* L., Coleoptera). Proc. Roy. Soc. London, Ser. B **111**, 376–390

Mellanby, K.: 1934. Effects of temperature and humidity on the clothes moth larva, *Tineola bisellliella* Hum. (Lepidoptera). Ann. Appl. Biol. **21**, 476–482

Mellanby, K.: 1938. Diapause and metamorphosis of the blowfly, *Lucilia sericata* Meig. Parasitology **30**, 392–402

Mellanby, K.: 1939. The functions of insect blood. Biol. Rev. **14**, 243–260

Mellanby, K.: 1942. Metabolic water and desiccation. Nature (London) **150**, 21

Mellanby, K.: 1958. Water content and insect metabolism. Nature (London) **181**, 1403

Mellanby, K., French, R. A.: 1958. The importance of drinking water to larval insects. Entomol. Exp. Appl. **1**, 116–124

Meredith, J., Kaufman, W. R.: 1973. A proposed site of fluid secretion in the salivary gland of the ixodid tick *Dermacentor andersoni*. Parasitology **67**, 205–217

Miller, P. L.: 1960a. Respiration in the desert locust. I. The control of ventilation. J. Exp. Biol. **37**, 224–236

Miller, P. L.: 1960b. Respiration in the desert locust. II. The control of the spiracles. J. Exp. Biol. **37**, 237–263

Miller, P. L.: 1960c. Respiration in the desert locust. III. Ventilation and the spiracles during flight. J. Exp. Biol. **37**, 264–278

Miller, P. L.: 1962. Spiracle control in adult dragonflies (Odonata). J. Exp. Biol. **39**, 513–535

Miller, P. L.: 1964a. Factors altering spiracle control in adult dragonflies: water balance. J. Exp. Biol. **41**, 331–343

Miller, P. L.: 1964b. Factors altering spiracle control in adult dragonflies: hypoxia and temperature. J. Exp. Biol. **41**, 345–357

Miller, P. L.: 1970. On the occurrence of some characteristics of *Cyrtopus fastuosus* Bigot and *Polypedilum* sp. from temporary habitats in Western Nigeria. Entomol. Mon. Mag. **105**, 233–238

Miller, P. L.: 1974. Respiration—aerial gas transport. In: The Physiology of Insecta. Rockstein, M. (ed.). New York: Academic Press, pp. 345–402

Mills, R. R.: 1967. Hormonal control of excretion in the American cockroach. I. Release of a diuretic hormone from the terminal abdominal ganglion. J. Exp. Biol. **46**, 35–41

Mills, R. R., Nielsen, D. J.: 1967. Changes in the diuretic and antidiuretic properties of the haemolymph during the six-day vitellogenic cycle in the American cockroach. Gen. Comp. Endocrinol. **9**, 380–382

Moffett, D. F.: 1975. Sodium and potassium transport across the isolated hindgut of the desert millipede, *Orthoporus ornatus* (Girard). Comp. Biochem. Physiol. **50 A**, 57–63

Moloo, S. K.: 1971. Some aspects of water absorption by the developing egg of *Schistocerca gregaria*. J. Insect Physiol. **17**, 1489–1495

Mordue, W.: 1969. Hormonal control of Malpighian tube and rectal function in the desert locust *Schistocerca gregaria*. J. Insect Physiol. **15**, 273–285

263

Moriarty, F.: 1969a. Water uptake and embryonic development in eggs of *Chorthippus brunneus* Thunberg (Saltatoria : Acrididae). J. Exp. Biol. **50**, 327–333

Moriarty, F.: 1969b. The sublethal effects of synthetic insecticides on insects. Biol. Rev. Cambridge Phil. Soc. **44**, 321–357

Moriarty, F.: 1970. The significance of water absorption by the developing eggs of five British Acrididae (Saltatoria). Comp. Biochem. Physiol. **34**, 657–669

Mullins, D. E.: 1974. Nitrogen metabolism in the American cockroach: an examination of whole body ammonium and other cations excreted in relation to water requirements. J. Exp. Biol. **61**, 541–556

Mullins, D. E., Cochran, D. G.: 1972. Nitrogen excretion in cockroaches: uric acid is not a major product. Science **177**, 699–701

Mullins, D. E., Cochran, D. G.: 1973. Nitrogenous excretory material from the American cockroach. J. Insect Physiol. **19**, 1007–1018

Mullins, D. E., Cochran, D. G.: 1974. Nitrogen metabolism in the American cockroach: an examination of whole body and fat body regulation of cations in response to nitrogen balance. J. Exp. Biol. **61**, 557–570

Mullins, D. E., Cochran, D. G.: 1976. A comparative study of nitrogen excretion in twenty three cockroach species. Comp. Biochem. Physiol. **53 A**, 393–399

Munk, R.: 1968. Autoradiografische Untersuchungen des Transportes einiger Nahrungsbestandteile im Darmtrakt zweier Kleinzikaden: *Euscelides variegatus* Klem. (Jassidae) und *Triecophora vulnerata* Germ. (Cercopidae). Z. Vergl. Physiol. **61**, 129–136

Munson, S. C., Yeager, J. F.: 1949. Blood volume and chloride normality in roaches (*Periplaneta americana* (L.)) injected with sodium chloride solutions. Ann. Entomol. Soc. Am. **42**, 165–173

Murray, D. R. P.: 1968. The importance of water in the normal growth of larvae of *Tenebrio molitor*. Entomol. Exp. Appl. **11**, 149–168

Nagy, K. A.: 1972. Water and electrolyte budgets of a free-living desert lizard, *Sauromalus obesus*. J. Comp. Physiol. **79**, 39–62

Nagy, K. A., Shoemaker, V. H.: 1975. Energy and nitrogen budgets of the free-living desert lizard *Sauromalus obesus*. Physiol. Zool. **48**, 252–262

Nauomoff, M., Jeuniaux, C.: 1970. Modification de la composante cationique inorganique de l'hemolymphe au cours de development et des metamorphoses de quelques Lepidopteres. Arch. Internat. Physiol. Biochimie **78**, 357–365

Navarro, S., Calderon, M.: 1973. Carbon dioxide and relative humidity—interrelated factors affecting the loss of water and mortality of *Ephestia cautella* (Lepidoptera, Phycitidae). Israel J. Entomol. **8**, 143–152

Nedergaard, S., Harvey, W. R.: 1968. Active transport by the cecropia midgut. IV. Specificity of the transport mechanism for potassium. J. Exp. Biol. **68**, 13–24

Needham, G. R., Sauer, J. R.: 1975. Control of fluid secretion by isolated salivary glands of the lone star tick. J. Insect Physiol. **21**, 1893–1898

Needham, J.: 1938. Contributions of chemical physiology to the problem of reversibility in evolution. Biol. Rev. **13**, 225–251

Nelson, E. V., Camin, J. G.: 1967. Cuticular critical temperature of the rabbit tick, *Haemaphysalis leporispalustris*. Am. Zool. **7**, 195

Nemenz, H.: 1954. Über den Wasserhaushalt einiger Spinnen mit besonderer Berücksichtigung der Transpiration. Öst. Zool. Z. **5**, 123–158

Neville, A. C.: 1970. Cuticle ultrastructure in relation to the whole insect. Symp. Roy. Entomol. Soc. London **5**, 17–39

Neville, A. C.: 1975. Biology of the Arthropod Cuticle. Heidelberg: Springer-Verlag, 448 pp

Neville, A. C., Thomas, M. G., Zelazny, B.: 1969. Pore canal shape related to molecular architecture of arthropod cuticle. Tissue Cell **1**, 183–200

Nicolson, S., Harsfield, P. M., Gardiner, B. O. C., Maddrell, S. H. P.: 1974. Effects of starvation and dehydration on osmotic and ionic balance in *Carausius morosus*. J. Insect Physiol. **20**, 2061–2069

Nicolson, S. W., Leader, J. P.: 1974. The permeability to water of the cuticle of the larva of *Opifex fuscus* (Hutton) (Diptera, Culicidae). J. Exp. Biol. **60**, 593–603

264

Nijhout, H. F.: 1975. Excretory role of the midgut in larvae of the tobacco hornworm, *Manduca sexta* (L.). J. Exp. Biol. **62**, 221–230

Nobel, P.: 1974. Introduction to Biophysical Plant Physiology. San Francisco: W. H. Freeman, 488 pp

Noble-Nesbitt, J.: 1963. The site of water and ionic exchange with the medium in *Podura aquatica* L. (Collembola, Isotomidae). J. Exp. Biol. **40**, 701–711

Noble-Nesbitt, J.: 1967. Aspects of the structure, formation and function of some insect cuticles. In: Insects and Physiology. Beament, J. W. L., Treherne, J. E. (eds.). Edinburgh: Oliver and Boyd, pp. 3–16

Noble-Nesbitt, J.: 1969. Water balance in the firebrat, *Thermobia domestica* (Packard). Exchanges of water with the atmosphere. J. Exp. Biol. **50**, 745–769

Noble-Nesbitt, J.: 1970a. Structural aspects of penetration through insect cuticles. Pesticid. Sci. **1**, 204–208

Noble-Nesbitt, J.: 1970b. Water balance in the firebrat, *Thermobia domestica* (Packard). The site of uptake of water from the atmosphere. J. Exp. Biol. **52**, 193–200

Noble-Nesbitt, J.: 1973. Rectal uptake of water in insects. In: Comparative Physiology. Bolis, L., Schmidt-Nielsen, K., Maddrell, S. H. P. (eds.). Amsterdam: Elsevier-North Holland, pp. 333–351

Noble-Nesbitt, J.: 1975. Reversible arrest of uptake of water from sub-saturated atmospheres by the firebrat, *Thermobia domestica* (Packard). J. Exp. Biol. **62**, 657–669

Noble-Nesbitt, J.: 1976. Active transport of water vapour. In: Transport of Ions and Water in Animals. Gupta, B. L., Oschman, J. L., Wall, B. J. (eds.). London: Academic Press, in preparation

Noirot, C.: 1973. Cytologie ultrastructurale et phylogenie des termites. Proc. 7th Congr. Intern. Union Study Social Insects. Southampton, U. K.: Univ. Press, pp. 292–296

Noirot, C., Noirot-Timothée.: 1971. Ultrastructure du proctodeum chez le *Lepismodes inquilinus* Newman (= *Thermobia domestica* Packard). II. Le sac anal. J. Ultrastruct. Res. 37, 335–350

Norgaard, E.: 1951. On the ecology of two lycosid spiders *(Pirata piraticus* and *Lycosa pullata)* from a Danish sphagnum bog. Oikos **3**, 1–21

Norris, M. J.: 1934. Contributions towards the study of insect fertility. Proc. Zool. Soc. London **3**, 333–360

Norris, M. J.: 1936. The feeding habits of the adult Lepidoptera Heteroneura. Trans. Roy. Entomol. Soc. London **85**, 61–90

Nuñez, J. A.: 1956. Über die Regelung des Wasserhaushaltes bei *Anisotarsus cupripennis* Germ. Z. Vergl. Physiol. **38**, 341–344

Nutman, S. R.: 1941. Function of the ventral tube in *Onychiurus armatus* (Collembola). Nature (London) **148**, 168–169

O'Donnell, M. J.: 1977. The site of water vapour absorption in *Arenivaga investigata*. In: How organisms regulate their internal and external environment. Schmidt-Nielsen, K., Bolis, L., Maddrell, D. H. P. (eds.) Proc. Itern. Conf. Comp. Physiol. Vol. 3. Cambridge Univ. Press (In press)

Okasha, A. Y. K.: 1971. Water relations of an insect, *Thermobia domestica*. I. Water uptake from sub-saturated atmospheres as a means of volume regulation. J. Exp. Biol. **55**, 435–448

Okasha, A. Y. K.: 1973. Water relations in the insect *Thermobia domestica*. III. Effects of desiccation and rehydration on the haemolymph. J. Exp. Biol. **58**, 385–400

O'Neill, R. V.: 1969. Adaptive responses to desiccation in the millipede, *Narceus americanus* (Beauvois). Am. Midl. Natur. **81**, 578–583

O'Riordan, A. M.: 1969. Electrolyte movement in the isolated midgut of the cockroach (*Periplaneta americana* L.). J. Exp. Biol. **51**, 699–714

Oschman, J. L., Berridge, M. J.: 1970. Structural and functional aspects of salivary fluid secretion in *Calliphora*. Tissue Cell **2**, 281–310

Oschman, J. L., Wall, B. J.: 1969. The structure of the rectal pads of *Periplaneta americana* L. with regard to fluid transport. J. Morph. **127**, 475–510

Oschman, J. L., Wall, B. J., Gupta, B. L.: 1974. Cellular basis of water transport. Symp. Soc. Exp. Biol. **28**, 305–350

Pal, R.: 1950. The wetting of insect cuticle. Bull. Entomol. Res. **41**, 121–139

Parry, D. A.: 1951. Factors determining the temperature of terrestrial arthropods in sunlight. J. Exp. Biol. **28**, 445–462

Parry, D. A.: 1954. On the drinking of soil capillary water by spiders. J. Exp. Biol. **31**, 218–227

265

Parry, G.: 1953. Osmotic and ionic regulation in the isopod crustacean *Ligia oceanica*. J. Exp. Biol. **30**, 567–574

Passano, L. M.: 1960. Molting and its control. In: The Physiology of Crustacea. Waterman, T. H. (ed.). New York: Academic Press, Vol. I, pp. 473–536

Patlack, C. S., Goldstein, P. A., Hoffman, J. F.: 1963. The flow of solute and solvent across a two-membrane system. J. Theor. Biol. **5**, 426–442

Pearse, A. S.: 1950. The Emigration of Animals from the Sea. Dryden: Sherwood Press, 210 pp

Perttunen, V.: 1953. Reactions of diplopods to the relative humidity of the air. Investigations on *Orthomorpha gracilis*, *Iulus terrestris*, and *Schizophyllum sabulosum*. Ann. Zool. Soc. Zool. Botan. Fenn. "Vanamo" **16**, 1–69

Perttunen, V., Hayrinen, T.: 1970. Effect of light intensity and air humidity on flight initiation in *Blastophagus piniperda* L. (Col. Scolytidae). Entomol. Scandinav. **1**, 41–46

Phillips, J. E.: 1964a. Rectal absorption in the desert locust, *Schistocerca gregaria* Forskål. I. Water. J. Exp. Biol. **41**, 15–38

Phillips, J. E.: 1964b. Rectal absorption in the desert locust, *Schistocerca gregaria* Forskål. II. Sodium, potassium and chloride. J. Exp. Biol. **41**, 39–67

Phillips, J. E.: 1964c. Rectal absorption in the desert locust, *Schistocerca gregaria* Forskål. III. The nature of the excretory process. J. Exp. Biol. **41**, 69–80

Phillips, J. E.: 1969. Osmotic regulation and rectal absorption in the blowfly *Calliphora erythrocephala*. Can. J. Zool. **47**, 851–863

Phillips, J. E., Dockrill, A. A.: 1968. Molecular sieving of hydrophilic molecules by the rectal intima of the desert locust *(Schistocerca gregaria)*. J. Exp. Biol. **48**, 521–532

Phillips, J. E., Maddrell, S. H. P.: 1975. Active transport of magnesium by the Malpighian tubules of the larvae of the mosquito, *Aedes campestris*. J. Exp. Biol. **61**, 761–771

Pickford, R.: 1975. Water uptake in eggs of *Camnula pellucida* (Orthoptera : Acrididae) and its relationship to embryogenesis. Can. Entomol. **107**, 533–542

Pierre, F.: 1958. Ecologie et peuplement entomologique des sable vifs du Sahara Nord-Occidental. Publ. Centre Rech. Sahariennes Ser. Biol. Paris. Vol. I, 332 pp

Pilcher, D. F. M.: 1970a. Hormonal control of the Malpighian tubules of the stick insect, *Carausius morosus*. J. Exp. Biol. **52**, 653–665

Pilcher, D. F. M.: 1970b. The influence of the diuretic hormone on the process of urine formation by the Malpighian tubules of *Carausius morosus*. J. Exp. Biol. **53**, 465–484

Pinchon, Y.: 1970. Ionic content of haemolymph in the cockroach *Periplaneta americana*. J. Exp. Biol. **53**, 195–209

Pinkston, K. N., Frick, J. H.: 1973. Determinations of osmolarity on the fresh unfixed hemolymph of four species of spiders. Ann. Entomol. Soc. Am. **66**, 696–697

Poinsot-Balaguer, N.: Dynamique des communautés de Collemboles en milieu xerique méditerranéen. Pedobiologia **16**, 1–17

Prosser, C. L. (ed.): 1973. Comparative Animal Physiology, 3rd ed. Philadelphia: Saunders, 966 pp

Prusch, R. D.: 1972. Secretion of NH_4Cl by the hindgut of *Sarcophaga bullata* larvae. Comp. Biochem. Physiol. **41A**, 215–223

Prusch, R. D.: 1973. Secretion of a hyperosmotic excreta by the blowfly larva, *Sarcophaga bullata*. Comp. Biochem. Physiol. **46A**, 691–698

Prusch, R. D.: 1974. Active ion transport in the larval hindgut of *Sarcophaga bullata* (Diptera : Sarcophagidae). J. Exp. Biol. **61**, 95–109

Prusch, R. D.: 1976. Unidirectional ion movements in the hindgut of larval *Sarcophaga bullata* (Diptera : Sarcophagidae). J. Exp. Biol. **64**, 89–100

Pryor, M. G. M.: 1940. On the hardening of the cuticle of insects. Proc. Roy. Soc. London, Ser. B **128**, 393–407

Punt, A.: 1950. The respiration of insects. Physiol. Comp. Oecol. **2**, 59–74

Punt, A.: 1956. The influence of carbon dioxide on the respiration of *Carabus nemoralis* Mull. Physiol. Comp. Oecol. **4**, 132–141

Punt, A., Parser, W. J., Kuchlein, J.: 1957. Oxygen uptake in insects with cyclic CO_2 release. Biol. Bull. Woods Hole **112**, 108–119

Rajulu, G. S.: 1969. Moult cycle of a millipede *Spirostreptus asthenes* (Diplopoda : Myriapoda). Sci. Cult. (Calcutta) **35**, 483–485

Rajulu, G. S.: 1970. A comparative study of the free amino acids in the haemolymph of a millipede, *Spirostreptus asthenes* and a centipede *Ethmostigmus spinosus* (Myriapoda). Comp. Biochem. Physiol. **37**, 339–344

Rajulu, G. S., Krishnan, G.: 1968. The epicuticle of millipedes belonging to the genera *Cingalobolus* and *Aulacobolus* with special reference to seasonal variation. Z. Naturforsch. **23**, 84–85

Ramsay, J. A.: 1935. The evaporation of water from the cockroach. J. Exp. Biol. **12**, 373–383

Ramsay, J. A.: 1952. The excretion of sodium and potassium by the Malpighian tubules of *Rhodnius*. J. Exp. Biol. **29**, 110–126

Ramsay, J. A.: 1953. Active transport of potassium by the Malpighian tubules of insects. J. Exp. Biol. **30**, 358–369

Ramsay, J. A.: 1954. Active transport of water by the Malpighian tubules of the stick insect, *Dixippus morosus* (Orthoptera : Phasmidae). J. Exp. Biol. **31**, 104–113

Ramsay, J. A.: 1955 a. The excretory system of the stick insect, *Dixippus morosus* (Orthoptera, Phasmidae). J. Exp. Biol. **32**, 183–199

Ramsay, J. A.: 1955 b. The excretion of sodium, potassium and water by the Malpighian tubules of the stick insect, *Dixippus morosus* (Orthoptera, Phasmidae). J. Exp. Biol. **32**, 200–216

Ramsay, J. A.: 1956. Excretion by the Malpighian tubules of the stick insect *Dixippus morosus* (Orthoptera, Phasmidae): calcium, magnesium, chloride, phosphate and hydrogen ions. J. Exp. Biol. **33**, 697–709

Ramsay, J. A.: 1958. Excretion by the Malpighian tubules of the stick insect, *Dixippus morosus* (Orthoptera, Phasmidae): amino acids, sugars, and urea. J. Exp. Biol. **35**, 871–891

Ramsay, J. A.: 1964. The rectal complex of the meal worm *Tenebrio molitor* L. (Coleoptera, Tenebrionidae). Phil. Trans. Roy. Soc., Ser. B **248**, 279–314

Ramsay, J. A.: 1976. The rectal complex in the larvae of Lepidoptera. Phil. Trans. Roy. Soc. London Ser. B, **274**, 203–226

Rao, K. R.: 1968. The pericardial sacs of *Ocypode* in relation to the conservation of water, molting and behavior. Am. Zool. **8**, 561–567

Rapoport, E. H., Tschapek, M.: 1967. Soil water and soil fauna. Rev. Ecol. Biol. Sol **4**, 1–58

Rayah, El Amin El.: 1970. Humidity responses of two desert beetles, *Adesmia antiqua* and *Pimelia grandis* (Coleoptera : Tenebrionidae). Entomol. Exp. Appl. **13**, 438–447

Razet, P.: 1966. Les elements terminaux du catabolisme azote chez les insectes. Année Biol. **5**, 43–73

Reichle, D. E., Goldstein, R. A., Van Hook, R. I. Jr., Dodson, G. J.: 1973. Analysis of insect consumption in a forest canopy. Ecology **54**, 1076–1084

Reichle, D. E., Nelson, D. J., Dunaway, P. B.: 1971. Biological concentration and turnover of radionuclides in food chains: a bibliography. Nuclear Safety Information Center, Oak Ridge National Laboratory NSIC-89

Reichle, D. E., Van Hook, R. I. Jr.: 1970. Radionuclide dynamics in insect food chains. Manitoba Entomol. **4**, 22–32

Reynolds, S. E.: 1974 a. Pharmacological induction of plasticization in the abdominal cuticle of *Rhodnius*. J. Exp. Biol. **61**, 705–718

Reynolds, S. E.: 1974 b. A post ecdysial plasticization of the abdominal cuticle in *Rhodnius*. J. Insect Physiol. **20**, 1957–1962

Reynolds, S. E.: 1975. The mechanical properties of the abdominal cuticle of *Rhodnius* larvae. J. Exp. Biol. **62**, 69–80

Richards, A. G.: 1951. The Integument of Arthropods. Minnesota Univ. Press, 411 pp

Richards, A. G.: 1953. Structure and development of the integument. In: Insect Physiology. Roeder, K. D. (ed.). New York: Wiley, pp. 1–22

Richards, A. G.: 1958. The cuticle of arthropods. Ergeb. Biol. **20**, 1–26

Richards, A. G., Clausen, B. B., Smith, M. N.: 1953. Studies on arthropod cuticle. X. The asymmetrical penetration of water. J. Cell. Comp. Physiol. **42**, 395–414

Richards, A. G., Korda, F. H.: 1950. Studies on arthropod cuticle. IV. An electron microscope survey of the intima of arthropod tracheae. Ann. Entomol. Soc. Am. **43**, 49–71

Riegel, J. A.: 1966. Analysis of formed bodies in urine removed from the crayfish antennal gland by micropuncture. J. Exp. Biol. **44**, 387–395

Riegel, J. A.: 1970. A new model of transepithelial fluid movement with detailed application to fluid movement in the crayfish antennal gland. Comp. Biochem. Physiol. **36**, 403–410

Riegel, J. A.: 1971. Excretion—Arthropoda. In: Chemical Zoology. Florkin, M., Scheer, B. T. (eds.). New York: Academic Press, Vol. VI, Part B, pp. 249–277

Rinterknecht, E., Levi, P.: 1966. Etude au microscope electronique du cycle cuticulaire au cours du 4eme stade larvaire chez *Locusta migratoria*. Z. Zellforsch. Mikrosk. Anat. **72**, 390–407

Roberts, J. A., Kitching, R. L.: 1974. Ingestion of sugar protein and water by adult *Lucilia cuprina* (Diptera : Calliphoridae). Bull. Entomol. Res. **64**, 81–88

Roberts, R. B., Miskus, R. P., Duckles, C. K., Sakai, T. T.: 1969. *In vivo* fate of the insecticide Zectran in spruce budworm, tobacco budworm and horsefly larvae. J. Agric. Fd. Chem. **17**, 107–111

Robertson, J. D.: 1960. Osmotic and ionic regulation. In: The Physiology of Crustacea. Waterman, T. H. (ed.). New York: Academic Press, Vol. I, pp. 317–339

Robinson, W.: 1928. Water conservation in insects. J. Econ. Entomol. **21**, 897–902

Robson, E. A.: 1964. The cuticle of *Peripatopsis moseleyi*. Quart J. Microsc. Sci. **105**, 281–299

Roth, L. M.: 1968a. Oothecae of the Blattaria. Ann. Entomol. Soc. Am. **61**, 83–111

Roth, L. M.: 1968b. Reproduction of some poorly known species of Blattaria. Ann. Entomol. Soc. Am. **61**, 571–579

Roth, L. M.: 1968c. Oviposition behavior and water changes in the oothecae of *Lophoblatta brevis* (Blattaria : Blattellidae : Plectopterinae). Psyche **75**, 99–106

Roth, L. M., Dateo, G. P.: 1965. Uric acid storage and excretion by accessory sex glands of male cockroaches. J. Insect Physiol. **11**, 1023–1029

Roth, L. M., Willis, E. R.: 1955a. Water relations of cockroach oothecae. J. Econ. Entomol. **48**, 33–36

Roth, L. M., Willis, E. R.: 1955b. Water content of cockroach eggs during embryogenesis in relation to oviposition behavior. J. Exp. Zool. **128**, 489–510

Roth, L. M., Willis, E. R.: 1958. An analysis of oviparity and viviparity in the Blattaria. Trans. Am. Entomol. Soc. **83**, 221–238

Rudolph, D., Knülle, W.: 1974. Site and mechanism of water vapour uptake from the atmosphere in ixodid ticks. Nature (London) **249**, 84–85

Saccharov, N. L.: 1930. Studies in cold resistance in insects. Ecology **11**, 505–517

Sackin, H., Boulpaep, E. L.: 1976. Models for coupling of salt and water transport. Proximal tubular reabsorption in *Necturus* kidney. J. Gen. Physiol. **66**, 671–733

Sahrhage, D.: 1953. Ökologische Untersuchungen an *Thermobia domestica* (Packard) und *Lepisma saccharina* L. Z. Wiss. Zool. **157**, 77–168

Saini, R. S.: 1964. Histology and physiology of the cryptonephridial system in insects. Trans. Roy. Entomol. Soc. London **116**, 347–392

Salminen, I., Lindqvist, O. V.: 1972. Cuticular water content and the rate of evaporation in the terrestrial isopod *Porcellio scaber* Latr. J. Exp. Biol. **57**, 659–674

Salt, R. W.: 1949. Water uptake in eggs of *Melanoplus bivittatus* (Say). Can. J. Res. **27**, 236–242

Salt, R. W.: 1952. Some aspects of moisture absorption and loss in eggs of *Melanoplus bivittatus* (Say). Can. J. Zool. **30**, 55–82

Salt, R. W.: 1961. Resistance of poikilothermic animals to cold. British Med. Bull. **17**, 5–8

Sauer, J. R., Frick, J. H., Hair, J. A.: 1974. Control of chlorine-36 uptake by isolated salivary glands of the lone star tick. J. Insect Physiol. **20**, 1771–1778

Sauer, J. R., Hair, J. A.: 1971. Water balance in the lone star tick (Acarina : Ixodidae): the effects of relative humidity and temperature on weight changes and total water content. J. Med. Entomol. **8**, 479–485

Sauer, J. R., Hair, J. A.: 1972. The quantity of blood ingestion by the lone star tick (Acarina : Ixodidae). Ann. Entomol. Soc. Am. **65**, 1065–1068

Sauer, J. R., Levy, J. J., Smith, D. W., Mills, R. R.: 1970. Effect of rectal lumen concentration on the reabsorption of ions and water by the American cockroach. Comp. Biochem. Physiol. **32**, 601–614

Scheurer, R., Leuthold, R.: 1969. Haemolymph proteins and water uptake in female *Leucophaea maderae* during the sexual cycle. J. Insect Physiol. **15**, 1067–1077

Schmidt, G.: 1955. Physiologische Untersuchungen zur Transpiration und zum Wassergehalt der Gattung *Carabus* (Ins. Coleopt.). Zool. Jb. **65**, 459–495

268

Schmidt-Nielsen, B., Gertz, K. H., Davis, L. E.: 1968. Excretion and ultrastructure of the antennal gland of the fiddler crab *Uca mordax*. J. Morph. **125**, 473–481

Schmidt-Nielsen, K.: 1964. Desert Animals. Oxford: Univ. Press, 277 pp

Schmidt-Nielsen, K.: 1975. Animal Physiology. Adaptation and Environment. Cambridge: Univ. Press, 699 pp

Schmitz, S. H., Schultz, T. W.: 1969. Digestive anatomy of terrestrial isopoda: *Armadillidium vulgare* and *Armadillidium nasatum*. Am. Midland Nat. **82**, 163–181

Schneider, F.: 1948. Beitrag zur Kenntnis der Generationsverhältnisse und Diapause räuberischer Schwebfliegen (Syrphidae, Dipt.). Mitt. Schweiz. Entomol. Ges. **21**, 249–285

Schneiderman, H. A.: 1953. The discontinuous release of carbon dioxide by diapausing pupal insects. Anat. Rec. **117**, 540

Schneiderman, H. A.: 1960. Discontinuous respiration in insects: role of spiracles. Biol. Bull. **119**, 494–528

Schneiderman, H. A., Schechter, A. N.: 1966. Discontinuous respiration in insects. V. Pressure and volume changes in the tracheal system of silkworm pupae. J. Insect Physiol. **12**, 1143–1170

Schneiderman, H. A., Williams, C. M.: 1953. Discontinuous carbon dioxide output by diapausing pupae of the giant silkworm, *Platysamia cecropia*. Biol. Bull. **105**, 382

Schneiderman, H. A., Williams, C. M.: 1955. An experimental analysis of the discontinuous respiration of the cecropia silkworm. Biol. Bull. **109**, 123–143

Schoffeniels, E.: 1960. Role des oxides amines dans la regulation de la pression osmotique du milieu interieur des insectes aquatiques. Arch. Int. Physiol. **68**, 507–508

Schoeffeniels, E., Gilles, R.: 1970. Nitrogenous constituents and nitrogen metabolism in arthropods. In: Chemical Zoology. Florkin, M., Scheer, B. T. S. (eds.). New York: Academic Press, Vol. V, pp. 199–227

Schulz, F. N.: 1930. Zur Biologie des Mehlwurms *(Tenebrio molitor)*. I. Der Wasserhaushalt. Biochem. Z. **227**, 340–353

Sedlag, U.: 1951. Untersuchungen über den Ventraltubus der Collembolen. Wiss. Z. Martin-Luther-Univ. Halle-Wittenb. **1**, 93–127

Seely, M. K., Hamilton, W. J. III.: 1976. Fog catchment sand trenches constructed by tenebrionid beetles, *Lepidochora*, from the Namib Desert. Science **193**, 484–486

Sewell, M. T.: 1955. The histology and histochemistry of the cuticle of a spider, *Tegenaria domestica* (L.). Ann. Entomol. Soc. Am. **48**, 107–118.

Seymour, R. S.: 1974. Convective and evaporative cooling in sawfly larvae. J. Insect Physiol. **20**, 2447–2457

Seymour, R. S., Vinegar, A.: 1973. Thermal relations, water loss and oxygen consumption of a north American tarantula. Comp. Biochem. Physiol. **44 A**, 83–96

Shaw, J.: 1955. Ionic regulation and water balance in the aquatic larva of *Sialis lutaria*. J. Exp. Biol. **32**, 353–382

Shaw, J., Stobbart, R. H.: 1972. The water balance and osmoregulatory physiology of the desert locust *(Schistocerca gregaria)* and other desert and xeric arthropods. Symp. Zool. Soc. London **31**, 15–38

Shulov, A.: 1952. The development of eggs of *Schistocerca gregaria* (Forskål) in relation to water. Bull. Entomol. Res. **43**, 469–476

Shulov, A., Pener, M. P.: 1963. Studies on the development of eggs of the desert locust (*Schistocerca gregaria* Forskål) and its interruption under particular conditions of humidity. Anti-Locust Res. Bull. **41**, 1–59

Sidorov, V. E.: 1960. The nature of the coxal fluid of the argassid ticks, *Ornithodorus lahorensis*, *O. papillipes* and *Argas persicus*. Dokl. Akad. Nauk. SSSR. Biol. Sci. **130**, 176–178

Sinoir, Y.: 1966. Interaction du deficit hydrique de l'insecte et de la teneur en eau de l'aliments dans la prise de nourriture chez le criquet migrateur, *Locusta migratoria migratorioides* (R. et F.) C. R. Acad. Sci., Paris **262**, 2480–2483

Sláma, K.: 1960. Physiology of sawfly metamorphosis. 1. Continuous respiration in diapausing prepupae and pupae. J. Insect Physiol. **5**, 341–348

Slifer, E. H.: 1938. The formation and structure of a special water absorbing area in the membranes covering the grasshopper egg. Quart. J. Microsc. Sci. **80**, 437–457

Slifer, E. H.: 1946. The effects of xylol and other solvents on diapause in the grasshopper egg, together with a possible explanation for the action of these agents. J. Exp. Zool. **102**, 333–356

269

Slifer, E. H.: 1948. Isolation of a wax-like material from the shell of the grasshopper egg. Disc. Faraday Soc. **3**, 182–187

Slifer, E. H.: 1958. Diapause in the eggs of *Melanoplus differentialis* (Orthoptera, Acrididae). J. Exp. Zool. **138**, 259–282

Slifer, E. H., Sekhon, S. S.: 1970. Sense organs of a thysanuran, *Ctenolepisma lineata pilifera* with special reference to those on the antennal flagellum (Thysanura, Lepismatidae). J. Morphol. **132**, 1–25

Small, D. M.: 1970. The physical state of lipids of biological importance: Cholesteryl esters, cholesterol, triglycerides. Adv. Exp. Medic. Biol. **7**, 55–83

Sohal, R. S.: 1974. Fine structure of the Malpighian tubules in the housefly, *Musca domestica*. Tissue Cell **6**, 719–728

Solomon, M. E.: 1966. Moisture gains, losses and equilibria of flour mites, *Acarus siro* L., in comparison with larger arthropods. Entomol. Exp. Appl. **9**, 25–41

Speeg, K. V., Campbell, J. W.: 1968. Formation and volatilization of ammonia gas by terrestrial snails. Am. J. Physiol. **214**, 1392–1402

Spencer, J. O., Edney, E. B.: 1954. The absorption of water by woodlice. J. Exp. Biol. **31**, 491–496

Stadden, B. W.: 1964. Water balance in *Corixa dentipes* (Thomas.) (Hemiptera, Heteroptera). J. Exp. Biol. **41**, 609–619

Steiner, A.: 1930. Die Temperaturregulierung im Nest der Feldwespe (*Polistes gallica* var. *biglumis* L.). Z. Vergl. Physiol. **11**, 461–502

Stewart, D. M., Martin, A. W.: 1970. Blood and fluid balance of the common tarantula, *Dugesiella hentzi*. Z. Vergl. Physiol. **70**, 223–246

Stewart, D. M., Martin, A. W.: 1974. Blood pressure in the tarantula, *Dugesiella hentzi*. J. Comp. Physiol. **88**, 141–172

Stewart, T. C., Woodring, J. P.: 1973. Anatomical and physiological studies of water balance in the millipedes *Pachydesmus crassicutis* (Polydesmida) and *Orthoporus texicolens* (Spirobolida). Comp. Biochem. Physiol. **44A**, 734–750

Stobbart, R. H., Shaw, J.: 1974. Salt and water balance; excretion. In: The Physiology of Insecta. Rockstein, M. (ed.). New York: Academic Press, Vol. V, pp. 361–446

Stoeckenius, W.: 1962. The molecular structure of lipid-water systems and cell membrane models studied with the electron microscope. Symp. Intern. Soc. Cell Biol. **1**, 349–367

Stower, W. J., Griffiths, J. F.: 1966. The body temperature of the desert locust (*Schistocerca gregaria*). Entomol. Exp. Appl. **9**, 127–178

Subramoniam, T.: 1974. A histochemical study on the cuticle of a millipede: *Spirostreptus asthenes* (Diplopoda: Myriapoda). Acta Histochem. **51**, 200–204

Sutcliffe, D. W.: 1962. The composition of haemolymph in aquatic insects. J. Exp. Biol. **39**, 325–343

Sutcliffe, D. W.: 1963. The chemical composition of haemolymph in insects and some other arthropods, in relation to their phylogeny. Comp. Biochem. Physiol. **9**, 121–135

Sutton, S. L.: 1972. Woodlice. London: Ginn, 144 pp

Sweetman, H. L.: 1931. Preliminary report on the physical ecology of certain *Phyllophaga* (Scarabaeidae, Coleoptera). Ecology **12**, 401–422

Takahashi, Y.: 1959. Studies on the structural differentiation of silkworm cuticle and its relations to the transpiration and the penetration of fungus hyphae. Bull. Nagano Sericultural Exp. Sta. **58**, 1–100

Tartivita, K., Jackson, L. L.: 1970. Cuticular lipids of insects: I. Hydrocarbons of *Leucophaea maderae* and *Blatta orientalis*. Lipids **5**, 35–37

Tatchell, R. J.: 1964. Digestion in the tick, *Argas persicus* Oken. Parasitology **54**, 423–440

Tatchell, R. J.: 1967. Salivary secretion in the cattle tick as a means of water elimination. Nature (London) **213**, 940–941

Tatchell, R. J.: 1969. Secretion of the cattle tick, *Boophilus microplus*. J. Insect Physiol. **15**, 1421–1430

Theodor, O.: 1936. On the relation of *Phlebotomus papatasii* to the temperature and humidity of the environment. Bull. Entomol. Res. **27**, 653–671

Thompson, T. E.: 1964. The properties of bimolecular phospholipid membranes. In: Cellular Membranes in Development. Locke, M. (ed.). New York: Academic Press, pp. 83–96

Thorpe, W. H.: 1950. Plastron respiration in aquatic insects. Biol. Rev. Cambridge Phil. Soc. **25**, 344–390

270

Tiegs, O. W.: 1947. The development and affinities of the Pauropoda, based on a study of *Pauropus silvaticus*. Part II. Quart. J. Microsc. Sci. **88**, 275–336

Tiegs, O. W., Manton, S. M.: 1958. The evolution of the Arthropoda. Biol. Rev. **33**, 255–337

Tobias, J. M.: 1948. Potassium, sodium and water interchange in irritable tissues and haemolymph of an omnivorous insect, *Periplaneta americana*. J. Cell. Comp. Physiol. **31**, 125–148

Tobe, S. S.: 1974. Water movement during diuresis in the tsetse-fly *Glossina austeni*. Experientia **30**, 517–518

Toye, S. A.: 1966. The effect of desiccation on the behaviour of three species of Nigerian millipedes: *Spirostreptus assiniensis*, *Oxydesmus* sp., and *Habrodesmus falx*. Entomol. Exp. Appl. **9**, 378–384

Toye, S. A.: 1970. Some aspects of biology of two common species of Nigerian scorpions. J. Zool. London **162**, 1–9

Travis, D. F.: 1960. The deposition of skeletal structures in the Crustacea. I. The histology of the gastrolith disc, skeletal tissue complex and the gastrolith in the crayfish, *Oreonectes (Cambarus) virilis* Hagen. Biol. Bull. Woods Hole **118**, 137–149

Travis, D. F.: 1963. Structural features of mineralization from tissue to macromolecular levels of organisation in the decapod Crustacea. Ann. N. Y. Acad. Sci. **109**, 177–245

Treherne, J. E., Willmer, P. G.: 1975a. Evidence for hormonal control of integumentary water loss in cockroaches. Nature (London) **254**, 437–439

Treherne, J. E., Willmer, P. G.: 1975b. Hormonal control of integumentary water loss: evidence for a novel neuroendocrine system in an insect *(Periplaneta americana)*. J. Exp. Biol. **63**, 143–159

Tyson, G. E.: 1968. The fine structure of the maxillary gland of the brine shrimp, *Artemia salina*: the end-sac. Z. Zellforsch. Mikrosk. Anat. **86**, 129–138

Ueda, Y.: 1974. A study on the function of the rectal bladder of spiders. Acta Arachnol. **25**, 47–52

Vannier, G.: 1974a. Variation in the flux of body evaporation and cuticular resistance in *Tetrodontophora bielanensis* (Collembola) living in an atmosphere with a variable hygroscopic regime. Rev. Ecol. Biol. Sol. **11**, 201–211

Vannier, G.: 1974b. Calcul de la resistance cuticulaire à la diffusion de vapeur d'eau chez un insecte collembole. C. R. Acad. Sci. Paris **278**, 625–628

Verrett, J. M., Mills, R. R.: 1973. Water balance during vitellogenesis by the American cockroach: translocation of water during the cycle. J. Insect Physiol. **19**, 1889–1901

Verrett, J., Mills, R. R.: 1975a. Water balance during vitellogenesis by the American cockroach: hydration of the oocytes. J. Insect Physiol. **21**, 1061–1064

Verrett, J. M., Mills, R. R.: 1975b. Water balance during vitellogenesis by the American cockroach: distribution of water during the six day cycle. J. Insect Physiol. **21**, 1841–1845

Vietinghoff, U.: 1966. Einfluß der Neurohormone C_1 und O_1 auf die Absorptionsleistung der Rektaldrüsen der Stabheuschrecke *(Carausius morosus* Br.). Naturwissenschaften **52**, 162–163

Vollmer, A. R., MacMahon, J. A.: 1974. Comparative water relations of five species of spiders from different habitats. Comp. Biochem. Physiol. **47A**, 753–765

Waggoner, P. E.: 1967. Moisture loss through the boundary layer. Biometeorology **3**, 41–52

Waku, Y.: 1974. Ultrastructure of Malpighian tubule cells in the silkworm *Bombyx mori* with special regard to metamorphosis. Zool. Mag. Tokyo **83**, 152–162

Wall, B. J.: 1967. Evidence for antidiuretic control of rectal water absorption in the cockroach *Periplaneta americana* L. J. Insect Physiol. **13**, 565–578

Wall, B. J.: 1970. Effects of dehydration and rehydration on *Periplaneta americana*. J. Insect Physiol. **16**, 1027–1042

Wall, B. J., Oschman, J. L.: 1970. Water and solute uptake by rectal pads of *Periplaneta americana*. Am. J. Physiol. **218**, 1208–1215

Wall, B. J., Oschman, J. L.: 1975. Structure and function of the rectum in insects. Fortschr. Zool. **23**, 193–222

Wall, B. J., Oschman, J. L., Schmidt-Nielsen, B.: 1970. Fluid transport: concentration of the intercellular compartment. Science **167**, 1497–1498

Wall, B. J., Ralph, C. L.: 1962. Responses of specific neurosecretory cells of the cockroach, *Blaberus giganteus*, to dehydration. Biol. Bull. Woods Hole **122**, 431–438

Wall, B. J., Ralph, C. L.: 1964. Evidence for hormonal regulation of Malpighian tubule excretion in the insect *Periplaneta americana* L. Gen. Comp. Endocrinol. **4**, 452–456

271

Wang, T. H., Wu, H. W.: 1948. On the structure of the Malpighian tubes of the centipede and their excretion of uric acid. Sinensia **18**, 1–11

Warburg, M. R.: 1965. The evaporative water loss of three isopods from semi-arid habitats in South Australia. Crustaceana **9**, 302–308

Warburg, R. M.: 1968a. Simultaneous measurement of body temperature and water loss in isopods. Crustaceana **14**, 39–44

Warburg, R. M.: 1968b. Behavioral adaptations of terrestrial isopods. Am. Zool. **8**, 545–559

Ward, R. D., Ready, P. A.: 1975. Chorionic sculpturing in some sandfly eggs (Diptera, Psychodidae). J. Entomol. **50A**, 127–134

Wardhaugh, K. G.: 1970. The development of eggs of the Australian plague locust *Chortoicetes terminifera* (Walk.) in relation to temperature and moisture. Proc. Int. Study Conf. Current Future Prob. Acridol. London, pp. 261–272

Wardhaugh, K. G.: 1973. A study of some factors affecting egg development in *Chortoicetes terminifera* (Walker) (Orthoptera : Acrididae). Thesis, Australian National University

Watson, J. A. L., Hewitt, P. H., Nel, J. J. C.: 1971. The water-sacs of *Hodotermes mossambicus*. J. Insect Physiol. **17**, 1705–1709

Weber, H.: 1931. Lebensweise und Umweltbeziehungen von *Trialeurodes vaporariorum* (Westwood) (Homoptera–Aleurodina). Z. Morph. Ökol. Tiere. **23**, 575–753

Weis-Fogh, T.: 1960. A rubber-like protein in insect cuticle. J. Exp. Biol. **37**, 889–907

Weis-Fogh, T.: 1964. Diffusion in insect wing muscle, the most active tissue known. J. Exp. Biol. **41**, 229–256

Weis-Fogh, T.: 1967. Respiration and tracheal ventilation in locusts and other flying insects. J. Exp. Biol. **47**, 561–587

Weis-Fogh, T.: 1970. Structure and formation of insect cuticle. In: Insect Ultrastructure. Neville, A. C. (ed.). London: Blackwell, pp. 165–185

Wharton, D. R. A., Wharton, M. L., Lola, J. E.: 1965a. Weight and blood volume changes induced by irradiation of the American cockroach. Radiat. Res. **25**, 514–525

Wharton, D. R. A., Wharton, M. L., Lola, J.: 1965b. Blood volume and water content of the male American cockroach, *Periplaneta americana* L.—methods and the influence of age and starvation. J. Insect Physiol. **11**, 391–404

Wharton, G. W., Arlian, L. G.: 1972. Utilization of water by terrestrial mites and insects. In: Insect and Mite Nutrition. Rodriguez, J. C. (ed.). Amsterdam: Elsevier—North Holland, pp. 154–165

Wharton, G. W., Devine, T. L.: 1968. Exchange of water between a mite, *Laelaps echidnina*, and the surrounding air under equilibrium conditions. J. Insect Physiol. **14**, 1303–1318

Wharton, G. W., Kanungo, K.: 1962. Some effects of temperature and relative humidity on water balance in females of the spiny rat mite, *Echinolaelaps echidninus* (Acarina : Laelaptidae). Ann. Entomol. Soc. Am. **55**, 483–492

Wharton, G. W., Parrish, W., Johnston, D. E.: 1968. Observations on the fine structure of the cuticle of the spiny rat mite, *Laelaps echidnina* (Acari, Mesostigmata). Acarologia **10**, 206–214

Wheeler, R. E.: 1963. Studies on the total haemocyte count and haemolymph volume in *Periplaneta americana* (L.) with special reference to the last moulting cycle. J. Insect Physiol. **9**, 223–235

Wheeler, W. M.: 1893. Contributions to insect embryology. J. Morphol. **8**, 1–160

Whitten, J.: 1968. Metamorphic changes in insects. In: Metamorphosis: A Problem in Developmental Biology. Etkin, W. E., Gilbert, L. I. (eds.). Amsterdam: North Holland, pp. 43–105

Whitten, J. M.: 1969. Coordinated development in the fly foot: sequential cuticle secretion. J. Morphol. **127**, 73–104

Wieser, W.: 1972a. O/N ratios of terrestrial isopods at two temperatures. Comp. Biochem. Physiol. **43A**, 859–868

Wieser, W.: 1972b. A glutaminase in the body wall of terrestrial isopods. Nature (London) **239**, 288–290

Wieser, W.: 1972c. Oxygen consumption and ammonia excretion in *Ligia beaudiana* M.-E. Comp. Biochem. Physiol. **43A**, 869–876

Wieser, W., Schweizer, G.: 1970. A re-examination of the excretion of nitrogen by terrestrial isopods. J. Exp. Biol. **52**, 267–274

272

Wieser, W., Schweizer, G.: 1972. Der Gehalt an Ammoniak und freien Aminosäuren, sowie die Eigenschaften einer Glutaminase bei *Porcellio scaber* (Isopoda). J. Comp. Physiol. **81**, 73–88

Wieser, W., Schweizer, G., Hartenstein, R.: 1969. Patterns in the release of gaseous ammonia by terrestrial isopods. Oecology **3**, 390–400

Wigglesworth, V. B.: 1931. The physiology of excretion in a blood sucking insect, *Rhodnius prolixus* (Hemiptera, Reduviidae). I–III. J. Exp. Biol. **8**, 411–451

Wigglesworth, V. B.: 1933. The physiology of the cuticle and of ecdysis in *Rhodnius prolixus* with special reference to the functions of the oenocytes and of the dermal glands. Quart. J. Microsc. Sci. **76**, 270–318

Wigglesworth, V. B.: 1938. The regulation of osmotic pressure and chloride concentration in the haemolymph of mosquito larvae. J. Exp. Biol. **15**, 235–247

Wigglesworth, V. B.: 1945. Transpiration through the cuticle of insects. J. Exp. Biol. **21**, 97–114

Wigglesworth, V. B.: 1947. The epicuticle in an insect, *Rhodnius prolixus* (Hemiptera). Proc. Roy. Soc. London, Ser. B **134**, 163–181

Wigglesworth, V. B.: 1948a. The insect cuticle as a living system. Discus. Faraday Soc. **3**, 172–177

Wigglesworth, V. B.: 1948b. The structure and deposition of the cuticle in the adult mealworm, *Tenebrio molitor* L. (Coleoptera). Quart. J. Microsc. Sci. **89**, 197–217

Wigglesworth, V. B.: 1948c. The insect cuticle. Biol. Rev. Cambridge Phil. Soc. **23**, 408–451

Wigglesworth, V. B.: 1954. Growth and regeneration in the tracheal system of an insect, *Rhodnius prolixus* (Hemiptera). Quart. J. Microsc. Sci. **95**, 115–137

Wigglesworth, V. B.: 1970. Structural lipids in the insect cuticle and the function of the oenocytes. Tissue Cell **2**, 155–179

Wigglesworth, V. B.: 1972. The Principles of Insect Physiology. London: Chapman and Hall, 827 pp

Wigglesworth, V. B.: 1973. The role of the epidermal cells in moulding the surface pattern of the cuticle in *Rhodnius* (Hemiptera). J. Cell Sci. **12**, 683–705

Wigglesworth, V. B.: 1975. Incorporation of lipid into the epicuticle of *Rhodnius* (Hemiptera). J. Cell Sci. **19**, 459–485

Wigglesworth, V. B., Gillett, J. D.: 1936. The loss of water during ecdysis in *Rhodnius prolixus* Stahl. (Hemiptera). Proc. Roy. Entomol. Soc. London A **11**, 104–107

Wilkins, M. B.: 1960. A temperature dependent endogenous rhythm in the rate of carbon dioxide output of *Periplaneta americana*. Nature (London) **185**, 481–482

Wilkinson, P. R.: 1953. Observations on the sensory physiology and behaviour of larvae of the cattle tick, *Boophilus microplus* (Ixodidae). Aust. J. Zool. **1**, 345–356

Willem, V.: 1924. Observations sur *Machilis maritima*. Bull. Biol. France Belge **58**, 306–320

Williams, C. B.: 1924. A third bioclimatic study in the Egyptian Desert. Tech. Sci. Service Bull. **50**, 1–32

Williams, C. B.: 1954. Some bioclimatic observations in the Egyptian Desert. In: Biology of Deserts. Cloudsley-Thompson, J. L. (ed.). London: Inst. Biology, pp. 18–27

Williams, R. T.: 1970. In vitro studies on the environmental biology of *Goniodes colchici* (Denny) (Mallophaga: Ischnocera). II. The effect of temperature and humidity on water loss. Aust. J. Zool. **18**, 391–398

Williams, R. T.: 1971. In vitro studies on the environmental biology of *Goniodes colchici* (Denny) (Mallophaga: Ischnocera). III. The effect of temperature and humidity on the uptake of water vapour. J. Exp. Biol. **55**, 553–568

Winston, P. W.: 1967. Cuticular water pump in insects. Nature (London) **214**, 383–384

Winston, P. W., Nelson, V. E.: 1965. Regulation of transpiration in the clover mite *Bryobia praetiosa* Koch (Acarina: Tetranychidae). J. Exp. Biol. **43**, 257–269

Wolbarsht, M. L.: 1957. Water taste in *Phormia*. Science **125**, 1248

Wolfe, L. S.: 1954a. The deposition of the third instar larval cuticle of *Calliphora erythrocephala*. Quart. J. Microsc. Sci. **95**, 49–66

Wolfe, L. S.: 1954b. Studies on the development of the imaginal cuticle of *Calliphora erythrocephala*. Quart. J. Microsc. Sci. **95**, 67–78

Wolfe, L. S.: 1955. Further studies of the third instar larval cuticle of *Calliphora erythrocephala*. Quart. J. Microsc. Sci. **96**, 181–191

Wood, D. W.: 1957. The effect of ions upon neuromuscular transmission in a herbivorous insect. J. Physiol. **138**, 119

Wood, J. L.: Farrand, P. S., Harvey, W. P.: 1969. Active transport of potassium by the cecropia midgut. VI. Microelectrode potential profile. J. Exp. Biol. **50**, 169–178

Woodring, J. P.: 1973. Comparative morphology, functions and homologies of the coxal glands of oribatid mites (Arachnida : Acari). J. Morphol. **139**, 407–430

Wynne-Edwards, V. C.: 1962. Animal Dispersion in Relation to Social Behaviour. Edinburgh: Oliver and Boyd, 653 pp.

Yeager, J. F., Munson, S. C.: 1950. Blood volume of the roach *Periplaneta americana* determined by several methods. Arthropoda **1**, 255–265

General Index

Zoophysiology and Ecology

Editors: D.S. Farner (coordinating editor); W.S. Hoar; B. Hoelldobler; H. Langer; M. Lindauer

Vol. 1: P.J. BENTLEY
Endocrines and Osmoregulation
A Comparative Account of the Regulation of Water and Salt in Vertebrates
Contents: Osmotic Problems of Vertebrates. – The Vertebrate Endocrine System. – The Mammals. – The Birds. – The Reptiles. – The Amphibia. – The Fishes.

Vol. 2: L. IRVING
Arctic Life of Birds and Mammals
Including Man
Contents: Environment of Arctic Life. – Mammals of the Arctic. – Arctic Land Birds and Their Migrations. – Maintenance of Arctic Populations: Birds. – Maintenance of Arctic Populations of Mammals. – Warm Temperature of Birds and Mammals. – Maintenance of Warmth by Variable Insulation. – Metabolic Supply of Heat. – Heterothermic Operation of Homeotherms. – Size and Seasonal Change in Dimensions. – Insulation of Man.

Vol. 3: A.E. NEEDHAM
The Significance of Zoochromes
Contents: General. – The Nature and Distribution of Zoochromes. – Physiological Functions of Zoochromes. – Biochemical Functions of Zoochromes. – The Significance of Zoochromes for Reproduction and Development. – Evidence from Chromogenesis in the Individual. – Evolutionary Evidence and General Assessment.

Vol. 4/5: A.C. NEVILLE
Biology of the Arthropod Cuticle
Contents: General Structure of Integument. – The Structural Macromolecules. – Molecular Cross-Linking. – Supermolecular Architecture. – Physiological Aspects. – Calcification. – Physical Properties. – Phylogenetical Aspects. – Outstanding Problems.

Vol. 6: K. SCHMIDT-KOENIG
Migration and Homing in Animals
Contents: Field Performance in Orientation and Experimental and Theoretical Analysis in Crustaceans and Spiders, Locusts, Bees, Butterflies, Fishes, Amphibians, Reptiles, Birds, Mammals: Bats, Whales and Terrestrial Mammals. – Conclusion. – Appendix: Some Statistical Methods for the Analysis of Animal Orientation Data.

Vol. 7: E. CURIO
The Ethology of Predation
Contents: Internal Factors. – Searching for Prey. – Prey Recognition. – Prey Selection. – Hunting for Prey.

Vol. 8: W. LEUTHOLD
African Ungulates
Contents: General Background. – Comparative Review of Non-Social Behavior. – Social Behavior. – Social Organization. – Behavior and Ecological Adaptation.

Springer-Verlag
Berlin Heidelberg
New York

Behavioral Ecology and Sociobiology

Managing Editor: H. Markl, Konstanz, Germany
Editors: J. H. Crook, B. Hölldobler, H. Kummer, E. O. Wilson

Advisory Editors:

S. A. Altmann	W. D. Hamilton	M. Lindauer
J. F. Eisenberg	D. von Holst	P. Marler
Th. Eisner	K. Immelmann	G. F. Oster
V. Geist	W. T. Keeton	R. L. Trivers
D. R. Griffin	W. E. Kerr	Ch. Vogel

Behavioral Ecology and Sociobiology
will publish original contributions and short communications
dealing with quantitative studies and with the experimental
analysis of animal behavior on the population. Special
emphasis will be given to the functions, mechanisms and
evolution of ecological adaptations of behavior.

Aspects of particular interest are:
- Orientation in space and time
- Communication and all other forms of social and inter-
  specific behavioral interaction, including predatory and
  antipredatory behavior
- Origins and mechanisms of behavioral preferences and
  aversions, e. g. with respect to food, locality, and social
  partners.
- Behavioral mechanism of competition and resource
  partitioning
- Population physiology
- Evolutionary theory of social behavior

Subscription information and sample copies upon request.

Springer-Verlag Berlin Heidelberg New York

DATE DUE

APR 28 1981		

DEMCO 38-297